W9-ATX-723

INTRODUCTION TO GEOGRAPHIC INFORMATION SYSTEMS

INTRODUCTION TO GEOGRAPHIC INFORMATION SYSTEMS

Kang-tsung Chang
University of Idaho

Boston   Burr Ridge, IL   Dubuque, IA   Madison, WI   New York   San Francisco   St. Louis
Bangkok   Bogotá   Caracas   Kuala Lumpur   Lisbon   London   Madrid   Mexico City
Milan   Montreal   New Delhi   Santiago   Seoul   Singapore   Sydney   Taipei   Toronto

McGraw-Hill Higher Education 🔀
A Division of The **McGraw-Hill** Companies

INTRODUCTION TO GEOGRAPHIC INFORMATION SYSTEMS

Published by McGraw-Hill, a business unit of The McGraw-Hill Companies, Inc., 1221
Avenue of the Americas, New York, NY 10020. Copyright © 2002 by The McGraw-Hill
Companies, Inc. All rights reserved. No part of this publication may be reproduced or distributed
in any form or by any means, or stored in a database or retrieval system, without the prior written
consent of The McGraw-Hill Companies, Inc., including, but not limited to, in any network or other
electronic storage or transmission, or broadcast for distance learning.

Some ancillaries, including electronic and print components, may not be available to customers
outside the United States.

This book is printed on acid-free paper.

International   1 2 3 4 5 6 7 8 9 0 QPF/QPF 0 9 8 7 6 5 4 3 2 1
Domestic   1 2 3 4 5 6 7 8 9 0 QPF/QPF 0 9 8 7 6 5 4 3 2 1

ISBN 0–07–238211–2
ISBN 0–07–112178–1 (ISE)

Publisher: *Margaret J. Kemp*
Developmental editor: *Lisa Leibold*
Associate marketing manager: *Tami Petsche*
Senior project manager: *Susan J. Brusch*
Senior production supervisor: *Sandy Ludovissy*
Coordinator of freelance design: *David W. Hash*
Cover designer: *Nathan Bahls*
Cover images: *©The Living Earth, Inc./©Corbis*
Senior photo research coordinator: *Lori Hancock*
Photo research: *Randall Nicholas*
Senior supplement producer: *Audrey A. Reiter*
Media technology producer: *Judi David*
Compositor: *GAC—Indianapolis*
Typeface: *10/12 Times Roman*
Printer: *Quebecor World Fairfield, PA*

Figure credit: Figure 2.10, By permission of National Geodetic Survey

Library of Congress Cataloging-in-Publication Data

Chang, Kang-tsung.
   Introduction to geographic information systems / Kang-tsung Chang.—1st ed.
      p.   cm.
   Includes bibliographical references (p.   ) and index.
   ISBN 0–07–238211–2– ISBN 0–07–112178–1 (ISE)
   1. Geographic information systems. I. Title.

G70.212 .C4735   2002
901'.285—dc21   2001044140
                                                     CIP

INTERNATIONAL EDITION ISBN 0–07–112178–1
Copyright © 2002. Exclusive rights by The McGraw-Hill Companies, Inc., for manufacture
and export. This book cannot be re-exported from the country to which it is sold by McGraw-Hill.
The International Edition is not available in North America.

www.mhhe.com

*This book is dedicated to
Rweichung Hsu Chang.*

BRIEF CONTENTS

TABLE OF CONTENTS

PREFACE

This book covers both the important concepts and practice of geographic information systems (GIS) and is intended for a first or second course in GIS. A geographic information system is a computer system for capturing, storing, querying, analyzing, and displaying geographically referenced data. For many years, GIS has been considered to be too difficult, expensive, and proprietary. The development of the user-friendly interface, powerful and affordable computer hardware and software, and public digital data has broadened the range of GIS applications and brought GIS to many disciplines. It is not unusual to find students from more than a dozen disciplines in an introductory GIS class.

Although the power of a GIS lies in computing, its design and application require an understanding of concepts from fields such as geography, cartography, spatial analysis, and database management. These concepts explain GIS operations in terms of their objectives, structures, and relationships. Our knowledge of concepts will neither change with GIS technology nor become outdated with new versions of a GIS package. We can learn GIS concepts from readings. But for GIS users, an efficient way to grasp concepts behind the menus and icons is through practice. As a tool, GIS has been compared to statistics. We learn the working of statistical theorems and concepts by solving practical problems. The same is true with GIS. Moreover, to be proficient in using GIS packages is a requirement in the current GIS job market.

THE APPROACH

This book is designed to provide students with a solid foundation in both GIS concepts and use of GIS. To achieve the goal, I have attempted a comprehensive coverage of GIS topics. The book is organized into three parts. Part 1 (Chapters 1 to 8) covers the fundamentals of GIS including coordinate systems, data models, data input, data management, and data display. Part 2 (Chapters 9 to 12) includes data exploration, analysis using vector and raster data, and terrain analysis. Part 3 (Chapters 13 to 16) covers spatial interpolation, GIS models and modeling, regions, and network and dynamic segmentation. Also included in the book are new developments in GIS, such as the object-oriented data model, and research-oriented questions such as the effect of spatial scale.

WHAT MAKES THIS BOOK UNIQUE?

I have included problem-solving tasks in each chapter, complete with data sets and instructions. This hands-on experience is intended to complement the discussions in the text and to enhance the working knowledge of GIS functions.

This book contains complimentary ArcView 3.2a software. I have chosen *ArcView 3.2* and *ArcInfo 8,* two GIS packages from Environmental Systems Research Institute, Inc. (ESRI), for the practice sections. They are among the most popular GIS packages on the market and, when combined, they offer an extensive suite of GIS functionalities. Most exercises use *ArcView 3.2* and the *Spatial Analyst* extension. *ArcInfo 8* (the *ArcInfo Workstation* component) is only used for those exercises that cannot be handled by *ArcView 3.2,* such as topological editing, regions, and dynamic segmentation. Although GIS vendors often advertise their products with special features, a surprising degree of overlap exists among commercial GIS packages. Whenever appropriate, I have referenced different GIS packages on specific GIS functions. It would not be difficult to use this book with other GIS packages.

I have cited in the book examples and references from a variety of disciplines. This is necessary because GIS is used in many disciplines. I have also used boxes throughout the book to provide additional information and worked examples of important computing algorithms. Although some may feel that computing algorithms are matters for GIS software developers, I have included these worked examples for students who want to know how results from such operations as geometric transformation and spatial interpolation are derived.

ACKNOWLEDGMENTS

I would like to thank Ken McGinty, who read and edited an earlier draft, and the following reviewers who provided many helpful comments on earlier drafts:
Sergei Andronikov, Austin Peay State University
Matthew Bampton, University of Southern Maine
W.B. Clapham, Jr., Cleveland State University
Ferko Csillag, University of Toronto
Greg Gaston, Oregon State University
John Heinrichs, Fort Hays State University
Helmut Kraenzle, James Madison University
Mahesh Rao, Oklahoma State University
V.B. Robinson, University of Toronto
May Yuan, University of Oklahoma
and Undral Batsukh, who tested the practice sections of the book. At McGraw-Hill, Lisa Leibold, Pat Forrest, Susan Brusch, and David Hash contributed guidance and assistance during various stages of the project. Finally, I am grateful to my sons, Gary and Mark, for their support.

INTRODUCTION

1.1 WHAT IS A GIS?

A **Geographic Information System** (**GIS**) is a computer system for capturing, storing, querying, analyzing, and displaying geographic data. Like any other information technology, GIS can be divided into the following four components:

- **Computer System.** The computer system includes the computer and the operating system to run GIS. Typically the choices are PCs that use the Windows operating system (e.g., Windows 2000, Windows NT) or workstations that use the UNIX operating system. Additional equipment may include monitors for display, digitizers and scanners for spatial data input, and printers and plotters for hard copy data display.
- **GIS Software.** The GIS software includes the program and the user interface for driving the hardware. Common user interfaces in GIS are menus, graphical icons, and commands.
- **Brainware.** Equally important as the computer hardware and software, the brainware refers to the purpose and objectives, and provides the reason and justification, for using GIS.
- **Infrastructure.** The infrastructure refers to the necessary physical, organizational, administrative, and cultural environments for GIS operations. The infrastructure includes requisite skills, data standards, data clearinghouses, and general organizational patterns.

GIS is not new. Since the late 1960s computers have been used to store and process geographic data. Early examples of GIS-related work include the following:

- Computer mapping at the University of Edinburgh (Coppock 1988), the Harvard Laboratory for Computer Graphics (Chrisman 1988), and the Experimental Cartography Unit (Rhind 1988).
- Canada Land Inventory and the subsequent development of the Canada Geographic Information System (Tomlinson 1984).

- Publication of Ian McHarg's *Design with Nature* (McHarg 1969) and the map overlay method.
- Introduction of an urban street network with topology in the U.S. Bureau of the Census' DIME (Dual Independent Map Encoding) system (Broome and Meixler 1990).

For many years, though, GIS has been considered to be too difficult, expensive, and proprietary. The advent of the graphical user interface (GUI), powerful and affordable hardware and software, and public digital data has broadened the range of GIS applications and brought GIS to mainstream use. The following is a list of GIS software producers and their main products.

- Environmental Systems Research Institute (ESRI), Inc. (http://www.esri.com/): **ArcInfo, ArcView**
- Autodesk Inc. (http://www3.autodesk.com/): **AutoCAD Map**
- Baylor University, Texas (http://www.baylor.edu/~grass/): **GRASS**
- Clark Labs (http://www.clarklabs.org/): **IDRISI**
- International Institute for Aerospace Survey and Earth Sciences, the Netherlands (http://www.itc.nl/ilwis/): **ILWIS**
- MapInfo Corporation (http://www.mapinfo.com/): **MapInfo**
- ThinkSpace Inc. (http://www.thinkspace.com/): **MFworks**
- Intergraph Corporation (http://www.intergraph.com/): **MGE, GeoMedia**
- Bentley Systems, Inc. (http://www.bentley.com/): **Microstation**
- PCI Geomatics (http://www.pcigeomatics.com/): **PAMAP**
- TYDAC Inc. (http://www.tydac.ch/): **SPANS**
- Caliper Corporation (http://www.caliper.com/): **TransCAD, Maptitude**
- Northwood Technologies Limited (http://www.northwoodtec.com/): **Vertical Mapper**

According to a published survey (Crockett 1997), ESRI Inc. and Intergraph Corp. have dominated the market for GIS software. Two main products from ESRI Inc. are ArcView and ArcInfo 8. A desktop GIS package, ArcView uses extensions and stand-alone utility programs to extend its capabilities. ArcInfo 8 consists of ArcInfo Workstation, ArcCatalog, ArcMap, and ArcToolbox. ArcInfo Workstation, which is basically the same as ARC/INFO version 7.x and referred to as ARC/INFO hereafter, operates on UNIX, Windows 2000, or Windows NT, whereas the other three are limited to Windows 2000 or Windows NT. Intergraph Corp. also has two main products: MGE and GeoMedia. MGE consists of a series of products and operates on Windows 2000, Windows NT, or UNIX. GeoMedia is a desktop GIS package and is compatible with standard Windows development tools.

Smaller GIS companies often deploy strategic partnerships to strengthen their roles on the GIS market (Box 1.1). GRASS is unique among GIS software packages because it is free. Originally developed by the U.S. Army Construction Engineering Research Laboratories, GRASS is currently maintained in both the United States (Baylor University) and Germany (University of Hannover). TransCAD is a GIS package specially designed for use by transportation professionals. Microsoft

and Oracle, two giant companies in information technology, have also entered the GIS industry. Microsoft MapPoint targets business analysts who need to analyze and map geographic data (http://www.microsoft.com/). Oracle Spatial provides the features to store, access, and manage spatial data in Oracle8*i*, a relational database management system (http://www.oracle.com/).

Along with the proliferation of GIS activities, numerous GIS textbooks have been published (see Appendix A), and several journals and trade magazines are now devoted to GIS and GIS applications (see Appendix B).

Since the beginning, GIS has been important in natural resource management, such as land use planning, timber management, wildlife habitat analysis, riparian zone monitoring, and natural hazard assessment. In more recent years GIS has been used in emergency planning, market analysis, facilities management, transportation planning, and military applications. Integration of GIS with other technologies such as Global Positioning System (GPS) and the Internet has introduced new applications (Box 1.2). In summary, GIS has become one type of information system, specializing in information that has a geographic or spatial component.

The ability of GIS to handle and process geographically referenced data distinguishes GIS from

---

Box 1.1   **Strategic Partnerships in the GIS Industry**

A strategic partnership is designed to increase the competitiveness of the companies involved on the GIS market. A partnership may involve marketing a package, providing data exchange functionalities, or jointly developing a new module or extension. The following is a list of such partnerships:

- MapInfo and Vertical Mapper, MapInfo and SPANS
- PC Geomatics and ILWIS
- GeoMedia and Mfworks

- ESRI and ERDAS (a raster imaging/image processing company)

The GIS industry has also witnessed acquisitions of companies by other companies. Examples include Atlas GIS by ESRI, Vision* Solutions by Autodesk, and SmallWorld by GE. SmallWorld used to be a GIS company, but GE-SmallWorld (http://smallworld-us.com/) no longer advertises GIS among its business operations.

Box 1.2   **Application Examples of Integrating GIS with Other Technologies**

Over the years, new applications have been found by integrating GIS with other technologies. Among the examples are precision farming, interactive mapping on the Internet, and location-based services. Precision farming refers to site-specific farming such as herbicide or fertilizer application. Technologies for precision farming include GPS to mark the location of weeds or plots, GIS to store and process data and to prepare maps, and advanced computer-controlled sprayers.

Interactive mapping lets Internet users select map layers for display and make their own maps. Websites set up for interactive mapping usually provide the functionalities of pan, zoom, and simple query. The U.S. Geological Survey (USGS) maintains a website for interactive mapping using the National Atlas of the United States (http://www.nationalatlas.gov/).

The U.S. Census Bureau also has a website called Tiger Mapping Service, where Internet users can map public geographic data of anywhere in the United States (http://tiger.census.gov/). Typically, GIS provides the database and the query and mapping functionalities for interactive mapping on the Internet.

A new application that has been promoted by GIS companies, especially MapInfo and Intergraph, is called location-based services (LBS). This application allows users to deliver and receive information that is related to a particular location through wireless telecommunication. For example, an LBS user may use a web-enabled mobile phone to be located and to access, through mobile location services, information that is relevant to the user's location such as nearby ATMs and Chinese restaurants.

---

other information systems. **Geographically referenced data** describe both the location and characteristics of spatial features on the Earth's surface. Roads and land use types are spatial features as are precipitation and elevation. In describing a road, for example, we need to refer to its location (i.e., where it is) and its characteristics (road classification, traffic volume, etc.). GIS therefore involves two geographic data components: **spatial data** relate to the geometry of spatial features and **attribute data** give the information about the spatial features.

1.1.1 Spatial Data

Mapmakers represent spatial features on the Earth's surface as map features on a plane surface. The locations of map features are based on a Cartesian coordinate system with intersecting perpendicular lines along the x- and y-axis, whereas the locations of spatial features are based on the geographic grid expressed in longitude and latitude values. The transformation from the spherical geographic grid to a plane coordinate system is called **map projection**. Hundreds of map projections have been developed for mapmaking. Every map projection preserves certain spatial properties while sacrificing other properties.

A basic principle in GIS is that map layers to be used together must be based on the same coordinate system. Otherwise, map features from different layers will not register with one another spatially.

Spatial features may be discrete or continuous. **Discrete features** are those that do not exist between observations, form separate entities, and are individually distinguishable (Dent 1999). Wells, roads, and land use types are examples of discrete features. **Continuous features** exist spatially between observations (Dent 1999). Precipitation and elevation are examples of continuous features.

GIS uses two basic data models to represent spatial features: vector and raster (Figure 1.1). The **vector data model** uses points and their x-, y- coordinates to construct spatial features of points,

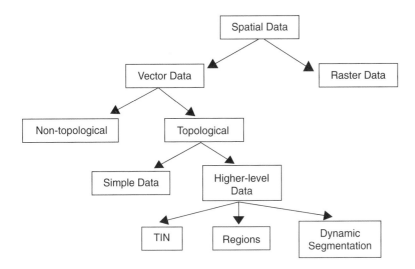

Figure 1.1

Data models for spatial data.

lines, and areas. Vector-based features are treated as discrete objects over the space. The **raster data model** uses a grid to represent the spatial variation of a feature. Each cell in the grid has a value that corresponds to the characteristic of the spatial feature at that location. Raster data are well-suited to the representation of continuous spatial features like precipitation and elevation.

The data model determines how the data are structured, stored, processed, and analyzed in a GIS. Many GIS functions are either vector-based or raster-based. So are different modules of a GIS package. ArcView provides its raster-based functionalities through the Spatial Analyst extension, and ARC/INFO offers its raster-based commands in the GRID module. Likewise, MGE include the Grid Analyst to work with raster data. Some GIS packages started as either raster-based or vector-based and later acquired capabilities of working with the other data model through strategic partnerships.

The data model is not really an issue to the GIS user nowadays. Raster and vector data can be displayed simultaneously. Raster data can be converted to vector data, and vice versa. A GIS has in fact become a useful tool for integrating raster and vector data.

Vector data are categorized as topological and non-topological (Figure 1.1). **Topology**, as used in GIS, expresses explicitly the spatial relationships between features, such as two lines meeting perfectly at a point and a directed line having an explicit left and right side. Topology is useful for detecting digitizing errors in digital maps and is necessary for some map overlay operations and network analysis. Non-topological data display faster and, more importantly, they can be used directly in different GIS software packages. GIS users should know whether their data are topological or not. ESRI, for example, uses the term **coverage** for topology-based digital maps and **shapefile** for non-topological vector data.

Topological vector data can be further divided into simple and higher-level data. Simple vector data consist of points, lines, and polygons. Higher-level data are built upon simple points, lines, and polygons. The **triangulated irregular network (TIN)** approximates the terrain with a set of non-overlapping triangles. Each triangle in a TIN consists of points and edges (lines) that connect these points to form triangles. A **region** is a collection of polygons, which may or may not be connected, and regions may overlap with one another or form a nested set. **Dynamic segmentation** is a data model that is built upon lines of a network and allows the use of real-world coordinates with linear measures such as mileposts.

Some GIS packages have recently introduced the object-oriented data model. The **object-oriented data model** uses objects to organize spatial data. An object has a set of built-in properties and can perform operations upon requests. To represent spatial features, objects may take the form of points, multipoints (a collections of points), polylines (a set of line segments), or polygons (including disjoint and nested polygons). As discussed in the next section, the object-oriented data model is different from the above data models because the geometries of objects are stored as an attribute called geometry.

1.1.2 Attribute Data

Attribute data describe the characteristics of spatial features. The amount of attribute data to be attached to a spatial feature can vary significantly depending on the feature type and the application. The Map Unit Interpretations Record (MUIR) attribute database, for example, contains about 88 estimated soil physical and chemical properties, interpretations, and performance data in a series of tables for each soil map unit surveyed by the Natural Resources Conservation Service (NRCS) in the United States (http://www.statlab.iastate.edu/soils/muir/).

Many commercial GIS packages store attribute data separate from spatial data in a split data system, often called the **georelational model**. Spatial data are stored in graphic files and managed by a file management system. Attribute data, on the other hand, are stored in a relational database. A **relational database** is a collection of tables, also called relations, which can be connected to each other by attributes whose values can uniquely identify a record in a table. Spatial data and attribute data in a GIS are typically linked through the feature IDs.

GIS users can use a relational database for data search, data retrieval, data editing, and creation of tabular reports. A relational database has two distinctive advantages in GIS applications. First, each table in the database can be prepared, maintained, and edited separately from other tables. Second, the tables can remain separate until a query or an analysis requires attribute data from different tables be linked together. Because the need for linking tables is often temporary, a relational database is efficient for both data management and data processing.

The relational database model is the standard model in GIS and management information system. With GIS increasingly becoming part of an organization's much larger information system, attribute data needed for a GIS project are likely to come from an enterprise-wide database. If spatial data distinguish GIS as a special type of information system, attribute data tend to connect GIS with other information systems.

Instead of storing spatial data and attribute data in a split data system, the object-oriented data model stores both data in a single database. A field called geometry is used to store spatial data. The object-oriented data model eliminates the complexity of coordinating and synchronizing two data file systems and actually brings GIS closer to other information systems.

1.2 GIS OPERATIONS

This section provides an overview of GIS operations. Although GIS activities no longer follow a set sequence, to explain what GIS users do, GIS activities can be grouped into spatial data input, attribute data management, data display, data exploration, data analysis, and GIS modeling (Figure 1.2).

1.2.1 Spatial Data Input

The most expensive part of a GIS project is database construction. Two basic options for database

Spatial Data Input	1. Data entry: use existing data, create new data 2. Data editing 3. Projection and re-projection 4. Geometric transformation
Attribute Data Management	1. Data entry and verification 2. Database management
Data Display	Use of maps, charts, and tables
Data Exploration	1. Attribute data query 2. Spatial data query 3. Geographic visualization
Data Analysis	1. Vector data analysis: buffering, overlay, distance measures, map manipulation 2. Raster data analysis: local, neighborhood, zonal, global 3. Terrain mapping and analysis 4. Spatial interpolation: global, local 5. Regions-based analysis 6. Network analysis
GIS Modeling	1. Binary models 2. Index models 3. Regression models 4. Process models

Figure 1.2

A classification of GIS activities.

construction are (1) use existing data and (2) create new data. Digital data clearinghouses have become commonplace on the Internet in recent years. The strategy for a GIS user nowadays is to look at what exists in the public domain before deciding either to create new data or to buy data from private companies.

New GIS data can be created from satellite images, GPS data, or paper maps. A variety of maps such as land use, land cover, and hydrography can be extracted from processing satellite images. Using satellites in space as reference points, a GPS receiver can determine its precise position on the Earth's surface, which can then be used to determine the location and shape of spatial features.

There are two methods for converting paper maps to digital maps: digitizing by using a digitizing table or a computer monitor, also called manual digitizing, and scanning. Scanning is preferred over manual digitizing in most cases, because scanning uses the machine and computer algorithm to do most of the work, thus avoiding human errors caused by fatigue or carelessness.

No matter how carefully it is prepared, a newly digitized coverage always has some errors. Digitizing errors can be removed through data editing, a part of database construction. One common type of digitizing error relates to the location accuracy of spatial data, such as missing lines or polygons or distorted lines. Another common type consists of topological errors, such as dangling arcs and unclosed polygons, which are caused by the failure of digitized features to follow the topological relationships among points, lines, and areas.

A newly digitized map has the same measurement unit (e.g., inches) as the source map used in digitizing or scanning. Therefore it must be converted to real-world coordinates by using a set of control points with known real-world coordinates and a process called geometric transformation.

1.2.2 Attribute Data Management

To complete database construction for a GIS project, attribute data must be entered, verified, and managed. Similar to spatial data, attribute data entry and verification also involve digitization and editing. Attribute data are usually managed in a relational database in GIS, whether attribute data are separate from spatial data or not. Because a relational database consists of separate tables, some of which may even reside at remote locations, the tables must be designed to facilitate data input, search, retrieval, manipulation, and output.

Two basic elements in the design of a relational database are the key and the type of data relationship. A key is a common field between two tables, which can establish a connection between corresponding records in the tables. The type of data relationship dictates how the tables are actually joined or linked. Expressed in the number of records, the type of relationship may be one-to-one, one-to-many, or many-to-one.

1.2.3 Data Display

Data display through maps, tables, and charts is a common part of a GIS project. As a visual tool, maps are most effective in communicating spatial data, whether the emphasis is on the location or the distribution pattern of spatial data. Either individually or grouped together, maps are important for visualization and query. Maps are also plotted to show results of GIS analysis.

A map consists of such elements as the title, subtitle, body, legend, north arrow, scale, acknowledgement, and border. These elements work together to bring spatial information to the map user. The first step in data display is to assemble map elements. Windows-based GIS packages have simplified the process of putting together a map. The GIS user can choose from program menus conventional symbols for qualitative and quantitative data, and can point and click the graphic icons to construct a map.

The second step in data display is map design. A well-designed map can help the mapmaker communicate spatial information to the map reader. A poorly designed map can confuse the map reader and even distort the information intended by the mapmaker. A creative process, map design cannot be easily performed by computer codes.

1.2.4 Data Exploration

Data exploration is data-centered query and analysis. Data exploration can be a GIS operation by itself or a precursor to formal data analysis. Data query allows the user to explore the general trends in the data, to take a close look at data subsets, and to focus on possible relationships between data sets. The purpose of data exploration is to better understand the data and to help formulate research questions and hypotheses.

An important component of effective data exploration consists of interactive and dynamically linked visual tools, including maps, graphs, and tables. The GIS user can perform data query using a map or a table. Following a query, the GIS user can view both the spatial and attribute components of the data subset, thus facilitating information processing and synthesis.

Geographic visualization (MacEachren 1995) is functionally similar to data exploration except that it is map-oriented. Tools for geographic visualization include data classification, data aggregation, and map comparison.

1.2.5 Data Analysis

Data analysis in GIS is closely related to the data model. The basic data models are vector and raster, each having its own set of analytical functions. Common vector functions include buffering, map overlay, distance measurement, and map manipulation. Raster data analysis can be conducted at the level of individual cells, or groups of cells, or cells within an entire grid. Based on the level of computation, raster data operations are commonly grouped into local, neighborhood, zonal, and global operations. Although some GIS concepts, such as map overlay and buffering, are the same for vector and raster data, the operational procedures differ.

The difference between vector and raster data is not an issue for terrain mapping and analysis because both types of data can be used. Raster data are ideal for the representation of surfaces, and the raster data structure is well-suited to intense computation often required for terrain analysis. But vector data, such as elevation points, contour lines, and streams and roads that represent changes of the terrain, are important inputs for terrain mapping and analysis. Moreover, the TIN is a basic data model in terrain analysis.

The terrain has been the object for mapping and analysis for hundreds of years. Mapping techniques such as contouring, profiling, hill shading, hypsometric tinting, and perspective views have been developed to portray the land surface. Measures of the land surface, such as slope, aspect (the directional measure of slope), and surface curvature, have also been developed. These land surface measures are important in studies of forest inventory, soil erosion, wildlife habitat suitability, site analysis, and many other fields. Terrain analysis also includes viewshed analysis (to determine areas of the land surface that are visible from an observation point) and watershed analysis (to derive flow direction, watershed boundaries, and stream networks).

The terrain with its undulating, continuous form is a familiar spatial feature to GIS users. Other surfaces used in GIS applications may not be physically present but can be visualized in the same way as the land surface. Cartographers called these types of surfaces statistical surfaces. Examples of statistical surfaces include precipitation, temperature, water table, and population density. Most studies of the statistical surface have been concentrated on developing spatial interpolation methods.

Spatial interpolation describes a process of using control points with known values to estimate values at other points. For example, from known temperature readings at nearby weather stations, interpolation can estimate the long-term average July temperature at locations that have no weather stations. GIS applications typically apply spatial interpolation to a grid (raster data) and make estimates for all cells in the grid. Spatial interpolation is therefore a means of converting point data to surface data. Like terrain analysis, spatial interpolation integrates vector and raster data.

Spatial interpolation methods can be grouped into global and local. A global method uses every control point available in estimating an unknown value. Trend surface analysis and regression modeling are two well-known global methods. A local method uses a sample of control points for estimation. GIS applications often use such local methods as Thiessen polygons, density estimation, inverse distance weighted, thin-plate splines, and kriging.

Higher-level vector data offer additional analytical functions that are not possible with simple points, lines, and polygons. Because regions may overlap and occupy the same area, they can be built as separate theme layers, each with its own attributes, in an integrated polygon coverage. An integrated coverage is therefore a collection of theme maps, which can be queried by different selection criteria. In this way, the regions data model allows map overlay (up to 32 coverages in ARC/INFO) to be combined with attribute data query in a single operation. With each region layer representing a spatial scale, the regions data model can also provide the frame for analyzing the effect of spatial scale on ecological systems and studies based on census data at the block group, census tract, and county levels.

A network is a connected line coverage with the appropriate attributes for the flow of objects such as traffic. Shortest path analysis is based on a network, with the objective of finding the path with the minimum cumulative cost in either time or distance between points on the network. Other network-related analyses include allocation, which studies the spatial distribution of resources, and location-allocation, which studies the spatial distribution of supply and demand.

Dynamic segmentation combines a network and a linear measurement system such as the milepost system to form higher-level objects. Because transportation planners and engineers use linear measures for events such as accident locations, speed limits, traffic volumes, and pavement conditions, dynamic segmentation allows the linking of those events to a road network for analysis.

1.2.6 GIS Modeling

A model is a simplified representation of a phenomenon or a system. **GIS modeling** refers to the use of GIS in building analytical models with spatial data. A very useful GIS operation for modeling is map overlay, which combines spatial and attribute data of different variables into a composite map. Because each map feature on the composite map represents a select set of data characteristics by location, the composite map can be further processed to extract new information for modeling purposes.

Analytical models differ in the degree of complexity. This book presents four types of GIS models: binary, index, regression, and process models.

A binary model turns a composite map into a binary map, separating map features that satisfy a set of selection criteria from map features that do not. An index model produces a ranked map with index values calculated using the attributes of a composite map. A regression model shows the statistical relationship between a dependent variable and independent variables from a composite map. A process model integrates existing knowledge about the environmental processes in the real world into a set of relationships and equations for quantifying the processes.

1.3 ORGANIZATION OF THIS BOOK

Chapters 2 through 8 of this book cover the basics of GIS. Chapter 2 discusses the concepts of map projection, coordinate system, and re-projection (projecting from one coordinate system to another). Common map projections such as transverse Mercator and Lambert conformal conic and coordinate systems such as the Universal Transverse Mercator (UTM) grid system and the State Plane Coordinate (SPC) system are discussed in the chapter along with their parameters.

Chapter 3 examines the vector data model, including simple geometric objects, higher-level objects, and the object-oriented data model. Spatial data concepts of map scale, accuracy, and precision are also included in the chapter. Chapter 4 describes use of existing digital data and creation of new data from satellite images, GPS readings, and paper maps for vector data input. The chapter also discusses metadata (information about data), data conversion, and geometric transformation. Chapter 5 deals with spatial data editing. Different methods for correcting topological errors and location errors are illustrated in the chapter. Continuing with the emphasis on vector data, Chapter 6 discusses attribute data input and management. The discussion includes the relational database model and its design, examples of the relational database model, and attribute data entry and verification.

Chapter 7 considers the raster data model and a large variety of data that use the raster format such as remotely sensed data, digital elevation data, scanned maps, and graphic files. The chapter also discusses different methods for raster data structure and compression, transformation of raster data, and conversion between vector and raster data. Chapter 8 focuses on data display and cartography. Data display, especially via maps, serves a range of purposes in a GIS project. Chapter 8 describes use of map symbol, color, and text in cartography as well as guidelines for map design.

Chapter 9 gives an overview of data exploration. The chapter describes different query methods that can take advantage of the capability of a Windows-based GIS in dynamically linking maps, graphs, and tables. Tools for map-based geographic visualization are also illustrated.

Chapters 10 through 13 constitute the core of data analysis. Chapter 10 discusses vector data analysis including buffering, map overlay, distance measurement, and map manipulation. Slivers and error propagation are also discussed in the context of map overlay. Raster data analysis using local, neighborhood, zonal, and global operations (including physical distance measures, cost distance measures, and spatial autocorrelation) is covered in Chapter 11. Chapter 12 examines terrain mapping and analysis using elevation grids or TINs. Different computational methods are explained, illustrated, and compared. Chapter 13 provides a summary of spatial interpolation methods including both global and local methods. A small data set is used throughout the chapter to illustrate the computing algorithm behind each of the local methods.

Chapter 14 provides an overview of GIS modeling. Many examples from different disciplines are used in the chapter to explain the basics of building binary, index, regression, and process models. Chapter 15 covers the higher-level object of regions. After discussing regions in real-world applications, the chapter describes how regions can be constructed in ARC/INFO and how regions can be used in data analysis. Chapter 16 covers network and dynamic segmentation. The first part of the chapter explains a network and its attributes and network analysis. The second part describes how the dynamic segmentation model can be built in ARC/INFO and how the model can be used for data query and analysis.

1.4 CONCEPTS AND PRACTICE

Each chapter in this book is divided into two main sections. The first section covers topics and concepts addressed in the chapter. The second section covers applications, usually with three to five problem-solving tasks. Each chapter also has boxes to provide either information relevant to topics covered in the chapter or the worked examples of the computing algorithms.

This book stresses both concept and practice. GIS concepts explain the purpose and objectives of GIS operations and the interrelationship among GIS operations. A basic understanding of map projection, for example, explains why we must project map layers to be used together to a common coordinate system and why we need to input numerous projection parameters. Our knowledge of map projection is long lasting: the knowledge will neither change with the technology nor become outdated with new updates of a GIS package.

For most GIS users, GIS is a problem-solving tool (Wright et al. 1997). To apply the tool correctly and efficiently, the GIS user must become proficient in using the tool. One constant complaint with command-driven GIS packages is that they are too difficult to learn. The graphical user interface has improved the human-computer interaction but it has also created a new problem of sorting out the functionalities of different documents, menus, buttons, and tools.

Practice, which is a regular feature in mathematics and statistics textbooks, is really the only way for us to become proficient in using the GIS tool. Practice can also help us grasp GIS concepts. For example, the root mean square (RMS) error, a measure of the goodness of control points in geometric transformation of a map or an image, is difficult to understand mathematically, but it becomes much easier to understand after a couple of runs of selecting (or digitizing) control points and executing geometric transformation. The change in RMS errors shows how the concept works.

Practice sections in a GIS textbook require data sets and GIS software. Many data sets used in this book are from GIS classes taught at the University of Idaho over the past 14 years. Most exercises in the practice sections use ArcView 3.2 and its extensions such as Spatial Analyst and Network Analyst. ARC/INFO (ArcInfo Workstation in ArcInfo 8) is used only when the functionalities to be used are not available in ArcView such as correcting topological errors, creating and analyzing regions, and using the dynamic segmentation data model. Notice that the practice sections do not use the Windows-based packages of ArcCatalog, ArcMap, and ArcToolbox in ArcInfo 8 because they cannot be used for some of the exercises such as correcting topological errors.

ArcView and ARC/INFO are among the most popular GIS packages on the market. When combined, they provide the most extensive suite of GIS functionalities. Besides adopting ArcView and ARC/INFO for the practice sections, this book also uses them, whenever necessary, to explain materials covered in each chapter. Other GIS packages are often referenced for comparison purposes.

KEY CONCEPTS AND TERMS

Attribute data: Data that describe the characteristics of spatial features.

Continuous features: Spatial features that exist between observations.

Coverage: The term used by ESRI for a vector-based digital map with topology.

Data exploration: Data-centered query and analysis.

Discrete features: Spatial features that do not exist between observations, form separate entities, and are individually distinguishable.

Dynamic segmentation: A data model that is built upon arcs of a line coverage and allows the use of real-world coordinates with linear measures.

Geographic information system (GIS): A computer system for capturing, storing, querying, analyzing, and displaying geographic data.

Geographic visualization: Use of maps for data exploration and visual information processing.

Geographically referenced data: Data that describe both the location and characteristics of spatial features on the Earth's surface.

Georelational model: A GIS package that stores spatial data and attribute data in two separate but related file systems.

GIS modeling: The process of using GIS in building models with spatial data.

Map projection: The process of transforming from the spherical geographic grid to a plane coordinate system.

Object-oriented data model: A data model that uses objects to organize spatial data. An object is an entity such as a land parcel that has a set of properties and can perform operations upon requests.

Raster data model: A spatial data model that uses a grid to represent the spatial variation of a feature.

Region: A collection of polygons, which may or may not be connected.

Relational database: A collection of tables, which can be connected to each other by attributes whose values can uniquely identify a record in a table.

Shapefile: The term used by ESRI for non-topological vector data.

Spatial data: Data that describe the geometry of spatial features.

Spatial interpolation: A process of using control points with known values to estimate values at other points.

Topology: A sub-field of mathematics that is applied in GIS to ensure that the spatial relationships between features are expressed explicitly.

Triangulated irregular network (TIN): A data model that approximates the terrain with a set of non-overlapping triangles.

Vector data model: A spatial data model that uses points and their x-, y-coordinates to construct spatial features of points, lines, and areas.

APPLICATIONS: INTRODUCTION

This applications section covers two tasks. Task 1 introduces ArcView and its use, and Task 2 ARC/INFO and its use.

Task 1: Introduction to ArcView

What you need: *emidalat*, an elevation grid; *emidastrm.shp*, a stream shapefile.

Task 1 is a simple exercise of adding an elevation grid and a shapefile to view. Because an elevation grid contains raster data, you need to use the Spatial Analyst extension to display it.

1. Start ArcView. ArcView creates an empty Project and opens the Project window. To load the Spatial Analyst extension, click File

and select Extensions. The Extensions dialog lists the extensions available on the computer, each with a check box. Click on Spatial Analyst, and a short message appears in the box below describing the function of Spatial Analyst. Check Spatial Analyst, and click OK.

2. ArcView organizes its objects in a hierarchical structure. A Project has access to the five documents of Views, Tables, Charts, Layouts, and Scripts. Views display maps, Tables display tabular data, Charts display data in charts and diagrams, Layouts are used to prepare map and data presentations, and Scripts work with macros (programs) written in Avenue. Each document has its own window(s). Task 1 lets you work with Views and Tables. Click on Views in the Project window, if it is not already highlighted, and then New. This opens a view window titled View1.

3. Each document in ArcView has its own menus, buttons, and tools, arranged in three rows at top of the window. They provide the user interface to ArcView. Buttons and tools are graphic icons, representing the commonly used menu items. A button does something as soon as you click it. A tool stays depressed when it is clicked, and nothing happens until you click or do something in the active document. When you hold the mouse point over a graphic icon, a short message called ToolTip appears in a floating yellow box to tell you the function of the icon.

4. Click the View menu, and a pull-down menu appears with menu items. Some items are clear, meaning that they are active, while others are fuzzy, meaning that they are not available. Select Add Theme from the View menu, or click on the Add Theme button. The Add Theme dialog lets you choose a theme or themes to add to view. ArcView uses the term theme for a set of map features linked to their attributes. *Emidalat* is a grid theme, and *emidastrm.shp* is a feature theme.

5. The Add Theme dialog shows Drives, Directory, and Data Source Types. First, use Drives and Directory to navigate to *emidalat* and *emidastrm.shp*. Then select Grid Data Source. Double click on *emidalat* to add it to view. Click on the Add Theme button again. This time select Feature Data Source and add *emidastrm.shp* to view.

6. *Emidastrm.shp* and *emidalat* now appear in the Table of Contents of the View1 window. To display them, check the boxes next to *emidastrm.shp* and *emidalat*. ArcView places the last theme added to view at the top of the Table of Contents. You can change the order by dragging a theme to a new position and dropping it.

7. Make *emidastrm.shp* active by clicking it in the Table of Contents. An active theme has a raised appearance. Select Table from the Theme menu. This opens a table called Attributes of *emidastrm.shp*, which shows attributes associated with *emidastrm.shp*. (*Emidalat* does not have a theme table; therefore, the menu item of Table is fuzzy.)

8. To exit ArcView, first close all documents and the project. Select Close All from the File menu to close all documents. Select Close Project from the File menu, and choose not to save changes made to the project. Then select Exit from the File menu to exit ArcView.

Task 2: Introduction to ARC/INFO (ArcInfo Workstation in ArcInfo 8)

What you need: *emidalat*, an elevation grid; *breakstrm*, a stream coverage.

As explained in the chapter, the practice sections of this book only use ArcInfo Workstation in ArcInfo 8, which is the same as ARC/INFO version 7.x and can operate on Windows 2000, Windows NT, or UNIX. The command-driven ARC/ INFO organizes commands into modules such as Arc, ArcPlot, ArcEdit, and Tables. Arc is the start up module; ArcPlot displays coverages, grids, and TINs; ArcEdit includes editing functions; and Tables works with tables or INFO files. While working in ARC/INFO, you need to move between modules, depending on the task you do.

Task 2 uses ARC/INFO to display *emidalat* and *breakstrm*. *Breakstrm* shows the same streams as *emidastrm.shp* for Task 1 except that *breakstrm* is an ARC/INFO coverage.

1. Start ARC/INFO or ArcInfo Workstation in ArcInfo 8. After the Arc prompt appears, use the WORKSPACE (or W) command to change the directory to where *emidalat* and *breakstrm* reside.
2. First, you need to go to Arcplot and set up the environment to display *emidalat* and *breakstrm*:
   Arc: arcplot
   Arcplot: display 9999 /*define the computer monitor as the display device
   Arcplot: mapextent emidalat /*define the map extent of *emidalat*

Note: To explain what a command does, this book includes the explanation following /.*

3. The next command display *emidalat* by using an item (value for this example) and a look-up table (emidalat.lut):
   Arcplot: gridpaint emidalat value emidalat.lut /*display *emidalat* by elevation zone
   Arcplot: arclines breakstrm 3 /*display *breakstrm* in green
   Note: A look-up table like emidalat.lut specifies the symbol for each elevation zone.
4. Use QUIT to exit Arcplot and Arc:
   Arcplot: quit
   Arc: quit

REFERENCES

Broome, F.R., and D.B. Meixler. 1990. The TIGER Data Base Structure. *Cartography and Geographic Information Systems* 17: 39–47.

Chrisman, N. 1988. The Risks of Software Innovation: A Case Study of the Harvard Lab. *The American Cartographer* 15: 291–300.

Coppock, J.T. 1988. The Analogue to Digital Revolution: A View from an Unreconstructed Geographer. *The American Cartographer* 15: 263–75.

Crockett, M. 1997. GIS Companies Race for Market Share. *GIS World* 10 (11): 54–57.

Dent, B.D. 1999. *Cartography: Thematic Map Design*, 5th ed. Dubuque, Iowa: WCB/McGraw-Hill.

McHarg, I.L. 1969. *Design with Nature.* New York: Natural History Press.

Rhind, D. 1988. Personality as a Factor in the Development of a Discipline: The Example of Computer-Assisted Cartography. *The American Cartographer* 15: 277–89.

Tomlinson, R.F. 1984. Geographic Information Systems: The New Frontier. *The Operational Geographer* 5:31–35.

Wright, D.J., M.F. Goodchild, and J.D. Proctor. 1997. Demystifying the Persistent Ambiguity of GIS as "Tool" versus "Science." *Annals of the Association of American Geographers* 87: 346–62.

MAP PROJECTION AND COORDINATE SYSTEM

2.1 INTRODUCTION

GIS users work with map features on a plane surface. These map features represent spatial features on the Earth's surface. The locations of map features are based on a coordinate system, whereas the locations of spatial features are based on the geographic grid expressed in longitude and latitude values. The transformation from the geographic grid to a coordinate system is referred to as **map projection**, which is the topic of this chapter.

A basic principle in GIS is that map layers to be used together must be based on the same coordinate system. The practice of this principle, however, is complicated by the fact that hundreds of coordinate systems are used in mapmaking and by different GIS data producers. For example, Figure 2.1 shows the road maps of Idaho and Montana downloaded from the Internet. The road maps

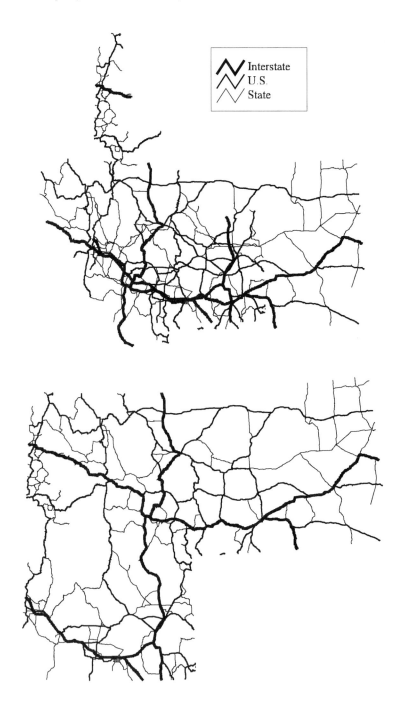

Figure 2.1
The top map shows two road coverages of Idaho and Montana based on different coordinate systems. The bottom map shows the road coverages based on the same coordinate system.

obviously cannot be used together because they are based on different coordinate systems, which result in the overlapping of the road networks. To make them usable, these two road maps must be converted to a common coordinate system.

Increasingly, GIS users download digital maps from the Internet, or get them from government agencies and private companies. Some digital maps are measured in longitude and latitude values, while others are in different coordinate systems from the one intended for the GIS project. Invariably these digital maps must be processed before they can be used together.

Processing in this case implies projection and re-projection. Projection means converting digital maps from longitude and latitude values to two-dimensional coordinates, and **re-projection** means converting from one coordinate system to another. Typically, projection and re-projection are among the initial tasks performed in a GIS project.

This chapter is divided into four sections. Section 1 describes the geographic grid. Section 2 discusses map projection and parameters required for different map projections. Section 3 covers coordinate systems, which are based on map projections. Section 4 uses ArcView as an example and discusses the projection function in a GIS package.

2.2 GEOGRAPHIC GRID

The **geographic grid** is the location reference system for spatial features on the Earth's surface (Figure 2.2). The geographic grid consists of meridians and parallels. **Meridians** are lines of longitude for the E-W direction. The Earth makes a complete, 360° rotation on its axis about every 24 hours. Using the meridian passing through Greenwich, England as the prime meridian or 0°, the longitude value of a point on the Earth's surface can be measured as 0° to 180°, east or west of the prime meridian. **Parallels** are lines of latitude for the N–S direction. Using the equator as 0° latitude, the latitude value of a point can be measured as 0° to 90° north or south of the equator.

Although applied to the spherical Earth's surface, the geographic grid is similar to a plane coordinate system. The origin of the geographic grid is where the prime meridian meets the equator. Longitude values are similar to x values in a coordinate

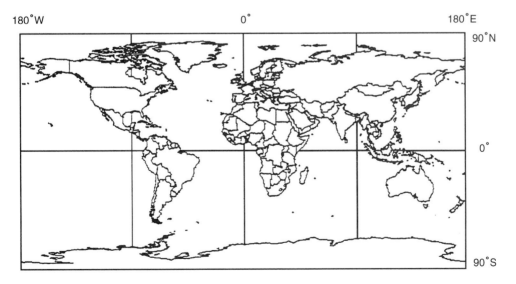

Figure 2.2
The geographic grid.

system and latitude values are similar to *y* values. Therefore, longitude and latitude values are placed in the *x* and *y* columns respectively.

In GIS, it is conventional to enter longitude and latitude values with positive (default) or negative signs. Latitude values are positive if north of the equator, and negative if south of the equator. Longitude values are positive in the eastern hemisphere and negative in the western hemisphere. Longitude and latitude values may be measured in **degree-minute-second (DMS)**, or **decimal degrees (DD)** system. One degree equals 60 minutes and one minute equals 60 seconds. Conversion between the two systems can be cumbersome. For example, a latitude value of 45°52'30" would be equal to 45.875°.

2.3 MAP PROJECTION

Map projection is the transformation of the spherical Earth's surface to a plane surface (Robinson et al. 1995; Dent 1999). The outcome of the transformation is a systematic construction of lines on a plane surface representing the geographic grid. Besides allowing the use of paper maps and digital maps, map projection enables map users to work with two-dimensional coordinates, rather than spherical or three-dimensional coordinates, which are much more complex for measurements and calculation (Box 2.1). But the transformation from the Earth's surface to a flat surface always involves

distortion and no map projection is perfect. This is why hundreds of map projections have been developed for mapmaking (Maling 1992; Snyder 1993). Every map projection preserves certain spatial properties while sacrificing other properties.

Cartographers typically group map projections into four classes by their preserved properties: conformal, equal area or equivalent, equidistant, and azimuthal or true direction. A **conformal** projection preserves local shapes. An **equivalent** projection represents areas in correct relative size. An **equidistant** projection maintains consistency of scale for certain distances. And an **azimuthal** projection retains certain accurate directions. The conformal and equivalent properties are mutually exclusive. Otherwise a map projection can have more than one preserved property such as conformal and azimuthal. The conformal and equivalent properties are global properties, meaning that they apply to the entire map projection. The equidistant and azimuthal properties are local properties and may be true only from or to the center of the map projection. The preserved property of a map projection is usually included in its name such as Lambert conformal conic or Albers equal-area conic.

Cartographers often use a geometric object to illustrate how a map projection can be constructed (Figure 2.3). For example, by placing a cylinder tangent to a lighted globe, a projection can be made by tracing the lines of longitude and latitude

---

Box 2.1   **How To Measure Distances on A Spherical Surface**

T he method for calculating distances between two points on a spherical surface uses trigonometric functions:

$$\cos D = \sin a \sin b + \cos a \cos b \cos c$$

where *D* is the distance between points A and B in degrees, *a* is the latitude of A, *b* is the latitude of B, and *c* is the difference in longitude between A and B. To convert *D* to a linear distance measure, one can

multiply *D* by the length of one degree at the equator, which is 111.32 kilometers or 69.17 miles.

The above method is more difficult than the method for measuring distances on a coordinate system, which uses the equation:

$$D = [(x_1 - x_2)^2 + (y_1 - y_2)^2]^{1/2}$$

where x_i and y_i are the coordinates of point *i*.

onto the cylinder. The cylinder in this case is the projection surface, and the globe is called the **reference globe**. Other common projection surfaces include a cone and a plane. A map projection is called a **cylindrical** projection if it can be constructed using a cylinder, a **conic** projection using a cone, and an **azimuthal** using a plane.

The use of a geometric object can also help explain two other concepts in map projection: case and aspect. Take the example of a conic projection: the cone can be placed so that it is tangent to the globe, or it can intersect the globe (Figure 2.3). The first is the simple case, which results in one line of tangency, and the second is the secant case, which results in two lines of tangency. A cylindrical projection behaves the same way as a conic projection in terms of case. An azimuthal projection, in contrast, has a point of tangency in the simple case and a line of tangency in the secant case. Aspect describes the placement of a geometric ob-

ject relative to a globe. A plane, for example, may be tangent at any point on a globe (Figure 2.4). A polar aspect refers to tangency at the pole, an equatorial aspect at the equator, and an oblique aspect anywhere between the equator and the pole.

The concept of case relates directly to the **standard line** in map projection, which refers to the line of tangency between the projection surface and the reference globe. For cylindrical and conic projections the simple case has one standard line whereas the secant case has two standard lines. The standard line is called the **standard parallel** if it follows a parallel, and the **standard meridian** if it follows a meridian. There is no projection distortion along the standard line because it has the same scale as that of the reference globe. Another way to describe the standard line is that it has the scale factor of 1. The **scale factor** is defined as the ratio of the local scale to the scale of the reference globe or the **principal scale**. Because a scale factor of 1 means that the standard line has the same scale as the scale of the reference globe, the standard parallel is sometimes called the latitude of true scale. The degree of projection distortion increases away from the standard line.

The standard line should not be confused with the **central line**. Whereas the standard line dictates the distribution pattern of projection distortion, the central lines (the central parallel and meridian) define the center or the origin of a map projection. The central parallel, sometimes called the latitude of origin, often differs from the standard parallel. Likewise, the central meridian often differs from the standard meridian. A good example showing the difference between the central meridian and the standard line is the transverse Mercator projection.

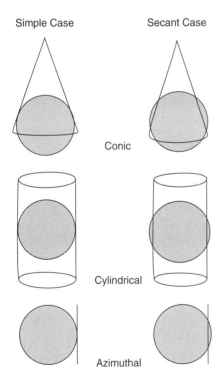

Figure 2.3
Use of a geometric object to construct a map projection.

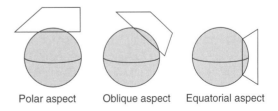

Polar aspect   Oblique aspect   Equatorial aspect

Figure 2.4
Aspect and map projection.

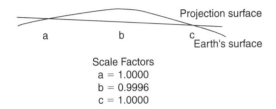

Scale Factors
a = 1.0000
b = 0.9996
c = 1.0000

Figure 2.5
The central meridian in this secant case transverse Mercator projection has a scale factor of 0.9996. The two standard lines on either side of the central meridian have a scale factor of 1.

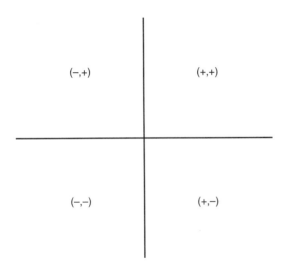

Figure 2.6
The central parallel and the central meridian divides a map projection into four quadrants. Points within the NE quadrant have positive x- and y-coordinates, points within the NW quadrant have negative x-coordinates and positive y-coordinates, points within the SE quadrant have positive x-coordinates and negative y-coordinates, and points within the SW quadrant have negative x- and y-coordinates.

Normally a secant projection, a transverse Mercator projection is defined by its central meridian and two standard lines on either side. The standard line has a scale factor of 1, and the central meridian has a scale factor of less than 1 (Figure 2.5).

A map projection can be used directly as a coordinate system in a GIS project. The center of the map projection, as defined by the central parallel and the central meridian, becomes the origin of the coordinate system and divides the coordinate system into four quadrants. The x-, y-coordinates of a point are either positive or negative, depending on where the point is located (Figure 2.6). To avoid having negative coordinates, GIS users can apply a **false easting** for x and a **false northing** for y to the origin of the map projection. Essentially, the false easting and false northing move the origin of the coordinate system to its SW corner so that all points will fall within the NE quadrant and have positive coordinates. False easting and false northing are optional parameters in map projection.

x-, y-coordinates based on a false origin often have very large numbers. For example, the NW corner of the Moscow East, Idaho quadrangle map (quad) has the UTM coordinates of 500,000 and 5,177,164 meters. To preserve data precision for computations with coordinates, **x-shift** and **y-shift** values can be applied to all coordinate readings to reduce the number of digits. Therefore, if the x-shift value is set as −500,000 meters and the y-shift value as −5,170,000 meters for the quad, the coordinates for its NW corner will be changed to 0 and 7164 meters. Because x-shift and y-shift, or false easting and false northing, change the values of x-, y-coordinates in a digital map, these changes should be documented in the metadata (information about data), especially if the map is to be shared with other users.

Finally, to help users choose from among hundreds of map projections, cartographers sometimes group map projections by how well they can be used to map the world, a hemisphere, a continent, a country, or a region. ArcView, for example, uses this approach with the predefined map projections.

2.3.1 Commonly Used Map Projections

2.3.1.1 Transverse Mercator
The **transverse Mercator** projection is a variation of the Mercator projection, probably the best-known projection of the world. Instead of using the standard parallel as in the case of the Mercator

Figure 2.7
The Lambert conformal conic projection of the United States. The central meridian is 96°W, the two standard parallels are 20°N and 60°N, and the latitude of projection's origin is 40°N.

Figure 2.8
The equidistance conic projection of the United States. The projection parameter values are the same as the Lambert conformal conic projection in Figure 2.7. The difference between the two projections does not show up at the printed scale.

projection, the transverse Mercator projection uses the standard meridian. As discussed later, the transverse Mercator is the basis for two common coordinate systems. The projection requires the following parameters: scale factor at central meridian, longitude of central meridian, latitude of origin (or central parallel), false easting, and false northing.

2.3.1.2 Lambert Conformal Conic

The **Lambert conformal conic** projection is a good choice for a mid-latitude area of greater east-west than north-south extent, such as the conterminous United States or the state of Montana (Figure 2.7). Typically used as a secant projection, the projection includes the parameters of the first and second standard parallels, central meridian, latitude of projection's origin, false easting, and false northing.

2.3.1.3 Albers Equal-Area Conic

The Albers equal-area conic projection requires the same parameters as the Lambert conformal

conic projection. The two projections in fact look similar except that one is equal area and the other is conformal.

2.3.1.4 Equidistant Conic

The equidistant conic projection is also called the simple conic projection. The projection preserves the distance property along all meridians and one or two standard parallels. It uses the same parameters as the Albers equal-area conic projection (Figure 2.8).

2.3.2 Datum

So far we have examined different map projections and their required parameters. We need to work with two other parameters in map projection: **spheroid** and **datum**. A spheroid is a model that approximates the Earth. Because the Earth is wider along the equator, a spheroid has its major axis along the equator and its minor axis connecting the poles. A spheroid is also called an **ellipsoid**, an

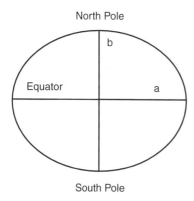

North Pole

Equator

South Pole

Figure 2.9

The semi-major axis is represented by *a* and the semi-minor axis is represented by *b*.

ellipse rotated about its minor (polar) axis. To uniquely describe the geographic coordinates of a location, the relationship between the Earth and the spheroid must be defined through a datum (Moffitt and Bossler 1998). In other words, the datum will change whenever a different spheroid is adopted to approximate the Earth.

Until recently, **Clarke 1866**, a ground-measured spheroid, was the standard spheroid for mapping in the United States. Clarke 1866's semi-major axis (equatorial radius) and semi-minor axis (polar radius) measure 6,378,206.4 meters and 6,356,583.8 meters respectively (Figure 2.9). The difference between the axes, although small on the global scale, affects large-scale mapping. **NAD27** (North American Datum of 1927) is based on the Clarke 1866 spheroid and has its center at Meades Ranch, Kansas.

Satellite-determined spheroids, however, have recently replaced ground-measured spheroids. **WGS84** (World Geodetic System 1984) is a spheroid determined from satellite orbital data. Its semi-major axis measures 6,378,137.0 meters and semi-minor axis, 6,356,752.3 meters. WGS84 is identical to the **GRS80** (Geodetic Reference System 1980) spheroid for mapping concerns. **NAD83** (North American Datum of 1983) is based on the WGS84, or the GRS80, spheroid, and is measured from the center of the spheroid.

GIS data providers in the United States are switching from NAD27 to NAD83. Not only is NAD83 more accurate, but it is also tied into a global network and GPS (Global Positioning System) measurements. The horizontal shift between NAD27 and NAD83 can be substantial (Figure 2.10). Positions of points can change between 10 and 100 meters in the conterminous United States, more than 200 meters in Alaska, and in excess of 400 meters in Hawaii. For example, for the Ozette quad from the Olympic Peninsula in Washington, the shift is 98 meters to the east and 26 meters to the north. Until the switch from NAD27 to NAD83 is complete, GIS users must keep watchful eyes on the datum because digital maps based on different datums will not register correctly.

2.4 COORDINATE SYSTEMS

Map projections provide the base map for small-scale mapping such as mapping of the world or a continent at the 1:1,000,000-scale or smaller. Because small-scale mapping accentuates map projection distortion, one should select map projections for their preserved properties. For example, to give the correct impression of population pressure, a population map of the world should be based on an equivalent projection, rather than a conformal projection.

Plane coordinate systems are typically used in large-scale mapping such as at a scale of 1:24,000 or larger. Coordinate systems are designed for detailed calculations and positioning. Therefore, accuracy in a feature's absolute position and its relative position to other features is more important than the preserved property of a map projection. To maintain the level of accuracy desired for measurements, a coordinate system is often divided into different zones, with each zone based on a separate map projection. Four coordinate systems are commonly used in the U.S.: the Universal Transverse Mercator (UTM) grid system, the Universal Polar Stereographic (UPS) grid system, the State Plane Coordinate (SPC) system, and the Public Land Survey System (PLSS).

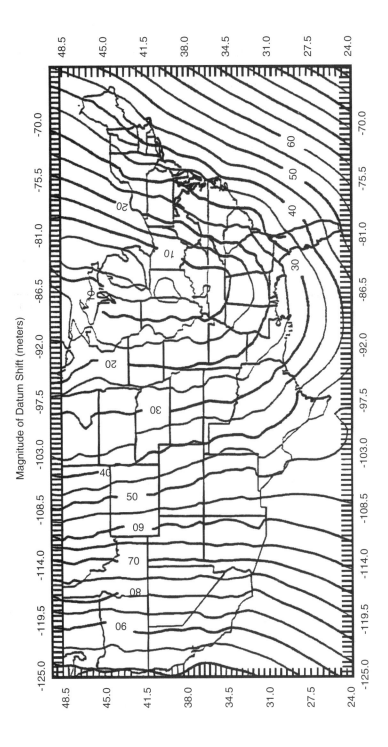

Figure 2.10
Magnitude of the horizontal shift from NAD27 to NAD83 in meters. (By permission of the National Geodetic Survey.)

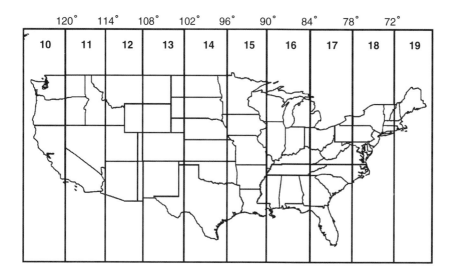

Figure 2.11
UTM zones in the conterminous United States.

2.4.1 The UTM Grid System

Used worldwide, the **UTM** grid system divides the Earth's suface between 84°N and 80°S into 60 zones. Each zone covers 6° of longitude, and is numbered sequentially with zone 1 beginning at 180°W. Figure 2.11 shows the UTM zones in the conterminous United States.

Each of the 60 UTM zones is mapped onto a transverse Mercator projection, with a scale factor of 0.9996 at the central meridian. The standard meridians are 180 kilometers to the east and the west of the central meridian. A false origin is assigned to each UTM zone. In the Northern Hemisphere, UTM coordinates are measured from a false origin at the equator and 500,000 meters west of the central meridian. In the Southern Hemisphere, UTM coordinates are measured from a false origin located at 10,000,000 meters south of the equator and 500,000 meters west of the central meridian. The UTM grid system maintains the accuracy of at least one part in 2500 (i.e., distance measured over a 2500-meter course on the UTM grid system would be accurate within a meter of the true measure).

2.4.2 The UPS Grid System

The **UPS** grid system covers the polar areas. The stereographic projection is centered on the pole and is used for dividing the polar area into a series of 100,000-meter squares, similar to the UTM grid system.

2.4.3 The SPC System

The **SPC** system was developed in the 1930s to permanently record original land survey monument locations in the United States. To maintain the required accuracy of one part in 10,000 or less, a state may have two or more SPC zones (Figure 2.12). As examples, Oregon has the North and South SPC zones and Idaho has the West, Central, and East SPC zones. Each SPC zone is mapped onto a map projection. The transverse Mercator projection is used for zones that are elongated in the north-south direction (e.g., Idaho's SPC zones) and the Lambert conformal conic projection for zones that are elongated in the east-west direction (e.g., Oregon's SPC zones). Point locations within each SPC zone are measured in feet from a false origin.

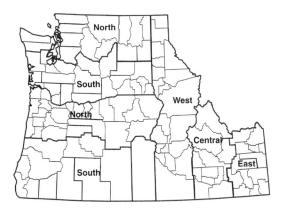

Figure 2.12
State Plane Coordinate Systems for Washington, Oregon, and Idaho. The Lambert conformal conic projection is the basis for the Washington and Oregon coordinate systems, whereas the transverse Mercator projection is the basis for the Idaho coordinate system.

Because of the switch from NAD27 to NAD83, there are SPC27 and SPC83. Besides the change of the datum, SPC83 has a few other changes. SPC83 coordinates are published in meters instead of feet. The states of Montana, Nebraska, and South Carolina have each replaced multiple zones by a single SPC zone. California has reduced from seven to six SPC zones. And Michigan has changed from transverse Mercator to Lambert conformal conic projections.

Some states in the United States have developed their own statewide coordinate system to maintain a desired level of accuracy. Idaho, for example, is divided into UTM zones 11 and 12 nearly in the center of the state. This division presents problems to GIS users in Idaho when their study area covers both zones. Because they must convert their database to a single zone so that it can be registered spatially, they cannot maintain the accuracy level designed for the UTM coordinate system. In 1994, the Idaho Geographic Information Advisory Committee adopted a statewide coordinate system (Box 2.2). The new system is still based on a transverse Mercator projection but its central meridian passes through the center of the state (114°W). Changing the location of the central meridian allows the Idaho system to use one zone for the entire state instead of two.

2.4.4 PLSS

The **PLSS** is a land partitioning system. Using the intersecting township and range lines, the system divided the lands mainly in the central and western states into 6 x 6 mile squares or townships. Each township was further partitioned into 36 square-mile parcels of 640 acres, called sections. A recent development in GIS is to use the PLSS for creation of the land parcel layer. The Bureau of Land Management (BLM) has been developing a Geographic Coordinate Data Base (**GCDB**) of the PLSS for the western United States (http://www.blm.gov/gcdb/). The GCDB contains longitude and latitude values and other descriptive information for section corners and monuments

---

Box **2.2**   **Parameters for the Idaho Statewide Transverse Mercator Coordinate System**

Based on a transverse Mercator projection, the Idaho statewide coordinate system has the following technical parameters:

- Measurement unit: Meter
- Central meridian: -114°

- Central meridian scale factor: 0.9996
- Datum: NAD27 (until NAD83 is adopted)
- Latitude of origin: 42°
- False northing: 100,000 meters
- False easting: 500,000 meters

recorded in the PLSS. Legal descriptions of a parcel layer can then be entered using, for example, bearing and distance readings originating from section corners.

2.5 PROJECTION IN ARCVIEW

All GIS packages have functions for projection and re-projection, although the way these functions are set up varies. Here we look at ArcView as an example. ArcView offers two methods for projection: Projection and Projection Utility. Projection is limited to projecting from geographic coordinates (i.e., longitude and latitude values) in decimal degrees. Projection Utility is a wizard-

based utility, which can project shapefiles from geographic coordinates to a coordinate system, or re-project one coordinate system to another, or convert from one datum to another. Projection Utility also allows the projected or re-projected shapefiles to be saved for future use.

Both Projection and Projection Utilities have the predefined or custom options for the projection method. The predefined systems cover the commonly used coordinate systems such as State Plane–1927, State Plane–1983, UTM, and national grids. Each category has a select list, and each projection on the list is predefined with its ellipsoid and parameter values. The custom options require the user to input the coordinate system and its parameter values.

KEY CONCEPTS AND TERMS

Azimuthal projection: One type of map projection that retains certain accurate directions. Azimuthal projection also refers to one type of map projection that uses a plane as the projection surface.

Central lines: The central parallel and the central meridian. Together, they define the center or the origin of a map projection.

Clarke 1866 spheroid: A ground-measured spheroid, which is the basis for North American Datum of 1927 (NAD27).

Conformal projection: One type of map projection that preserves local shapes.

Conic projection: One type of map projection that uses a cone as the projection surface.

Cylindrical projection: One type of map projection that uses a cylinder as the projection surface.

Datum: The basis for a coordinate system. A datum is derived from a spheroid.

Decimal degrees (DD) system: A measurement system for longitude and latitude values such as 42.5°.

Degree-minute-second (DMS) system: A measuring system for longitude and latitude values

such as 42°30'00". One degree equals 60 minutes and one minute equals 60 seconds.

Ellipsoid: A model that approximates the Earth. Also called *spheroid*.

Equidistant projection: One type of map projection that maintains consistency of scale for certain distances.

Equivalent projection: One type of map projection that represents areas in correct relative size.

False easting: A value applied to the origin of the map projection to change the x-coordinate readings.

False northing: A value applied to the origin of the map projection to change the y-coordinate readings.

Geographic Coordinate Data Base (GCDB): A database that is being developed by the U.S. Bureau of Land Management (BLM) to include longitude and latitude values and other descriptive information for section corners and monuments recorded in the PLSS.

Geographic grid: The location reference system for spatial features on the Earth's surface.

GRS80 spheroid: Geodetic Reference System reference spheroid of 1980, which is a satellite-determined spheroid.

Lambert conformal conic projection: A common map projection, which is the basis for the State Plane Coordinate (SPC) system for many states.

Map projection: The process of transforming the spatial relationship of map features on the Earth's surface to a flat map.

Meridians: Lines of longitude on the geographic grid for the E-W direction.

NAD27: North American Datum of 1927, which is based on the Clarke 1866 spheroid and has its center at Meades Ranch, Kansas.

NAD83: North American Datum of 1983, which is based on the WGS84, or the GRS80, spheroid, and is measured from the center of the spheroid.

Parallels: Lines of latitude on the geographic grid for the N-S direction.

Principal scale: Same as the scale of the reference globe.

Public Land Survey System (PLSS): A land partitioning system used in the United States.

Reference globe: The reduced model of the Earth from which map projections are constructed. Also called a *nominal* or *generating globe.*

Re-projection: Projection of spatial data from one coordinate system to another.

Scale factor: Ratio of the local scale to the scale of the reference globe. The scale factor is 1.0 along a standard line.

Spheroid: A model that approximates the Earth. Also called *ellipsoid.*

Standard line: Line of tangency between the projection surface and the reference globe. A standard line has no projection distortion and has the same scale as that of the reference globe.

Standard meridian: A standard line that follows a meridian or a line of longitude.

Standard parallel: A standard line that follows a parallel or a line of latitude.

State Plane Coordinate (SPC) system: A coordinate system developed in the 1930s to permanently record original land survey monument locations in the United States. Most states have more than one zone based on the SPC27 or SPC83 system.

Transverse Mercator projection: A common map projection, which is the basis for the Universal Transverse Mercator (UTM) grid system and the State Plane Coordinate (SPC) system for many states.

UPS grid system: Universal Polar Stereographic grid system, which divides the polar area into a series of 100,000-meter squares, similar to the UTM grid system.

UTM coordinate system: Universal Transverse Mercator coordinate system, which divides the Earth's surface between 84°N and 80°S into 60 zones, with each zone covering 6° of longitude.

WGS84 spheroid: World Geodetic System reference spheroid of 1984, which is a satellite-determined spheroid.

x-shift: A value applied to x-coordinate readings to reduce the number of digits.

y-shift: A value applied to y-coordinate readings to reduce the number of digits.

APPLICATIONS: PROJECTION AND COORDINATE SYSTEM

This applications section consists of three tasks. Task 1 uses ArcView to project two maps, both measured in longitude and latitude values and in decimal degrees, into real-world coordinates. ArcView's Projection Utility Wizard is used for Tasks 2 and 3. Task 2 lets you create a shapefile from a text file containing longitude and latitude values and then project the shapefile to real-world coordinates. Task 3 asks you to re-project a shapefile from one coordinate system to another.

Task 1: Use Projection in ArcView

What you need: *stationsll.shp* and *idll.shp*, two shapefiles measured in longitude and latitude values and in decimal degrees. *Stationsll.shp* contains snow courses, and *idll.shp* is an outline map of Idaho.

In this task, you will project *stationsll.shp* and *idll.shp* to the Idaho Transverse Mercator coordinate system (IDTM). Because IDTM is not one of the predefined systems, you need to choose the custom option. The custom option lists 18 systems and 12 ellipsoids. Each of the custom systems requires the user to input a set of parameters. IDTM has the following parameters:

Projection Transverse Mercator
Datum NAD27 (based on the Clarke 1866 ellipsoid)
Units meters
Parameters
  scale factor: 0.9996
  central meridian: -114.0
  reference latitude: 42.0
  false easting: 500000
  false northing: 100000

1. Start ArcView, and open a new view. Add *stationll.shp* and *idll.shp* to view.
2. Select Properties from the View menu. Select meters as map units in the View Properties dialog and click on Projection.
3. In the Projection Properties menu, click on Custom and set the projection properties according to those of IDTM. Click OK in the Projection Properties menu and the View Properties menu. The themes are now projected into IDTM.

Task 2: Create a shapefile from a text file and project the shapefile using ArcView's Projection Utility

What you need: ArcView 3.2 with Projection Utility; *snow.txt*, a text file containing the longitude and latitude values of 40 snow courses in Idaho; *stations.shp*, a projected shapefile to verify the result of Task 2.

Task 2 shows you how you can create a shapefile from a delimited text file and then project the shapefile from the geographic coordinate system to the Lambert conformal conic projection. ArcView's Projection Utility involves four steps, and each step comes with a dialog for you to input data.

1. Start ArcView, and load the extension of Projection Utility Wizard.
2. Open a new view. To create a shapefile from *snow.txt*, click on Tables in the Project window and then Add. In the Add Table dialog, first change the file type to Delimited Text, select *snow.txt*, and click OK. The file shows the longitude and latitude values, in decimal degrees, of 40 snow courses.
3. Activate the view window. Select Add Event Theme from the View menu. In the Add Event Theme dialog, specify Longitude as the *x* field and Latitude as the *y* field. Click OK. A new theme called *snow.txt* is now added to the Table of Contents. Check the box next to *snow.txt* to view the event theme.
4. To use ArcView's Projection Utility, the event theme must be converted to a shapefile first. Make the event theme active, and select Convert to Shapefile from the Theme menu. Name the shapefile *trial.shp* and add it to view. Activate *trial.shp*.
5. *Trial.shp* is measured in longitude and latitude values. The next step is to project *trial.shp* to the Lambert conformal conic projection, with the following parameters:
Units: meters
Datum: NAD27
Ellipsoid: Clarke1866
Central meridian: -114
Latitude of origin: 42
1st standard parallel: 33
2nd standard parallel: 45
6. Select ArcView Projection Utility from the File menu. After the wizard is loaded, *trail.shp* should appear in the Step 1 dialog; if not, click Browse to navigate to *trial.shp*. Highlight *trial.shp* and click Next.
7. You need to define the coordinate system for *trial.shp* in Step 2. Make sure the box next to Show Advanced Option is checked. The

definition consists of four parts: Name, Parameters, Datum, and Ellipsoid. Start with the name. Select Geographic as the Coordinate System Type, GCS_North_ American_1927 as the Name, and Degree as the Units. Now click Parameters, and the dialog should show the Name as GCS_North_American_1927. Next click Datum. The Datum dialog should show the Name as D_North_American_1927; you do not have to input or change any values in the dialog. Click Ellipsoid, and the dialog should show the Name as Clarke_1866. Click Next and proceed to Step 3. If asked, press Yes to save the input coordinate system information. In Step 3, you will define the coordinate system for the new shapefile, which is the Lambert conformal conic. Again, you need to go through the definition in four parts. First is the Name. Select Projected as the Coordinate System Type. Because the Lambert conformal conic is not one of the predefined projections, you will select Custom (at the bottom of the scroll list) for the Name. Choose Meter for the Units. Now click Parameters. In the Parameters dialog, first select GCS_North_American_1927 as the Geographic Coordinate System. Then select Lambert_Conformal_Conic as the Base Projection, and enter −114, 42, 33, and 45 for the Central_Meridian, Central_Parallel, Standard_Parallel_1, and Standard_Parallel_2 respectively. Click Next and proceed to Step 4.

8. Step 4 lets you name the new shapefile and its path. Name the new shapefile *trial2.shp*. Then click Next.

9. ArcView takes the information you have provided above and displays a summary sheet. If you find any mistakes, you can click Back to correct the mistakes; otherwise, click Finish.

10. ArcView is now projecting *trial.shp* to *trial2.shp*. A progress dialog will pop up, and the projection will take a while. When it is completed, click OK in the Complete dialog.

You can then add *trial2.shp* to view. To check if *trial2.shp* is correctly projected, you can add *stations.shp* to view. The two shapefiles, *trail2.shp* and *stations.shp*, should be exactly the same.

11. After completing the projection, ArcView's Projection Utility creates a project file called *trial2.prj*. You can open *trial2.prj*, a text file, and read the parameter values of the projection.

Task 3: Use ArcView Projection Utility to re-project a shapefile

What you need: ArcView 3.2 with Projection Utility, *idoutl.shp*, and *stations.shp*, which is the same as *trial2.shp* from Task 2.

Task 3 lets you work with ArcView's Projection Utility one more time. This time you will re-project *stations.shp*, which is based on the Lambert conformal conic projection, to the Idaho transverse Mercator coordinate system. The parameter values for the two coordinate systems are the same as in Tasks 1 and 2.

1. Start ArcView and load Projection Utility Wizard.

2. Open a new view and add *stations.shp* to view.

3. Select ArcView Projection Utility from the File menu. You should see *stations.shp* as the shapefile to be re-projected in the Step 1 dialog. If not, click on Browse and find the path to *stations.shp*. Click Next.

4. The coordinate system of *stations.shp* is defined in Step 2. Because *stations.prj* already has the parameter values of the coordinate system, all you have to do is go through the Name, Parameters, Datum, and Ellipsoid and make sure the definition is correct. When you are ready, click Next to go to Step 3.

5. The Step 3 dialog asks you to define the new coordinate system that *stations.shp* will be converted to. Start with the Name: select Projected for the Coordinate System Type, Custom for the Name, and Meter for the

Units. Next is the Parameters dialog. For Base Projection, choose Transverse_Mercator and enter −114, 42, and 0.9996 for Central_Meridian, Central_Parallel, and Scale_Factor respectively. Then select GCS_North_American_1927 for the Geographic Coordinate System, and set False Easting as 500000 and False Northing as 100000. Make sure the Datum is D_North_American_1927 and the Ellipsoid is Clarke_1866, before clicking Next to go to Step 4.

6. Specify in the Step 4 dialog the name of the re-projected shapefile as *idstations.shp* and its path. Then click Next.

7. The Summary dialog displays the input coordinate system and the output coordinate system. If you find incorrect information, you can click Back to the previous dialog(s) to make corrections. If the information is correct, click Finish.

8. The re-projection takes a while. Click OK in the completion dialog and add the new shapefile to a new view. To check *idstations.shp* is correctly projected, you can add to view *idoutl.shp*, an outline map of Idaho based on the Idaho transverse Mercator coordinate system. Most snow courses should fall within Idaho.

REFERENCES

Dent, B.D. 1999. *Cartography: Thematic Map Design,* 5th ed. Dubuque, Iowa: Wm C. Brown Publishers.

Maling, D.H. 1992. *Coordinate Systems and Map Projections,* 2d ed. Oxford: Pergamon Press.

Moffitt, F.H., and J.D. Bossler. 1998. *Surveying,* 10th ed. Menlo Park, CA: Addison-Wesley.

Robinson, A.H., J.L. Morrison, P.C. Muehrcke, A.J. Kimerling, and S.C. Guptill. 1995. *Elements of Cartography,* 6th ed. New York: John Wiley & Sons.

Snyder, J.P. 1993. *Flattening the Earth: Two Thousand Years of Map Projections.* Chicago: University of Chicago Press.

VECTOR DATA MODEL

3.1 INTRODUCTION

Looking at a paper map, we can see what map features are like and how they are spatially related to one another. Figure 3.1 is a reference map, which shows that Idaho borders Montana, Wyoming, Utah, Nevada, Oregon, Washington, and Canada, and contains the Indian reservations. The map communicates

Figure 3.1
A reference map showing Idaho and land parcels held in trust by the United States for Native Americans.

to us through its symbols and text. We can easily see the map features and their spatial relationships, but how can the computer see these features and relationships? That is the basic question we ask in this chapter about the vector data model.

The **vector data model** uses points and their *x*-, *y*-coordinates to construct spatial features. Vector-based features are treated as discrete geometric objects over the space. The process of developing the vector data model consists of several steps. First, spatial features are represented as simple geometric objects of points, lines, and areas. Second, for some GIS applications, the spatial relationships between features must be expressed explicitly. Third, a logical structure of data files must be in place so that the computer can efficiently process data for spatial features and their spatial relationships. Fourth, land surface data, overlapping spatial features, and road networks are better represented as composites of simple geometric objects.

This chapter is divided into six sections. Section 1 covers the representation of vector data as geometric objects. Section 2 describes the data structure for computer processing. Section 3 discusses non-topological vector data. Section 4 covers spatial features that are better represented as composites of points, lines, and areas. Section 5 introduces the new object-oriented data model. The final section discusses the spatial data concepts of data accuracy, scale, and precision.

3.2 VECTOR DATA REPRESENTATION

3.2.1 Geometric Objects

The vector data model uses *x*-, *y*-coordinates and the simple geometric objects of **point**, **line**, and **area** to represent spatial features. A point may represent a well, a benchmark, or a gravel pit. A line may represent a road, a stream, or an administrative boundary. And an area may represent a vegetated land, a water body, or a sinkhole.

Dimensionality and property distinguish the three types of geometric objects. A point has 0 dimension and has only the property of location. A line is one-dimensional and has the property of length. An area is two-dimensional and has the properties of area and boundary. In the GIS literature, a point may also be called a node, vertex, or 0-cell; a line, an edge, link, chain, or 1-cell; and an area, a polygon, face, zone, or 2-cell (Laurini and Thompson 1992).

The basic units of the vector data model are points and their coordinates. A line feature is made of points (Figure 3.2). Between two end points a line consists of a series of points marking the shape of the line, which may be a smooth curve or a connection of straight-line segments. Smooth curves are typically fitted by mathematical equations. Straight-line segments may represent human-made features or approximations of curves. Line features may intersect or join with other lines and may form a network.

An area feature is defined by lines, which are in turn defined by points (Figure 3.3). The boundary of an area feature separates the interior area from the exterior area. Area features may be isolated or connected. An isolated area feature typically has a point serving as both the beginning and end point of its boundary. Area features may form holes within other areas, such as the Indian reservations surrounded by the state of Idaho. Area features may overlap one another and create overlapped areas. For example, the burned areas from previous forest fires may overlap each other. Holes and overlapped areas are treated as separate objects in the simple data model.

Vector data representation using points, lines, and areas is not always straightforward because it depends on map scale and, occasionally, criteria established by government mapping agencies (Robinson et al. 1995). **Map scale** is the ratio of the map distance to the ground distance. On a 1:24,000 scale map, for example, a map distance of 1 centimeter would represent 24,000 centimeters, or 240 meters. A 1:24,000 scale map shows a smaller area but contains more details than a 1:1,000,000 scale map. This is why a city on a 1:1,000,000 scale map may be represented as a point, but the same city may be shown as an area on a 1:24,000 scale map.

A stream may be shown as a single line near its headwaters but as an area along its lower reaches. In this case, the width of the stream determines

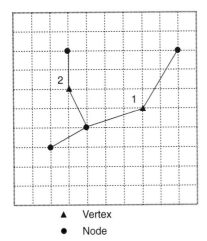

▲   Vertex

●   Node

Figure 3.2
Line objects: two lines with beginning node, end node, and vertex.

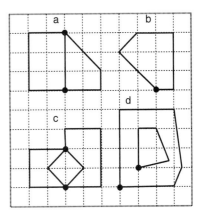

Figure 3.3
Area objects: a—contiguous areas; b—isolated area; c—three areas formed by two overlapped objects; d—a hole within an area.

how it should be represented on a map. The U.S. Geological Survey (USGS) uses single lines to represent streams less than 40 feet wide on 1:24,000 scale topographic maps and double lines for larger streams. Therefore, a stream may appear as a line or an area depending on its width and the criterion used by the government agency.

3.2.2 Topology

The conceptual representation of spatial features as points, lines, and areas is the first step in building the vector data model. For some GIS applications, the next step is to turn to topology to express explicitly the spatial relationships between features. **Topology** studies properties of geometric objects that remain invariant under certain transformations such as bending or stretching (Massey 1967). For example, a rubber band can be stretched and bent without losing its intrinsic property of being a closed circuit, as long as the transformation is within its elastic limits.

Topology is often explained through **graph theory**, a field of mathematics that uses diagrams or graphs to study the arrangements of geometric objects and the relationships between objects (Wilson and Watkins 1990). Important to the vector data model are digraphs (directed graphs), which include points and directed lines (also called arcs). Adjacency and incidence are two relationships that can be established between the point and line objects in digraphs (Box 3.1).

A good example of use of topology is the TIGER (Topologically Integrated Geographic Encoding and Referencing) database from the U.S. Bureau of the Census (Broome and Meixler 1990). In the TIGER database, points are called 0-cells, lines 1-cells, and areas 2-cells (Figure 3.5). Each 1-cell in a TIGER file is a directed line, meaning that the line is directed from a starting point toward an ending point with an explicit left and right side. Each 2-cell and 0-cell has knowledge of the 1-cells associated with it. The USGS' Digital Line Graph (DLG) products, which include such features as roads, streams, boundaries, and contours, also use topology to define the spatial relationship

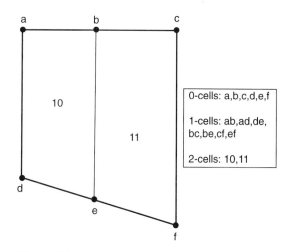

Figure 3.5
Topology in the TIGER database involves 0-cells or points, 1-cells or lines, and 2-cells or areas.

between features. Commercial GIS vendors such as ESRI and Intergraph have proprietary topological data structures.

ESRI defines the standard, topological vector data format used in ARC/INFO as **coverage** and groups coverages by point, line, and polygon. A coverage supports three basic topological relationships:

- **Connectivity**: Arcs connect to each other at nodes.
- **Area definition**: An area is defined by a series of connected arcs.
- **Contiguity**: Arcs have directions and left and right polygons.

Other than the use of terms, the above three topological relationships are similar to the topological relationships in the TIGER database. Topology is useful for detection of errors on digital maps. Topology can detect lines that do not meet correctly, polygons that are not closed properly, and other digitizing errors in a digital map. These kinds of errors must be corrected to avoid incomplete features and problems in data analysis. For example, a shortest path analysis requires roads to meet correctly. If a gap exists on a supposedly continuous

Box 3.1 Adjacency and Incidence

If a line joins two points, the points are said to be adjacent and incident with the line, and the adjacency and incidence relationships can be expressed explicitly in matrices. Figure 3.4 shows an adjacency matrix and an incidence matrix for a digraph. The row and column numbers of the adjacency matrix correspond to the node numbers, and the numbers within the matrix refer to the number of arcs joining the corresponding nodes in the digraph. For example, 1 in (11,12) means one arc joint from Node 11 to Node 12, and 0 in (12,11) means no arc joint from Node 12 to Node 11. The direction of the arc determines if 1 or 0 should be assigned.

The row numbers of the incidence matrix correspond to the node numbers in Figure 3.4, and the column numbers correspond to the arc numbers. The number 1 in the matrix means an arc is incident from a node, -1 means an arc is incident to a node, and 0 means an arc is not incident from or to a node. Take the example of Arc 1. It is incident from Node 13, incident to Node 11, and not incident to all the other nodes. What the matrices have accomplished is to express the adjacency and incidence relationships mathematically.

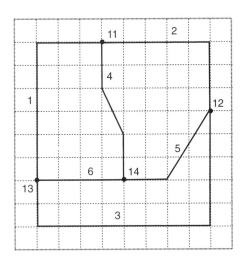

Adjacency matrix

	11	12	13	14
11	0	1	0	1
12	0	0	1	0
13	1	0	0	0
14	0	1	1	0

Incidence matrix

	1	2	3	4	5	6
11	-1	1	0	1	0	0
12	0	-1	1	0	-1	0
13	1	0	-1	0	0	-1
14	0	0	0	-1	1	1

Figure 3.4
The adjacency matrix and incidence matrix for a simple digraph.

road, the analysis will take a circuitous route to avoid the gap.

Topology is also important for some types of GIS analysis. One example is traffic volume analysis, in which traffic volume measurements must follow lines with the same traffic direction. Another example is deer habitat analysis, which often involves edges between habitat types, especially edges between old growth and clear-cuts (Chang et al. 1995). Because edges are coded with left and right polygons in a topologically structured data model, specific habitat types along edges can be easily tabulated and analyzed.

3.3 TOPOLOGICAL DATA STRUCTURE

So far we have examined the concepts involving the topology-based vector data model. The application of the concepts falls within the domain of data structure, that is, the structure of digital data files and the relationship between files.

Point features are simple: they can be coded with their identification numbers (IDs) and pairs of *x*- and *y*-coordinates (Figure 3.6). Topology does

not apply to points because points are separate from one another.

Figure 3.7 shows the data structure of a line feature. Using the ARC/INFO terminology, a line segment is called an **arc**, which is connected to two end points called **nodes**. The starting point is called the from-node and the ending point the to-node. The arc-node list sorts out the arc-node relationship. For example, Arc 2 has 12 as the from-node and 13 as the to-node. The arc-coordinate list shows the *x*-, *y*-coordinates that make up each arc. For example, Arc 3 consists of three line segments connected at the points of (2, 6) and (4, 4).

Figure 3.8 shows the data structure of an area feature. The polygon/arc list shows the relationship between polygons and arcs. For example, Arcs 1, 4, and 6 connect to define Polygon 101. Polygon 104 differs from the other polygons in being surrounded by Polygon 102. To show that Polygon 104 is a hole within Polygon 102, the arc list for Polygon 102 contains a zero to separate the external and internal boundaries. Polygon 104 is an isolated polygon consisting of only one arc (7). A node (15) is placed along the arc to be the beginning and end node. Polygon 100, which is outside the map area, is called the external or universe polygon.

The left/right list in Figure 3.8 shows the relationship between arcs and their left and right polygons. For example, Arc 1 is a directed line from Node 13 to Node 11 and has Polygon 100 as the polygon on the left and Polygon 101 as the polygon on the right. Finally, each polygon is assigned a label point to link the polygon to its attribute data.

The topology-based data structure facilitates the organization of data files and reduces data redundancy. The shared boundary between two polygons is listed once, not twice, in the arc-coordinate list. Moreover, because the shared boundary defines both polygons, updating the polygons is made easier. For example, if Arc 4 in Figure 3.8 is changed to a straight line between two nodes, only the coordinate list for Arc 4 needs to be changed.

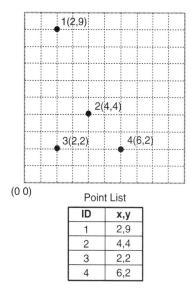

(0 0)

Point List

ID	x,y
1	2,9
2	4,4
3	2,2
4	6,2

Figure 3.6
Points with *x*-, *y*-coordinates.

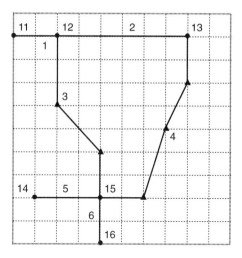

Arc-node list

Arc#	Fnode	Tnode
1	11	12
2	12	13
3	12	15
4	13	15
5	15	14
6	15	16

Arc-coordinate list

Arc#	x,y Coordinates
1	(0,9) (2,9)
2	(2,9) (8,9)
3	(2,9) (2,6) (4,4) (4,2)
4	(8,9) (8,7) (7,5) (6,2) (4,2)
5	(4,2) (1,2)
6	(4,2) (4,0)

Figure 3.7
The data structure of a line data model.

3.4 NON-TOPOLOGICAL VECTOR DATA

GIS developers introduced topology two decades ago to separate GIS from CAD (Computer Aided Design). AutoCAD by Autodesk was, and still is, the leading CAD package. A data format used by AutoCAD for transfer of data files is called DXF (digital exchange format). DXF maintains data in separate layers and allows the user to draw each layer using different line symbols, colors, and text. But DXF does not support topology.

The issue of topology has again returned to GIS (Box 3.2). The main advantage of using non-topological vector data is that they display more rapidly on the computer monitor than topology-based data. In recent years, non-topological data format has become one of the standard, non-proprietary data formats. Several commercial GIS packages, such as ArcView, MapInfo, and GeoMedia, have in fact adopted non-topological data formats, which can be used directly in different GIS software packages.

The standard non-topological data format used in ArcView is called the **shapefile**. Although the shapefile saves a point as a pair of x-, y-coordinates, a line as a series of points, and a polygon as a series of lines, no files describe the spatial relationship among geometric objects. Shapefile polygons actually have duplicate arcs for the shared boundaries and can overlap one another. Rather than having multiple files as for an ARC/INFO

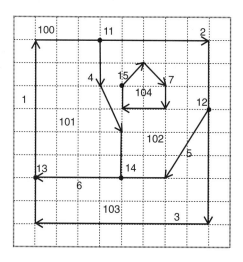

Left/Right list

Arc#	Lpoly	Rpoly
1	100	101
2	100	102
3	100	103
4	102	101
5	103	102
6	103	101
7	102	104

Polygon/Arc list

Polygon #	Arc#
101	1,4,6
102	4,2,5,0,7
103	6,5,3
104	7

Figure 3.8
The data structure of an area data model.

Box 3.2   Topology or No Topology

After two decades, topology has again become an issue. GIS users may wonder about the importance of topology. As described in this chapter, topology is useful for data editing and some types of spatial analysis. The topological capabilities of GIS software can differ extensively. ARC/INFO uses three basic topological relationships, which are sufficient for its needs, whereas sophisticated GIS operations may require more topological relationships to be present between map features.

The decision on topology therefore depends on the GIS project. For some projects, topological functions are not necessary, while for others they are a must. For example, a producer of GIS data will find it absolutely necessary to use topology for error checking and for ensuring that lines meet correctly and polygons are closed properly. For spatial analysis that requires topology, GIS users can actually build topology for non-topological vector data on the fly.

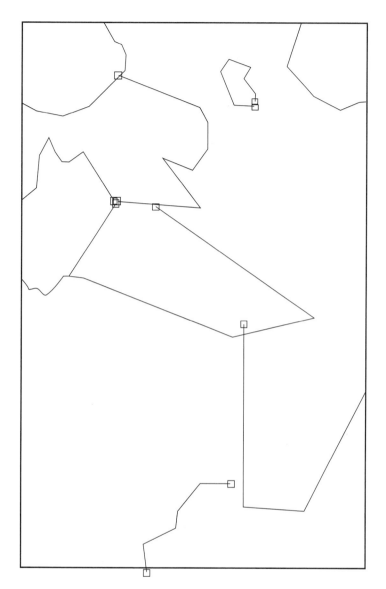

Figure 3.9a
A polygon coverage with topological errors.

coverage, the geometry of a shapefile is stored in two basic files: the .shp file stores the feature geometry, and the .shx file maintains the index of the feature geometry.

Shapefiles can be converted to coverages, and vice versa. The conversion from a shapefile to a coverage requires the building of topological relationships and the removal of duplicate arcs. The conversion from a coverage to a shapefile is simpler. But if a coverage has topological errors—such as lines not joined perfectly—the errors can lead to problems of missing features in the shapefile. Figure 3.9a shows a coverage that has errors in line joining. After converting to a shapefile,

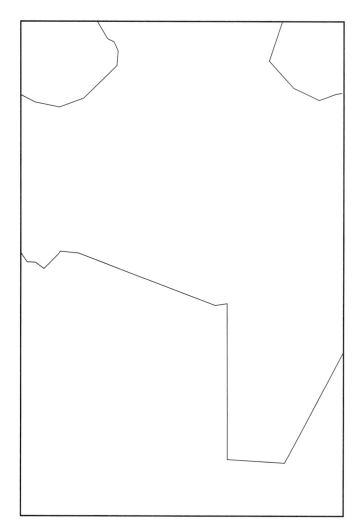

Figure 3.9b
A shapefile converted from the coverage shown in Figure 3.9a.

all lines that have errors disappear in the shapefile (Figure 3.9b).

3.5 HIGHER-LEVEL OBJECTS

To complete the discussion of the vector data model, this section discusses the higher-level objects of TIN, regions, and dynamic segmentation. The creation and applications of TIN, regions, and

dynamic segmentation are covered in Chapters 12, 15, and 16, respectively.

3.5.1 TIN

A vector data structure for terrain mapping and analysis is called the **triangulated irregular network (TIN)**. A TIN approximates the surface with a set of non-overlapping triangles (Figure 3.10). Each triangle in the TIN is assumed to have a constant

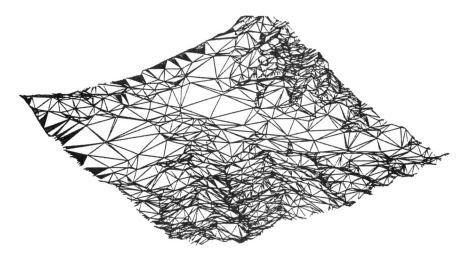

Figure 3.10
TIN in a perspective view.

gradient. These triangles are constructed using **Delaunay triangulation,** which is an iterative process of connecting points with their two nearest neighbors to form triangles as equi-angular as possible. Different computing algorithms have been proposed for Delaunay triangulation, and these algorithms are continually being improved (Watson and Philip 1984; Tsai 1993).

A TIN consists of two basic data elements: elevation points with x, y, and z values, and edges (lines) that connect these points to form triangles. The x, y values represent the location of a point, and the z value represents the elevation at the point. The slope and aspect of each triangle are computed from the x, y, and z values at the three points that make up the triangle. These elevation points in a TIN are actually sample points selected to represent the surface. Flat areas of the surface are represented by a small number of sample points and large triangles. Areas with high variability in elevation need a large number of sample points and small triangles.

The TIN data structure includes the triangle number, the number of each adjacent triangle, and data files showing the lists of points, edges, as well as x, y, and z values of each elevation point.

3.5.2 Regions

Built on simple lines and areas, the **regions data model** consists of region layers and regions (Figure 3.11). A region layer is made of regions of the same attribute. The regions data model has two important characteristics. First, region layers may overlap or cover the same area. For example, a region layer representing areas burned in a 1917 forest fire may overlap with another region layer representing areas burned in a 1930 fire. When different region layers cover the same area, they form a hierarchical region structure with one layer nested within another layer.

Second, a region may have disconnected or disjoint components. The state of Hawaii may therefore be a region, including several islands. In contrast, the simple data model would require each island to be a separate polygon. This characteristic also applies to the void or empty area of a region. Therefore, private land parcels scattered within a national forest can be grouped as one void area.

ARC/INFO treats region layers as subclasses in a polygon coverage. Each region subclass has its own attributes. In this way a series of region layers can be based on the same polygon coverage, which is called an integrated coverage. The Idaho

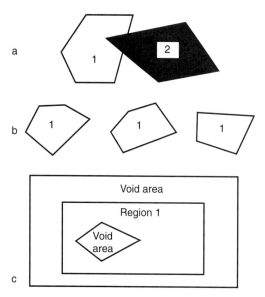

Figure 3.11

Properties of the regions data model: a—overlapped regions; b—a region with three components or rings; c—void area made of an area within the region and the external area.

Panhandle National Forest has built an integrated fire coverage with fire records spanning from the late 1900s to the 1960s (Box 3.3). Each fire year or period is handled as a region subclass.

The regions data structure consists of two basic elements: a file on the region-arc relationship and another on the region-polygon relationship.

Figure 3.12 shows the file structure for an example with four polygons, five arcs, and three regions. The region-polygon list relates the regions to the polygons. Region 101 consists of Polygons 11 and 12. Region 102 has two components: one includes Polygons 12 and 13, and the other consists of Polygon 14. Region 101 overlaps Region 102 in Polygon 12. The region-arc list links the regions to the arcs. Region 101 has only one ring, which connects Arcs 1 and 2. Region 102 has two rings: one connects Arcs 3 and 4, and the other consists of Arc 5.

3.5.3 Dynamic Segmentation

The **dynamic segmentation** model combines a line coverage and a linear measurement system such as the milepost system to form a higher-level object. ARC/INFO uses the basic elements of sections, routes, and events for the model. **Sections** refer directly to arcs of a line coverage and positions along arcs. Because arcs of a line coverage are made of a series of *x*-, *y*-coordinates based on a real-world coordinate system, sections are also measured in real-world coordinates. **Routes** are a collection of sections that represent linear phenomena such as highways, bike paths, or streams. Attribute data to be associated with routes are called **events** in dynamic segmentation. Events such as pavement conditions, accidents, and speed limits are measured in a linear system such as the milepost. But as long as events are accompanied by

---

Box **3.3**   **A Regions-based Fire Coverage**

The regions-based fire coverage prepared by the Idaho Panhandle National Forest includes 31 region subclasses: Pre1886, C1889, C1894, C1900, C1905, F1908-09, F1910, F1911-13, F1914-15, F1917, F1918, F1919, F1920-21, F1922-23, F1924, F1925, F1926, F1927-28, F1929, F1930, F1931, F1932-33, F1934, F1935-39, F1940-49, F1950-59, F1960-69, All_Fires, Doubles, Triples, and Quad_burns. Forest

fires prior to 1905 are less certain and thus carry letter C for circa. Doubles, triples, and quad_burns show areas burned twice, three times, and four times, respectively. Each region subclass has its attribute table, which includes acres of burned areas. This regions-based fire coverage can be displayed and queried in ARC/INFO or ArcView.

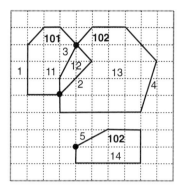

Region-polygon list

Region #	Polygon #
101	11
101	12
102	12
102	13
102	14

Region-arc list

Region #	Ring #	Arc#
101	1	1
101	1	2
102	1	3
102	1	4
102	2	5

Figure 3.12
The file structure for the regions data model.

measures of their locations, the dynamic segmentation model can relate events to route systems.

Figure 3.13 shows how a route, Route 109, is coded for a route system called BIKEPATH in a line coverage called ROADS. The coverage ROADS contains a number of topologically structured arcs, shown in thin lines. Route 109 in a thick shaded line is a collection of three sections, numbered 1, 2, and 3 with BIKEPATH# and BIKEPATH-ID. Using the ARC/INFO terminology, BIKEPATH# is called the machine ID, which is the ID assigned and used internally by ARC/INFO, and BIKEPATH-ID is called the user ID, which can be assigned and used by the user.

The section table relates the sections to the arcs in Roads. Section 1 covers the entire length of Arc 7; therefore, the FROM-POSITION is 0% and the TO-POSITION is 100%. The FROM-MEASURE of Section 1 is 0 because it is the beginning point of Route 109. The TO-MEASURE of Section 1 is 40 units (meters or feet, depending on the measurement unit of ROADS), which is measured from the line coverage. Similar to Section 1, Section 2 also covers the entire length of Arc 8. Its FROM-MEASURE and TO-MEASURE continues from Section 1. Section 3 covers 80% of Arc 9, thus its TO-POSITION is coded 80. The TO-MEASURE of Section 3 is computed by adding 80% of the length of Arc 9 to its FROM-MEASURE.

The route table in Figure 3.13 shows the route has a machine ID of 1 and a user ID of 109. The route table is linked to the section table using the ID value of BIKEPATH# and ROUTELINK#, which is 1 in the example. Attribute data for the route system can be directly added to the route table. Although attribute data can also be included in the section table, they are usually prepared as event tables. Events can be point events such as signs along a bike path, continuous events such as gradients along a bike path, or linear events such as stretches of poor visibility. A point event is

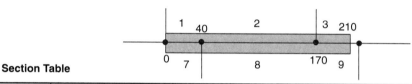

Section Table

Route Link #	Arc Link #	F-MEAS	T-MEAS	F-POS	T-POS	Bikepath #	Bikepath-ID
1	7	0	40	0	100	1	1
1	8	40	170	0	100	2	2
1	9	170	210	0	80	3	3

Route Table

BIKEPATH #	BIKEPATH-ID
1	109

Point Event Table

BIKEPATH-ID	LOCATION	ATTRIBUTE
109	40	Stop sign

Linear Event Table

BIKEPATH-ID	FROM	TO	ATTRIBUTE
109	100	120	Steep

Figure 3.13
The dynamic segmentation data model.

measured by its location, a continuous event by its to measure, and a linear event by its from and to measures (Figure 3.13). Event tables include the route ID, location measures, and attributes.

ARC/INFO has the built-in commands to develop the dynamic segmentation data model. The "measured polyline shapefile" used in ArcView has a measured value in addition to x- and y-coordinates. If the measured value represents the linear distance along a line, the shapefile can be used as an alternative to the dynamic segmentation data model in ARC/INFO. But the process of adding the measured values to a shapefile is tedious.

The hierarchical structure of arcs, sections, and routes in the dynamic segmentation data model has several advantages over the arc-node data model. First, the hierarchical structure allows one-to-many relationships. For example, routes and sections can efficiently represent different bus routes using the same downtown streets. In con-

trast, using the arc-node data model to deal with one-to-many relationships would be cumbersome and difficult.

Second, dynamic segmentation combines real-world locations with linear measures in the section table. The FROM-MEASURE and TO-MEASURE function as linear measures, like mileposts, in a linear reference system. But because the FROM-MEASURE and TO-MEASURE are based on a line coverage, they are also geo-referenced, meaning that they are referenced to real-world coordinates. This geo-referencing makes it easier to register route systems with other geo-referenced coverages in the database.

Third, dynamic segmentation offers a much more efficient means for working with segmented data, such as pavement conditions along highways. Using the arc-node data model would require adding and removing nodes to store segmented data.

3.6 OBJECT-ORIENTED DATA MODEL

The **object-oriented data model** uses objects to organize spatial data. Unlike a geometric object of a point, line, or area, an **object** is defined here as something that has a set of properties and can perform operations upon requests. By this definition, almost everything one uses in a GIS is an object. For example, a land use map is an object, which has properties such as its coordinate system and feature type and can respond to requests such as zoom in, zoom out, and query.

3.6.1 Structural Aspects of Objects

One area of concern in applying the object-oriented data model to GIS is the structural aspects of objects (Worboys et al. 1990; Worboys 1995). The principles that can be used to group objects include association, aggregation, generalization, instantiation, and specialization (Larman 1997; Zeiler 1999).

Association describes the relationships between objects of two types. If owner and land parcel represent two types of objects, the relationships between them can follow the rules that (1) an owner can own one or more parcels, and (2) a parcel can be owned by one or more owners. Aggregation is an asymmetric association in a whole-part relationship. For example, block groups are connected to form a census tract and census tracts are connected to form a county. Generalization identifies the commonality among objects, and groups objects of similar types into a higher-order type. For example, parcel, zoning, and census tract maps may be grouped into a higher-order class called boundary. The grouping of objects forms a hierarchical structure, which organizes objects into classes and classes into superclasses and subclasses. Instantiation means that an object of a class can be created from an object of another class. For example, a high-density residential area object may be created from a residential area object. Specialization differentiates objects of a given class by a set of rules. For example, roads may be separated by average daily traffic volume.

3.6.2 Behavioral Aspects of Objects

Another area of concern in applying the object-oriented data model to GIS is the behavioral aspects of objects (Egenhofer and Frank 1992). Inheritance is the basic principle in explaining the behaviors of objects: subclasses inherit properties and operations from a superclass, and objects inherit properties and operations from a subclass. Suppose that residential area is a superclass and low-density area and high-density area are the subclasses. All properties of the class residential area are inherited by its two subclasses. Through inheritance, properties need only be defined once in the class hierarchy. Additional properties (i.e., not inherited) may be defined for a particular subclass to separate it from other subclasses. For example, the lot size may be added as a property to separate low-density and high-density areas.

Encapsulation and polymorphism are two other principles closely related to the behavioral aspects of objects. Encapsulation refers to the mechanism to hide the properties and operations of an object so that the object can perform an operation by responding to a predefined message or request. For example, a polygon object can respond to a request called ReturnCenter (return the center of) by returning the physical center of the polygon. Polymorphism allows the same operation to be implemented in different ways in different objects. For example, the same request called GetDimension (get the dimension of) can be sent to a point, a line, or a polygon but the result differs depending on the feature type. The number 0 is returned if the object is a point, 1 if the object is a line, and 2 if the object is a polygon.

3.6.3 Applications of the Object-oriented Data Model in GIS

SmallWorld (now GE-SmallWorld) is thought by many to be the first commercial GIS package to implement the object-oriented data model, but the use of SmallWorld has been limited to utilities and communications (http://www.smallworld-us.com/). Two developments relative to the object-oriented data model have emerged during the past several years. The first is the introduction of object-oriented macro languages such as Avenue for ArcView and MapX for MapInfo (Box 3.4). These macros allow users to change the user interface or develop specialized functionality. The second development is the adoption of object-oriented programming languages such as Visual Basic and C++ by GIS vendors such as Intergraph, MapInfo, and ESRI. These languages can be used to customize GIS applications.

The geodatabase model, offered alongside the georelational data model in ArcInfo 8, is one of the newest entries in object-oriented GIS. Directed toward personal or small workgroup use, a geodatabase may be built from imported coverages, shapefiles, and tables. Geodatabase users can add behavior, properties, and relationships to objects in the database and construct a class hierarchy.

The geodatabase model uses the geometries of point, line, and polygon to represent vector-based spatial features. A point feature may be a simple feature with a point or a multipoint feature with a set of points. A line feature is a set of line segments, which may or may not be connected. A polygon feature may be made of one or many rings. A ring is defined as a set of connected, closed, nonintersecting line segments (Zeiler 1999). A polygon feature with many rings would

be the same as a region with disjoint components in the regions data model.

The geometries of spatial features are stored in a field called geometry in the geodatabase model. Instead of storing spatial data and attribute data in a split data system, the geodatabase model stores both data in a single database. This data structure, which is the same as Oracle Spatial, not only eliminates the complexity of coordinating disparate data stores but also reduces the processing overhead.

A general adoption of the geodatabase model is difficult because it requires universal agreement among the users on the structure and behavior of objects. One approach is to develop geodatabase models for specific industries and applications (http://www.esri.com/software/arcgisdatamodels/). Even if the object-oriented data model is not used to its full extent in the near future, GIS users may find some of its concepts appealing. For example, ArcInfo 8 provides four general validation rules for the grouping of objects: attribute domains, relationship rules, connectivity rules, and custom rules (Zeiler 1999). Attribute domains group objects into classes, or subtypes, by a valid range of values or a valid set of values for an attribute. Relationship rules connect objects that are associated. Connectivity rules let users build geometric

---

Box 3.4   Object-Oriented User Interface and Programming

The object-oriented data model is one application of object-oriented technology. GIS users are probably more familiar with two other applications, the user interface and the programming system. An object-oriented user interface uses icons, dialog boxes, and glyphs (i.e., graphical objects) in place of command lines. ArcView, for example, has an object-oriented user interface, which allows the user to communicate with the GIS package by point and click. Object-oriented user interface can restrict user choice of operations to those defined for the selected data set, thus avoiding operational errors. This is why ArcView

users often find menu selections differ depending on the input themes.

An object-oriented computer language uses the principles of inheritance, encapsulation, and polymorphism, and allows a programmer to write code by flowcharting a program's logic. Visual Basic and C++ are examples of the object-oriented computer language. Avenue, the macro language for ArcView, is also object-oriented. Avenue scripts can modify and extend ArcView's user interface as well as its analytical capabilities.

networks such as streams, roads, and water and electric utilities. And custom rules allow users to create custom features for advanced applications. These validation rules can circumvent data entry errors, thus ensuring data integrity.

3.7 SPATIAL DATA CONCEPTS

3.7.1 Map Scale, Spatial Resolution, and Spatial Data Accuracy

As discussed earlier, map scale can influence the representation of spatial features such as the line or area representation of streams. Map scale is also an indicator of map accuracy—how close the apparent location of a map feature is to its true ground location. The accuracy of a map feature is less reliable on a 1:100,000 scale map than on a 1:24,000 scale map. The standards for horizontal accuracy of USGS topographic maps state that no more than 10% of the well-defined map points tested can be more than 0.02 inch (0.5 millimeters) out of correct position at publication scales of 1:20,000 or smaller (Box 3.5). This tolerance of 0.02 inch corresponds to 40 feet on the ground for 1:24,000 scale maps and about 167 feet on the ground for 1:100,000 scale maps.

Map scale influences the level of detail on a map. As the map scale becomes smaller, the amount of map details decreases, and the degree of line generalization increases. For example, a meandering stream on a large-scale map becomes less sinuous on a small-scale map. Therefore, to obtain a high level of detail, GIS users must look for the largest scale map possible for the study area.

Although maps are still the most common source for spatial data entry, new data entry methods using Global Positioning Systems (GPS) and remote sensing imagery can bypass maps and the practice of various methods of map generalization (Monmonier 1996). The accuracy of spatial data collected by GPS or satellite images is directly related to the resolution of the measuring instrument. The spatial resolution of satellite images can range from 1 meter to 1 kilometer. Similarly, the spatial resolution of GPS readings can range from several millimeters to 100 meters.

3.7.2 Location Accuracy and Topological Accuracy

Location accuracy refers to the accuracy of spatial feature locations. Topological accuracy refers to how well the topological relationships between spatial features are maintained. Map scale and the

---

Box 3.5   Map Accuracy Standards

Revised and adopted in 1947, the U.S. National Map Accuracy Standard (NMAS) sets the accuracy standard for published maps such as topographic maps from the U.S. Geological Survey. The standards for horizontal accuracy require that no more than 10% of the well-defined map points tested shall be more than 1/30 inch at scales larger than 1:20,000, and 1/50 inch at smaller scales. This means that the tolerance is 40 feet on the ground for 1:24,000 scale maps.

NMAS defines the location accuracy of paper maps by map scale. Because digital spatial data is not constrained by scale, the National Standard for Spatial Data Accuracy (NSSDA) has been developed in the U.S. to replace the NMAS (http://www.fgdc.gov/standards/). The NSSDA omits accuracy metrics or threshold values that digital spatial data must achieve. Instead, agencies are encouraged to establish criteria for their product standards. Federal agencies typically adopt 6–12 meters (20–39 feet) as the tolerance for 1:24,000 scale maps.

data entry process largely determine the location accuracy of spatial features. Topological accuracy, on the other hand, depends on data entry, the capability of a GIS package to detect errors, and the ability of a GIS data producer to remove errors.

3.7.3 Location Data Accuracy and Precision

Location data accuracy measures how close the recorded location of a spatial feature comes to its ground location, whereas precision measures how exactly the location is recorded. The number of significant digits used in recording expresses the precision of a recorded location. For example, distances may be measured with decimal digits or rounded off to the nearest meter or foot. The precision of location data stored in the computer is dictated by the hardware's word length (single or double precision) and by whether the data are stored as integers or floating points. Coverage coordinates in ARC/INFO, for example, are stored as either single-precision real numbers with 6 to 7 significant digits or double precision with 13 to 14 significant digits.

KEY CONCEPTS AND TERMS

Arc: A line segment with two end points.

Area definition: An area is defined by a series of connected arcs.

Area: A spatial feature that is represented by a series of lines and has the geometric properties of size and perimeter. Also called polygon, face, or zone.

Connectivity: A topological relationship that stipulates that arcs connect to each other at nodes.

Contiguity: Arcs have directions and left and right polygons.

Coverage: The standard, topological vector data format used in ARC/INFO.

Delaunay triangulation: An iterative process of connecting points with their two nearest neighbors to form triangles as equi-angular as possible in a triangulated irregular network (TIN).

Dynamic segmentation: A data model that is built upon arcs of a line coverage and allows the use of real-world coordinates with linear measures.

Event: A basic element of the dynamic segmentation model that associates attribute data such as pavement conditions, accidents, and speed limits with routes.

Graph theory: A sub-field of mathematics that uses diagrams or graphs to study the arrangements of objects and the relationships between objects.

Line: A spatial feature that is represented by a series of points and has the geometric properties of location and length. Also called *arc, edge, link,* or *chain.*

Map Scale: A ratio of map distance to ground distance.

Node: The beginning or end point of a line.

Object: An entity such as a land parcel that has a set of properties and can perform operations upon requests.

Object-oriented data model: A data model that uses objects to organize spatial data. An object is an entity such as a land parcel that has a set of properties and can perform operations upon requests.

Point: A spatial feature that is represented by a single coordinate and has only the geometric property of location. Also called *node* or *vertex.*

Regions data model: A data model that is built upon lines and polygons and allows disjoint components and overlapped areas.

Route: A basic element of the dynamic segmentation model that uses a collection of sections to represent a linear phenomenon such as a highway, bike path, or stream.

Section: A basic element of the dynamic segmentation model that refers directly to arcs and positions along arcs.

Shapefile: The standard, non-topological vector data format used in ArcView.

Topology: A sub-field of mathematics that studies properties of geometric objects, which

remain invariant under certain transformations such as bending or stretching.

Triangulated irregular network (TIN): A vector data structure that approximates the terrain with a set of non-overlapping triangles.

Vector data model: A data model that uses points and their *x*-, *y*-coordinates to construct spatial features.

APPLICATIONS: THE VECTOR DATA MODEL

This applications section consists of three tasks. The first task reviews the topological relationships and the data structure for the simple arc-node data model. Tasks 2 and 3 examine TIN and the higher-level data models of regions and dynamic segmentation.

Task 1: The Data File Structure of ARC/INFO Coverages and Shapefile

What you need: *land.e00*, an ARC/INFO interchange file.

Task 1 asks you to import the interchange file *land.e00* into an ARC/INFO coverage and then convert the coverage to a shapefile. You can examine the data structure of the coverage and the shapefile by opening the folder(s) where the data sets reside.

1. Import71 is a utility program in the ArcView program group that can convert an interchange file into an ARC/INFO coverage. Start the program by double clicking Import71. The next dialog asks you to enter the name of the interchange file and its path, and the name of the coverage and its path. For example, the input to the dialog may be as follows:
   [Export Filename]
   d:\bookdata\chapter3\land.e00
   [Output Data Source]
   d:\bookdata\chapter3\land_arc

Data files associated with *land_arc* reside in two folders under chapter3: land_arc and INFO. Land_arc contains such binary data files as arc, pal, and lab, whereas INFO contains the INFO files of *land_arc*. INFO is the database management program for ARC/INFO, and INFO files all have generic file names such as arc000dat and arc000nit. If there are other ARC/INFO coverages under chapter3, their INFO files are also placed in the INFO folder. This makes it difficult to copy individual ARC/INFO coverages using the drag and drop method. One option to deal with this problem is to convert ARC/INFO coverages to shapefiles, as shown in the following steps.

2. Start ArcView and open a new view. Add *land_arc* to view. Make *land_arc* active. Select Convert to Shapefile from the Theme menu. In the next dialog, name the shapefile *land.shp* and provide its path. Add *land.shp* to view. *Land.shp* is actually one of several files created from the conversion. Other files, which also reside in the chapter3 folder, are land.dbf, land.shx, land.sbn, and land.sbx.

3. Shapefiles can be copied, renamed, or deleted in ArcView. But you cannot use these functions with shapefiles that are currently in use. Therefore, to copy, rename, or delete *land.shp*, you must first close the project and

then open a new project. Open a new view, and select Manage Data Sources from the File menu. The Shapefile Manager dialog lets you specify the action you want to take with *land.shp*.

Task 2: TIN, Regions, and Dynamic Segmentation using ArcView

What you need: *emidatin*, a TIN prepared from a digital elevation model; *fire*, a polygon coverage with region subclasses; *highway*, a network coverage based on the dynamic segmentation model. You need the 3D Analyst extension to display *emidatin*.

Task 2 lets you view the data models of TIN, regions, and dynamic segmentation using Arc-View. Of the three data models, TIN is the only one that can be created in ArcView with the 3-D Analyst extension (Chapter 12). You must use the Import 71 utility program to import coverages that are based on the regions or dynamic segmentation data model.

1. TIN

1. Start ArcView, and load the 3-D Analyst extension.
2. Open a new view, and click the AddTheme button. In the AddTheme dialog, select TIN Data Source for the Data Source Type, and navigate to *emidatin*. Click OK.
3. Check the box next to *emidatin* in the Table of Contents. The default area symbols show elevation ranges. These area symbols do not follow the triangular boundaries. To see triangles in *emidatin*, do the following: double click *emidatin* to open its legend editor, change the legend for lines to single symbol, and click Apply.
4. Now click the Zoom In button, and select a small area to zoom in. Each triangle represents a surface and has a constant slope and aspect. To verify these TIN characteristics, you can click the Identify tool and click several points within a triangle. The

Identify Results table should show constant slope and aspect values but different elevation values depending on where points are clicked.

2. Regions

1. Start ArcView and open a new view.
2. Click the AddTheme button. In the Add-Theme dialog, navigate to *fire* and click on the folder icon next to *fire* to open its content. The region subclasses of *fire1* (burned once), *fire12* (burned twice), and *fire123* (burned three times) should be listed along with polygon. Double-click on polygon to first display the *fire* coverage. Then double-click on *region.fire123* to display the region subclass.
3. Make the theme *region.fire123* active and select Table from the Theme menu. The fields of fire1, fire2, and fire3 and their values should appear in the table.

3. Dynamic Segmentation

1. Start ArcView and open a new view.
2. Click the AddTheme button. In the AddTheme dialog, navigate to *highway* and click on the folder icon next to *highway*. The list should include *route.fastroute* and *arc*. Double-click on *arc* to first display the highway network. Repeat the same procedure to add *route.fastroute* to view. You can open the table of *route.fastroute*, but the table only lists the ID value and does not contain attribute data.

Task 3: TIN, Regions, and Dynamic Segmentation using ARC/INFO

What you need: *emidatin*, a TIN prepared from a digital elevation model; *fire*, a polygon coverage with regions subclasses; and *highway*, a network coverage.

Task 3 lets you work with the data models of TIN, regions, and dynamic segmentation using ARC/INFO. You will examine the data structure of these data models. You will also create a route system on a network.

1. TIN

Like a coverage, a TIN is stored as a directory in ARC/INFO. Change the workspace to *emidatin* and list data files in the directory. You should see the edge list, node list, node/xy list, and node/z list. Change the workspace to the level above *emidatin*. Go to Arcplot and type the following commands:

Arcplot: display 9999
Arcplot: mapextent emidatin
Arcplot: tin emidatin

The TIN, *emidatin*, is displayed as a network of triangles. Triangles near the border are elongated rather than compact like those inside the border. This is caused by a lack of data points near the border. Unlike a coverage, a TIN does not have INFO files associated with it. Therefore you cannot further examine the TIN data model.

2. Regions

The coverage *fire* contains areas that have been burned once, and some twice or even three times. The regions data model is ideal for working with overlapped polygons, such as overlapped polygons of past forest fires. Change the workspace to *fire* and list data files in the directory. Three region subclasses reside in the directory. *Fire1* consists of areas burned at least once, *fire12* consists of areas burned at least twice, and *fire123* consists of areas burned three times. Each region subclass has a set of data files including the region-arc list (pal) and the region-polygon list (rxp). Change the workspace to the level above *fire*.

To display the three region subclasses, go to Arcplot and type the following commands:
Arcplot: mapextent fire

Arcplot: arcs fire /*shows polygon boundaries of the coverage *fire*
Arcplot: regionshades fire fire1 3 /*shade areas burned at least once in green
Arcplot: regionshades fire fire12 4 /*shade areas burned at least twice in blue

Arcplot: regionshades fire fire123 2 /*shade areas burned three times in red

The three subclasses form nested regions on the computer monitor: *fire123* is nested within *fire12*, and *fire12* is nested within *fire1*. Exit ArcPlot.

Next, go to Tables and type the following commands:

Enter Command: select fire.patfire1
Enter Command: list

The list shows those areas that were burned at least once and the year of the fire.

Enter Command: select fire.patfire12
Enter Command: list

The list shows those areas that were burned at least twice and the years (under fire1 and fire2) of the fires.

Enter Command: select fire.patfire123
Enter Command: list

Finally, the display shows areas that were burned three times and the years of those fires. The DROPFEATURES command in Arc can remove region subclasses.

3. Dynamic Segmentation

For this part of Task 3, you will define a route system on a network first and then examine the data structure for the route system. The coverage *highway* shows the interstate and state highways in Idaho. To define a route system, go to ArcPlot and view the transportation network:

Arcplot: mapextent highway
Arcplot: arcs highway /* plot the network
Arcplot: mapextent * /* zoom in a portion of the network
Arcplot: clear
Arcplot: arcs highway
Arcplot: markercolor 2 /* set the node color as red
Arcplot: nodes highway

Now you can define a route system on *highway* by typing the following commands:

Arcplot: netcover highway fastroute /* called the route system on *highway* as *fastroute*
Arcplot: path * /* path is the command to define a route system interactively

Enter point
Enter point
9 to exit
/* Enter two nodes as the beginning and end nodes of the route system. Enter 9 to exit.
Arcplot: routelines highway fastroute 3 /* draw fastroute in green.
Arcplot: sectionlines highway fastroute /* draw sections of fastroute in different colors
The display shows the relationships between arcs, sections, and the route system: Sections are based on arcs in *highway*, and *fastroute* is a collection of sections.

To view the data structure of the route system, exit ArcPlot and change the workspace to highway at the Arc prompt. List data files in the *highway* directory. You should see the route (rat) and section (sec) data files for *fastroute*. Change the workspace to the level above *highway*. Go to Tables and type the following commands:

Enter Command: select highway.secfastroute
Enter Command: items

You should see items that were discussed under the dynamic segmentation data model in the chapter.

Enter Command: list

The first record in the listing should have the F-MEAS value of 0 because this is the beginning point of *fastroute*. The T-MEAS value in the first record represents the length of the first section or arc in real-world measurement units (feet, in this case). The second record is for the second section in *fastroute*. Its T-MEAS value should be the sum of the lengths of the first two sections. This continues until the last record whose T-MEAS value should be the total length of the route system. The F-POS value of each record should be either 0 or 100, depending on the direction of the arc. But in either case each section should cover 100% of the arc because the route system was defined between two nodes on a network.

Enter Command: select highway.ratfastroute
Enter Command: items

You should see the items of FASTROUTE# and FASTROUTE-ID. Use the LIST command to see the item values. The DROPFEATURES command in Arc can remove the route system and its sections.

REFERENCES

Broome, F.R., and D.B. Meixler. 1990. The TIGER Data Base Structure. *Cartography and Geographic Information Systems* 17: 39–47.

Chang, K., D.L. Verbyla, and J.J. Yeo. 1995. Spatial Analysis of Habitat Selection by Sitka Black-tailed Deer in Southeast Alaska. *Environmental Management* 19: 579–89.

Egenhofer, M.F., and A. Frank. 1992. Object-Oriented Modeling for GIS. *Journal of the Urban and Regional Information System Association* 4: 3–19.

Larman, C. 1997. *Applying UML and Patterns: An Introduction to Object-Oriented Analysis and Design*. Upper Saddle River, NJ: Prentice Hall PTR.

Laurini, R., and D. Thompson. 1992. *Fundamentals of Spatial Information Systems*. London: Academic Press.

Massey, W.S. 1967. *Algebraic Topology: An Introduction*. New York: Harcourt, Brace & World, Inc.

Monmonier, M. 1996. *How to Lie with Maps* 2^d ed. Chicago: Chicago University Press.

Robinson, A.H., J.L. Morrison, P.C. Muehrcke, A.J. Kimerling, and S.C. Guptill. 1995. *Elements of Cartography* 6th ed. New York: John Wiley & Sons.

Tsai, V.J.D. 1993. Delaunay Triangulations in TIN Creation: An Overview and A Linear Time Algorithm. *International Journal of Geographical Information Systems* 7: 501–24.

Watson, D.F., and G.M. Philip. 1984. Systematic Triangulations. *Computer Vision, Graphics, and Image Processing* 26: 217–23.

Wilson, R.J., and J.J. Watkins. 1990. *Graphs: An Introductory Approach*. New York: John Wiley & Sons.

Worboys, M.F., H.M. Hearnshaw, and D.J. Maguire. 1990. Object-Oriented Data Modelling for Spatial Databases. *International Journal of Geographical Information Systems* 4: 369–83.

Worboys, M.F. 1995. *GIS: A Computing Perspective*. London: Taylor & Francis.

Zeiler, M. 1999. *Modeling Our World: The ESRI Guide to Geodatabase Design*. Redlands, CA: ESRI Press.

4

VECTOR DATA INPUT

4.1 INTRODUCTION

The most expensive part of a GIS project is database construction. Converting from paper maps to digital maps used to be the first step in constructing a database. But in recent years this situation has changed as digital data clearinghouses have become commonplace on the Internet. Now the strategy for a GIS user is to look at what exists in the public domain before deciding to create new data. Private companies have also entered the GIS market. Some of them produce new GIS data for their customers, while others produce value-added GIS data from public data.

The proliferation of available GIS data has made it easier to organize a GIS project but has not reduced the importance of data input in GIS. GIS data must still be produced to be put on the Internet or to be sold to customers. GIS users may not find the digital data they want and may have to create their own data. New GIS data can be created from satellite images, GPS (Global Positioning System) data, or paper maps. GIS users can also choose manual digitizing or scanning for converting paper maps to digital maps.

A newly digitized map has the same measurement unit as the source map used in digitizing or scanning. To make the digitized map useful in a GIS project, it must be converted to real-world coordinates through geometric transformation. Using a set of control points in geometric transformation of a digitized map makes the process somewhat uncertain. A measure called the root mean square (RMS) error helps determine the quality of transformation.

This chapter is divided into five sections. Section 1 discusses existing GIS data on the Internet, including examples from different levels of government and private companies. Sections 2 and 3 cover metadata and the data exchange methods, topics important to the use of existing data. Section 4 provides an overview of creating new GIS data from satellite images, GPS, and paper maps. Both manual digitizing and scanning are included in the section. This chapter concludes with geometric transformation and the interpretation of RMS error.

4.2 EXISTING GIS DATA

To find existing GIS data for a project is often a matter of knowledge, experience, and luck. Government agencies at different levels have set up websites for sharing public data (Onsrud and Rushton 1995) and for directing users to the source of the desired information. The Internet is also a medium for finding existing data from nonprofit organizations and private companies. But searching for GIS data on the Internet can be difficult. A keyword search with GIS will probably result in thousands of matches, but most hits are irrelevant to the user. Internet addresses may be changed or discontinued. Data on the Internet may be in a format that is incompatible with the GIS package used for a project, or to be usable for a project, the data may need extensive processing such as clipping the study area from a large data set.

Most GIS data on the Internet, especially from government agencies, belong to **framework data**, that is, data that most organizations regularly use for GIS activities. Framework data typically include seven basic layers: geodetic control (accurate positional framework for surveying and mapping), orthoimagery (rectified imagery such as ortho photos), elevation, transportation, hydrography, governmental units, and cadastral information (http://www.fgdc.gov/framework/framework.html/). This section discusses existing GIS data from public agencies and private companies as of 2000.

4.2.1 Public Data

4.2.1.1 Federal Geographic Data Committee

Public data are often free and downloadable from the Internet. All levels of government let GIS users access their public data through clearinghouses. To GIS users looking for data, the website maintained by the **Federal Geographic Data Committee** (**FGDC**) (http://www.fgdc.gov/) is a good start. Consisting of 16 federal agencies, FGDC coordinates the development of the National Spatial Data Infrastructure (NSDI), which is aimed at the sharing of geospatial data throughout all levels of government, the private and nonprofit sectors, and the academic community. The FGDC website provides a link to the Geospatial Data Clearinghouse, a collection of more than 100 spatial data nodes in the U.S. and overseas. Also available is a listing of websites that provide mostly free U.S. geospatial and attribute data.

4.2.1.2 U.S. Geological Survey

Through its national mapping program, the U.S. Geological Survey (USGS) is the major provider of GIS data in the U.S. Its website (http://mapping.usgs.gov/) offers pathways to USGS national mapping and remotely sensed data and to thematic data clearinghouses on national biological information, geologic information, and water resources information. Vector data that can be downloaded from the USGS website include **Digital Line Graphs** (**DLG**s) and land use/land cover data. DLGs are digital representations of point, line, and area features from USGS quadrangle maps (quads) at the scales of 1:24,000, 1:100,000, and 1:2,000,000. DLGs include such data categories as hypsography (i.e., contour lines and spot elevations), hydrography, boundaries, transportation, and the U.S. Public Land Survey System. DLGs contain attribute data and are topologically structured.

USGS has compiled land use and land cover (LULC) data from aerial photographs and earlier land use maps and field surveys. Level 1 of the data includes the categories of urban or built-up land, agricultural land, rangeland, forest land, water, wetland, barren land, tundra, and perennial snow and ice. Each level-1 category is subdivided by a modified Anderson level II scheme (Anderson et al. 1976). For example, water at level 1 is subdivided into streams and canals, lakes, reservoirs, and bays and estuaries at level 2. The LULC data use a minimum-mapping unit of 10 acres or 4 hectares. These data are available at scales of 1:250,000 and 1:100,000, sorted by state, and also in vector and raster formats.

4.2.1.3 U.S. Bureau of the Census

The U.S. Bureau of the Census offers the **TIGER** (**Topologically Integrated Geographic Encoding and Referencing**) database for the 1990 Census (Sperling 1995). The TIGER database contains legal and statistical area boundaries that can be linked to the census data. The database also includes roads, railroads, streams, water bodies, power lines, and pipelines. An earlier database prepared by the Census Bureau, the GBF/DIME (Geographic Base File/Dual Independent Map Encoding), formed the cartographic base for 345 urban centers in TIGER. This cartographic base was then joined with USGS data compiled from the 1:100,000 scale quads. The TIGER database is attribute-rich. For example, it contains the range of address numbers located along each side of each street segment for most urban areas of the country.

4.2.1.4 Statewide Public Data: An Example

Most states in the U.S. have designated websites for statewide GIS data (Appendix C). The Montana State Library, for example, maintains a website on GIS data in Montana (http://www.nris.state.mt.us/gis/gis.html/). This website contains both statewide and regional data. Statewide data include such categories as people/places, land/water, administration, map sheets/grids, and environmental monitoring. Most data are available in ARC/INFO export files and ArcView shapefiles for downloading.

4.2.1.5 Regional Public Data: An Example

The Greater Yellowstone Area Data Clearinghouse (GYADC) (http://www.sdvc.uwyo.edu/gya/) is a FGDC data node sponsored by a group of federal agencies, state agencies, universities, and nonprofit organizations. This data clearinghouse is unique in that it focuses on basic framework data for Yellowstone and Grand Teton national parks.

4.2.1.6 Metropolitan Public Data: An Example

Sponsored by 18 cities and the county government in the San Diego area, the San Diego Association of Governments (SANDAG) (http://www.sandag.cog.ca.us/index2.html/) is an example of a metropolitan data clearinghouse. Data that can be downloaded free from SANDAG's website include administrative boundaries, base map features, district boundaries, land cover and activity centers, transportation, and sensitive lands/natural resources.

4.2.2 GIS Data from Private Companies

Most GIS companies are engaged in software development, technical service, consulting, and data production. Some also provide data on the Internet or can direct GIS users to suitable sources. ESRI (Environmental Systems Research Institute, Inc.), for example, maintains a website called ArcData Online (http://www.esri.com/data/online/). From the site GIS users can download free data or samples of commercial data. Data selection can be based on geographic area, data type, or data provider.

Some companies provide specialized GIS data for their customers. GDT (Geographic Data Technology) offers street, address, postal, and census databases at its website (http://www.geographic.com/). Etak (now Tele Atlas North America, http://www.etak.com/) offers road and address databases for urban centers and rural areas.

4.3 METADATA

Metadata provide information about spatial data. They are particularly important to GIS users who want to use public data for their projects. First, metadata let GIS users know if public data meet their specific needs for area coverage, data quality, and data currency. Second, metadata show GIS users how to process and interpret spatial data. Third, metadata include the contact information for GIS users who need more information.

FGDC has developed the content standards for metadata. These standards have been adopted by federal agencies in developing their public data. FGDC metadata standards describe a data set based on the following:

- Identification information—basic information about the data set, including title, geographic data covered, and currency.

- Data quality information—information about the quality of the data set, including positional and attribute accuracy, completeness, consistency, sources of information, and methods used to produce the data.
- Spatial data organization information— information about the data representation in the data set such as method for data representation (e.g., raster or vector) and number of spatial objects.
- Spatial reference information—description of the reference frame for and means of encoding coordinates in the data set such as the parameters for map projections or coordinate systems, horizontal and vertical datums, and the coordinate system resolution.
- Entity and attribute information— information about the content of the data set, such as the entity types and their attributes and the domains from which attribute values may be assigned.
- Distribution information—information about obtaining the data set.
- Metadata reference information— information on the currency of the metadata information and the responsible party.

The FGDC website (http://www.fgdc.gov/metadata/metadata.html/) contains detailed information about the above metadata standards. Appendix D shows a complete example of metadata that describe a city and town coverage from a statewide clearinghouse. Each of the entries in Appendix D must be collected from the appropriate information resources and entered. USGS and NASA (National Aeronautical and Space Administration) have recently developed software tools that can assist the entry of metadata (Guptill 1999).

4.4 CONVERSION OF EXISTING DATA

Existing data from government or private sources often require **data conversion** to convert the data to a format compatible with the GIS package. Existing data cannot be easily converted because of a large variety of GIS packages and data formats in use. The choice of a conversion method basically depends upon the specificity of the data format. Proprietary data formats require special translators for data conversion, while neutral or public formats require a GIS package to have translators that can work with the formats (Robinson et al. 1995).

4.4.1 Direct Translation

Direct translation uses a translator or algorithm in a GIS package to directly convert spatial data from one format to another (Figure 4.1). Direct translation used to be the only method for data conversion before the development of data standards and open GIS. Direct translation is still the dominant method, which GIS users prefer because it is easier to use than other methods. ArcView, for example, has translators to convert ARC/INFO's interchange files, MGE and Microstation's DGN files, Auto-CAD's DXF and DWG files, and MapInfo files into shapefiles. Likewise, GeoMedia can access and integrate data from ARC/INFO, ArcView, AutoCAD, MapInfo, MGE, and Microstation.

4.4.2 Neutral Format

A **neutral format** is a public or de facto format for data exchange. DLG is a familiar neutral format to GIS users. USGS provides DLG files, and other government agencies such as the Natural Resources Conservation Service (NRCS, formerly the Soil Conservation Service) distribute digital soil maps in the DLG format. This is because most

GIS packages have translators to work with the DLG format (Figure 4.2).

A new entry in neutral format is the **Spatial Data Transfer Standard** (**SDTS**) (http://mcm-cweb.er.usgs.gov/sdts/). Approved by the Federal Information Processing Standards (FIPS) Program in 1992, SDTS is now mandatory for federal agencies. In practice, SDTS uses "profiles" to transfer spatial data. Each profile is targeted at a particular type of spatial data. Two profiles are currently available. The Topological Vector Profile (TVP) covers DLG, TIGER, and other vector data. The Raster Profile covers digital orthophotos, DEM, and other raster data, which are discussed in Chapter 7. From all indications, SDTS with its Topological Vector Profile and Raster Profile will eventually become the standard format for public data from federal agencies and probably from state and local agencies as well. GIS vendors have already developed translators for SDTS data.

The Vector Product Format (VPF) is a standard format, structure, and organization for large geographic databases that are based on a georelational data model. The National Imagery and Mapping Agency (NIMA) uses VPF for digital vector products developed at a variety of scales (http://www.nima.mil/). NIMA's vector products for drainage system, transportation, political

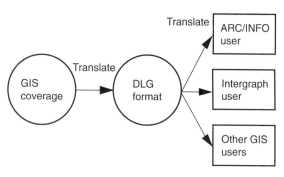

Figure 4.2
To accommodate users of different GIS packages, a government agency can translate public data into a neutral format such as the DLG format. Using the translator in the GIS package, the user can convert the neutral data file into the format used in the GIS.

Figure 4.1
The IGDSARC command in ARC/INFO converts a DGN file prepared by an Intergraph user to an ARC/INFO coverage.

boundaries, and populated places are also part of the global database that is being developed by the International Steering Committee for Global Mapping (ISCGM) (http://www.iscgm.org/).

Although neutral format is typically used for public data from government agencies, it can also be found with "industry standards" in the private sector. A good example is the DXF (Drawing Interchange File) format of AutoCAD. Another example is the ASCII format. Many GIS packages can import ASCII files, which have point data with x-, y-coordinates, into digital maps.

4.5 CREATING NEW DATA

Data sources for creating new data include remotely sensed data, GPS data, and paper maps. Remotely sensed data and GPS data are primary data sources in that they are raw data and have not been manipulated and processed as with paper maps. Paper maps, such as USGS quads, are secondary data sources. Maps have gone through the cartographic processes of compilation, generalization, and symbolization. Each of these processes

can affect the accuracy of the mapped data. Although high-resolution remotely sensed data and GPS data have become important sources for vector data input, paper maps are still the predominant source for creating new GIS data.

4.5.1 Remotely Sensed Data

Remotely sensed data, such as digital orthophotos (DOQs) and satellite images, are data acquired by a sensor from a distance. Although they are raster data, DOQs and satellite images are useful for vector data input. DOQs are digitized aerial photographs that have been differentially rectified or corrected to remove image displacements by camera tilt and terrain relief. Black-and-white DOQs have a 1-meter ground resolution (i.e., each pixel in the image measures 1 by 1 meter on the ground) and have pixel values representing 256 gray levels. Because DOQs combine the image characteristics of a photograph with the geometric qualities of a map, they can be effectively used as a background for digitizing or updating of existing digital maps (Figure 4.3). GIS users can process satellite images and extract data for a variety of maps in vector format

Figure 4.3
A digital orthophoto (DOQ) can be used as the background for digitizing or updating of existing coverages.

The following printout is an example of GPS data. The header information shows that the datum used is NAD27 (North American Datum 1927) and the coordinate system is UTM (Universal Transverse Mercator). The GPS data include seven point locations. The record for each point location includes the UTM zone number (i.e., 11), easting (*x*-coordinate), and northing (*y*-coordinate). The GPS data do not include the height or ALT value.

H R DATUM
M G NAD27 CONUS

H Coordinate System
U UTM UPS

H	IDNT	Zone	Easting	Northing	Alt	Description
W	001	11T	0498884	5174889	-9999	09-SEP-98
W	002	11T	0498093	5187334	-9999	09-SEP-98
W	003	11T	0509786	5209401	-9999	09-SEP-98
W	004	11T	0505955	5222740	-9999	09-SEP-98
W	005	11T	0504529	5228746	-9999	09-SEP-98
W	006	11T	0505287	5230364	-9999	09-SEP-98
W	007	11T	0501167	5252492	-9999	09-SEP-98

such as land use and land cover, hydrography, water quality, and areas of eroded soils.

4.5.2 GPS Data

Using satellites in space as reference points, a GPS receiver can determine its precise position on the Earth's surface (Moffitt and Bossler 1998). **GPS data** include the horizontal location based on the geographic grid or a coordinate system and, if chosen, the elevation of the point location (Box 4.1). When taking GPS positions along a line such as a road, one can determine the linear feature by the collection of positional readings. By extension, a polygon can be determined by a series of lines measured by GPS. This is why GPS has become a useful tool for spatial data input (Kennedy 1996).

The method by which a GPS receiver determines its position on the Earth's surface is similar to triangulation in surveying. The GPS receiver measures its distance (range) from a satellite using the travel time of radio signals it receives from the satellite. With three satellites simultaneously available, the receiver can determine its position in space (*x*, *y*, *z*), relative to the center of mass of the Earth. But to correct timing errors, a fourth satellite is required to get precise positioning (Figure 4.4). The receiver's position in space can then be converted to latitude, longitude, and height based on a reference ellipsoid.

The U.S. military maintains a constellation of 24 NAVSTAR (Navigation Satellite Timing and Ranging) satellites in space, and each satellite follows a precise orbit. This constellation gives GPS users between five and eight satellites visible from any point on the Earth's surface. A good practice is to pick satellites that are widely separated to minimize errors in GPS readings.

An important aspect of using GPS for spatial data entry is the need to correct errors in GPS data. The first type of error may be described as noise, which can come from different sources. Although a satellite orbit is precisely defined, it may involve slight position or ephemeris (orbital) errors between monitoring times. Distance between a GPS receiver and a satellite is calculated by multiplying the travel time of a radio signal by the speed of light, but the speed of light is not constant as the signal passes through the outer space and the atmosphere to the Earth's surface. When it does reach the surface, the signal may bounce off obstructions before reaching the receiver, thus creating a problem called multi-path interference.

The second type of error is intentional. For example, to make sure that no hostile force can get precise GPS readings, the U.S. military used to

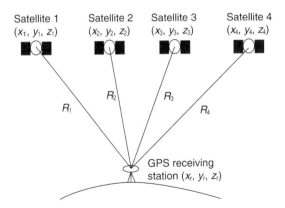

Figure 4.4
Use four GPS satellites to determine the coordinates of a receiving station. x_i, y_i, and z_i are coordinates relative to the center of mass of the Earth. R_i represents the distance (range) from a satellite to the receiving station.

Figure 4.5
A portable GPS receiver.
(Courtesy of Trimble.)

degrade their accuracy under the policy called "Selective Availability" or "SA" by introducing noise into the satellite's clock and orbital data. On May 2, 2000, SA was switched off. Without the effect of SA, the horizontal accuracy of GPS readings is improved by almost a factor of 10.

With the aid of a reference or base station, **differential correction** can significantly reduce noise errors. A reference station is located at a point that has been accurately surveyed. Using its known position, the reference receiver can calculate what the travel time of the GPS signals should be. The difference between the predicted and actual travel times thus becomes an error correction factor.

The reference receiver computes error correction factors for all visible satellites. These correction factors are then available to GPS receivers within about 300 miles (500 kilometers) of the reference station. GIS applications usually do not need real-time transmission of error correction factors. Differential correction can be made later as long as records are kept of measured positions and the time each position is measured. Reference stations are operated by private companies and by public agencies such as the U.S. Coast Guard. GIS users considering use of differential GPS data should find out their closest reference stations.

Equally important as correcting errors in GPS data is the type of GPS receiver. Most GIS users use code-based receivers (Figure 4.5). With differential correction, code-based GPS readings can easily achieve the accuracy of 3 to 5 meters, and some newer receivers are even capable of submeter accuracy. Carrier phase receivers, the other type of GPS receivers, are mainly used in geodetic control and precise surveying and are capable of sub-centimeter differential accuracy (Lange and Gilbert 1999).

4.5.3 Digitizing Using A Digitizing Table

Digitizing is the process of converting data from analog to digital format. Digitizing using a digitizing table is also called manual digitizing (Figure 4.6). A **digitizing table** has a built-in electronic mesh, which can sense the position of the cursor. To transmit the x-, y-coordinates of a point to the connected computer, the operator simply clicks on a button on the cursor after lining up the cursor's cross hair with the point. Large-size digitizing tables typically have an absolute accuracy of 0.001 inch (0.003 centimeter).

GIS packages typically have a built-in digitizing module for manual digitizing. A topology-

Figure 4.6
A large digitizing table and a cursor with a 16-button keypad.
(Courtesy of GTCO Calcomp, Inc.)

based GIS package has functionalities to ensure the topology of the digitized map. ARC/INFO, for example, offers several commands to help users in digitizing. The ARCSNAP command allows the user to specify a distance within which an arc will be snapped to an existing arc (Figure 4.7). The NODESNAP command works the same way as ARCSNAP except it snaps a node to an existing node (Figure 4.8). The INTERSECTARCS command calculates arc intersections and adds nodes at intersections (Figure 4.9).

Digitizing usually begins with a set of control points, which are later used for converting the digitized map to real-world coordinates. Digitizing point features is simple: each point is clicked once to record its location. Digitizing line or area features can follow either point mode or stream mode. The operator selects points to digitize in point mode. In stream mode, lines are digitized at a preset time or distance interval. For example, lines can be automatically digitized at a 0.01-inch interval. Most GIS users prefer point mode because point mode creates a smaller data file than stream mode and is more efficient in digitizing simple line features with straight-line segments.

Digitizing line or polygon features can also follow either discrete or continuous mode. In

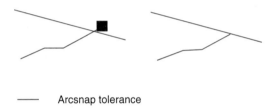

Arcsnap tolerance

Figure 4.7
The arcsnap tolerance allows the end of an arc to be snapped to an existing arc. The left diagram shows the overextension of a digitized arc. Because the overextension is smaller than the arcsnap tolerance, the end of the digitized arc is snapped to an existing arc, as shown in the right diagram.

discrete mode the operator observes the arc-node topological relationship. Points are digitized as nodes if they are where lines meet or intersect. In continuous mode, which is also called spaghetti digitizing, the operator digitizes long, continuous lines and lets the GIS package build the arc-node relationship in processing.

Because the vector data model treats a polygon as a series of lines, digitizing polygon features is the same as digitizing line features except that each polygon requires a label in addition to its boundary. The label is needed to link the polygon with its attribute data.

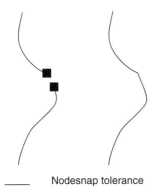

_____ Nodesnap tolerance

Figure 4.8
The nodesnap tolerance allows the snapping of nodes. The left diagram shows the digitized arc not reaching its intended end point. Because the gap between the nodes are smaller than the nodesnap tolerance, the end node of the arc is snapped to the other node.

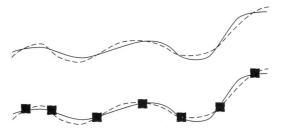

Figure 4.10
Duplicate lines create problems for editing. The top diagram shows the same arc was digitized twice. After the topology of the coverage is built, nodes are created at the intersections of the duplicate lines and a series of small polygons are formed between the lines in the bottom diagram. The extra arcs must be removed in editing.

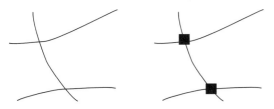

Figure 4.9
The intersectarcs option calculates arc intersections and adds nodes at the intersections. The left diagram shows the arcs were digitized without adding nodes at the intersections. The right diagram shows the creation of nodes with the intersectarcs option.

Although digitizing itself is mostly manual, the quality of digitizing can be improved with planning and checking. An integrated approach is useful in digitizing different map layers of a GIS database that share common boundaries. For example, soils, vegetation types, and land use types may share some common boundaries in a study area. Digitizing these boundaries only once and using them on each of the map layers not only saves time in digitizing but also ensures the matching of the layers.

A rule of thumb in digitizing line or polygon features is to digitize each line once and only once

to avoid duplicate lines. Duplicate lines are seldom on top of one another because of the high accuracy of a digitizing table. In fact, duplicate lines form a series of polygons between them and are difficult to correct in editing (Figure 4.10). One way to reduce the number of duplicate lines is to put a transparent sheet on top of the source map and to mark off each line on the transparent sheet after the line is digitized. This method can also reduce the number of missing lines.

4.5.4 Scanning

Scanning is a digitizing method that converts an analog map into a scanned file, which is then converted back to vector format through tracing (Verbyla and Chang 1997). A scanner (Figure 4.11) converts an analog map into a scanned image file in raster format. The map to be scanned is typically a black-and-white map: black lines represent map features, and white areas represent the background. The map may be a paper map or a Mylar map, and inked or penciled. Scanning converts the map into a binary scanned file in raster format; each pixel has a value of either 1 (map feature) or 0 (background). Map features are shown as raster lines, a series of connected pixels on the scanned file

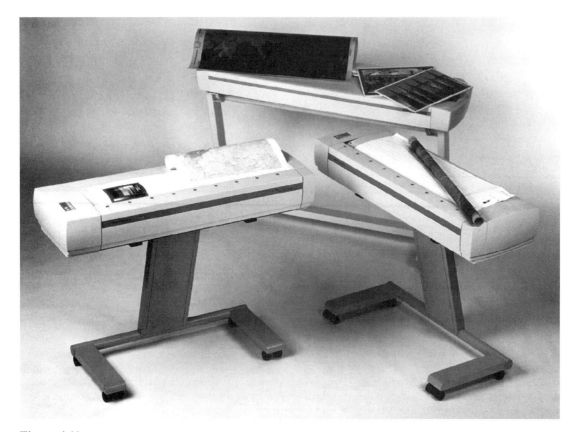

Figure 4.11
Large format drum-scanners.
(Courtesy of GTCO Calcomp, Inc.)

(Figure 4.12). The pixel size depends on the scanning resolution, which is often set at 300 dots per inch (dpi) or 400 dpi for digitizing. A raster line representing a thin inked line on the source map may have a width of 5 to 7 pixels (Figure 4.13).

Scribe sheets and color maps can also be scanned for digitizing. Because a scribed sheet is essentially a negative of a black-and-white map, the values of 1 and 0 need to be reversed for the scanned file. Expensive scanners can recognize colors. The Digital Raster Graph (DRG) from the USGS is a raster image scanned from the USGS quad. The DRG has 13 different colors, each representing one type of map feature on the USGS quad.

To complete the digitizing process, a scanned file must be vectorized. **Vectorization** turns raster lines into vector lines in a process called tracing. Commercial GIS packages such as ARC/INFO offer raster-to-vector algorithms for tracing. Tracing involves three basic elements: line thinning, line extraction, and topological reconstruction (Clarke 1995). Lines in the vector data model have length but no width. Lines in a scanned file (raster lines), however, usually occupy several pixels in width. Raster lines must be thinned, ideally to a one-cell width, for vectorization. Line extraction is the process of determining where individual lines begin and end. Finally, topology is built between lines extracted from the raster model. Results of raster-to-vector conversion often show step-like features along diagonal lines. A line smoothing operation can eliminate those artifacts from raster data.

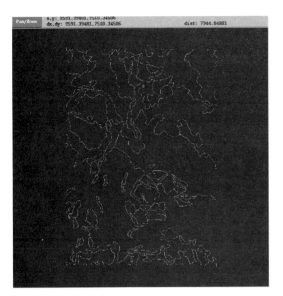

Figure 4.12
A binary scanned file: the lines are soil lines, and the black areas are the background.

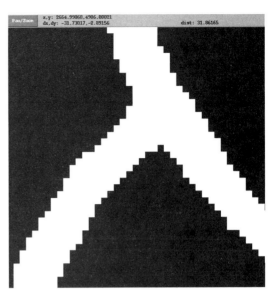

Figure 4.13
A raster line on a scanned file has a width of several pixels.

Tracing can be semi-automatic or manual. In semi-automatic mode, the user selects a starting point on the image map and lets the computer trace all the connecting raster lines (Figure 4.14). In manual mode, the user determines the raster line to be traced and the direction of tracing.

Scanning is preferred over manual digitizing for several reasons. First, scanning uses the machine and computer algorithm to do most of the work, thus avoiding human error caused by fatigue or carelessness. Second, tracing has both the scanned image and vector lines on the same screen, making tracing more flexible than manual digitizing. With tracing, the operator can zoom in or out and can move around the image map with ease. Manual digitizing requires the operator's attention on both the digitizing table and the computer monitor, and the operator can easily get tired. Third, scanning has been reported to be more cost effective than manual digitizing. The cost of scanning by a service company has dropped significantly in recent years. Scanning a black-and-white map the size of a USGS quad costs less than $10.

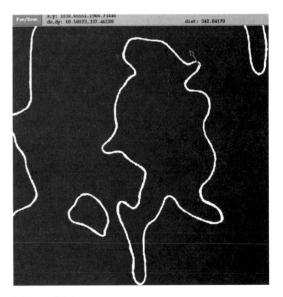

Figure 4.14
Semi-automatic tracing starts at a point (shown with an arrow) and traces all lines connected to the point.

The tracing algorithm for raster-to-vector conversion plays a key role in digitizing by scanning. No single tracing algorithm will work satisfactorily with different types of maps under different conditions. The algorithm must decide how to trace a vector line within a raster line that encompasses five to seven pixels. The width of a raster line actually doubles or triples when lines meet or intersect (Figure 4.15). The algorithm must also decide where to continue when a raster line is broken or when two raster lines are close together. Small mistakes are common in tracing. ARC/INFO users can adjust parameter values used in tracing. For example, line-straightening properties, which apply to lines at intersections or corners, may be adjusted depending on if the map contains human-made features (i.e., more straight lines) or natural features (i.e., more smooth, continuous lines). ARC/INFO users can also specify the maximum raster line width to be traced and the maximum distance between raster line segments to be jumped.

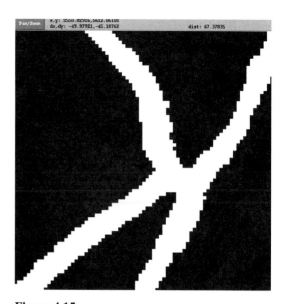

Figure 4.15
The width of a raster line doubles or triples when lines meet or intersect.

4.5.5 On-screen Digitizing

On-screen digitizing is an alternative to manual digitizing and scanning for limited digitizing work such as editing or updating an existing map. **On-screen digitizing** is manual digitizing on the computer monitor using a data source such as a DOQ as the background. This is an efficient method for digitizing, for example, new trails or roads that are not on an existing map but are on a new DOQ. Likewise, the method can be used for editing a vegetation map based on new information from a new DOQ that shows recent clear-cuts or burned areas. The resolution of the computer monitor can be a problem for on-screen digitizing.

4.5.6 Importance of Source Maps

Digitizing converts a source map to its digital format. The accuracy of the digital map is therefore directly related to the accuracy of the source map. The digital map can be only as good or as accurate as its source map. If the compilation of the source map contains errors, the errors will be passed on to the digital map. Besides the accuracy of the source map, other factors can also affect the accuracy of the digital map.

Paper maps generally are not good source maps because they tend to shrink and expand with changes in temperature and humidity. In even worse scenarios, GIS users may use copies of paper maps or mosaics of paper maps for digitizing. Such source maps will not yield good results. Their plastic backing makes Mylar maps much more stable than paper maps and better for digitizing.

The quality of line work on the source map will determine not only the accuracy of the digital map but also the GIS user's time and effort in digitizing and editing. The line work should be thin, continuous, and uniform, as expected from inking or scribing. Felt-tip markers should never be used to prepare the line work. Penciled source maps may be adequate for manual digitizing but are not recommended for scanning. Scanned files are binary data files, separating only the map feature from the background. Because the contrast between penciled lines and the background (i.e., surface of paper or Mylar) is not as sharp as inked lines, scanning parameters need to be adjusted to increase the contrast. The result of the adjustment is that any erased lines and smudges are also scanned, thus messing up the scanned file. For words or symbols on the source map to give supplemental information, they can be drawn in orange color, which will not be scanned.

4.6 GEOMETRIC TRANSFORMATION

A newly digitized map is measured in the same measurement unit as the source map used for digitizing or scanning. Before the digitized map can be used in a GIS project, it must be converted to real-world coordinates such as UTM coordinates. **Geometric transformation** is the process of converting a digital map or an image from one coordinate system to another using a set of control points and transformation equations.

Figure 4.16 illustrates a typical scenario of geometric transformation. The process begins with digitizing a set of control points, which are points with known longitude and latitude values or real-world coordinates. For example, a USGS 1:24,000 scale quad has 16 points with known longitude and latitude values: 12 points along the border of the quad, and four additional points within the quad. These 16 points divide the quad into 2.5 minutes in longitude and latitude. The next step is to convert the longitude and latitude readings of the control points to real-world coordinates. This step can be saved if the real-world coordinates of the control points are already known. For example, a state agency in charge of GIS services may have already compiled a master tic (control point) file that lists the UTM coordinates of quad corner points in a state. A map without map features is then copied from the digitized map, and its control points are updated to their real-world coordinates. Finally, the digitized map is converted into real-world coordinates through the use of a transformation method and the control points.

Different methods are available for geometric transformation (Taylor 1977). Each method is distinguished by the geometric properties it can preserve and by the operations or changes it allows on an object such as a rectangle corresponding to a USGS quad (Figure 4.17).

- Equi-area (congruence) transformation: The method allows rotation of the rectangle and preserves its shape and size.
- Similarity transformation: The method allows rotation of the rectangle and preserves its shape but not size.
- **Affine transformation**: The method allows angular distortion but preserves the parallelism of lines.
- Projective transformation: The method allows angular and length distortion, thus allowing the rectangle to be transformed into an irregular quadrilateral.
- Topological transformation: The method preserves the topological properties of an object but not the shape, thus allowing the rectangle to be transformed into a circle.

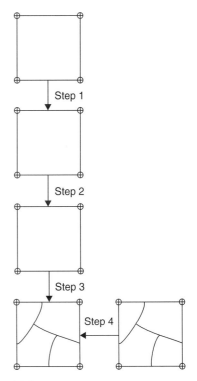

Figure 4.16
A typical scenario of geometric transformation. Step 1 updates the tics (control point) from digitizer units to longitude and latitude values. Step 2 projects the tics to real-world coordinates. Step 3 creates a coverage with tics only. Step 4 converts the digitized map features to real-world coordinates and places them in the output coverage.

4.6.1 Affine Transformation

Although GIS packages such as ARC/INFO and MGE offer similarity, affine, and projective transformation methods, most GIS users use the affine transformation on digital maps and satellite images and the projective transformation on aerial photographs with displacement. The following discussion thus focuses on the operations allowed in affine transformations.

While preserving line parallelism, the affine transformation allows rotation, translation, skew, and differential scaling on the rectangular object (Pettofrezzo 1978). Rotation rotates its x and y axes from the origin. Translation shifts its origin to a new location. Skew changes its shape to a parallelogram with a slanted direction. And differential scaling changes the scale by expanding or reducing in the x and/or y direction (Figure 4.18).

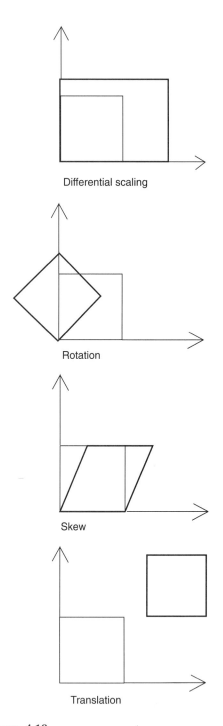

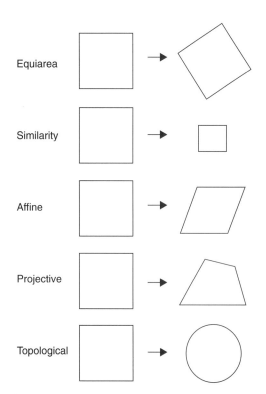

Figure 4.17
Different types of geometric transformation between coordinate systems.

Figure 4.18
Differential scaling, rotation, skew, and translation in affine transformation.

Because geometric transformation of a map is based on the control points, the affine transformation and its operations are first applied to the control points. In other words, the control points are transformed from their locations in the digitized map (also called the input or estimated control points) to their locations in real-world coordinates (also called the output or actual control points). Mathematically, the affine transformation is expressed as a pair of linear equations:

$$x' = Ax + By + C \qquad (4.1)$$
$$y' = Dx + Ey + F \qquad (4.2)$$

where x and y are in digitizer units, x' and y' are in real-world coordinates, and $A, B, C, D, E,$ and F are the transformation coefficients. The six coefficients can be estimated by using the digitized location and the real-world coordinates of the control points. A minimum of three control points is sufficient for the estimation. But often four or more control points are used to reduce problems in measurement errors. With four or more control points, a least squares solution, similar to that of a regression analysis, is used to estimate the transformation coefficients (Box 4.2). After the coefficients are estimated, the transformation equations are applied to the map features in the digitized map and convert their coordinates from digitizer units to real-world coordinates.

4.6.2 Geometric Interpretation of Transformation Coefficients

The geometric properties of the affine transformation can be interpreted from the transformation coefficients. The coefficient C represents the translation in the x direction, and F the translation in the y direction. The coefficients of $A, B, D,$ and E are related to rotation, skew, and scaling in the following equations:

Box 4.2   Estimation of Transformation Coefficients

The example used here and later in the chapter is a third-quad soil map (one third of a USGS 1:24,000 scale quad map), which has been scanned at 300 dpi. The map has four control points marked at the corners: Tic 1 at the NW corner, Tic 2 at the NE corner, Tic 3 at the SE corner, and Tic 4 at the SW corner. X and Y denote the control points' real-world (output) coordinates in meters based on the UTM coordinate system, and x and y denote the control points' digitized (input) locations. The measurement unit of the digitized locations is (1/300) of an inch, corresponding to the scanning resolution. The following table shows the input and output coordinates of the control points:

Tic-ID	x	y	X	Y
1	465.403	2733.558	518843.844	5255910.5
2	5102.342	2744.195	528265.750	5255948.5
3	5108.498	465.302	528288.063	5251318.0
4	468.303	455.048	518858.719	5251280.0

One method for deriving the transformation coefficients is to run two multiple regression analyses: the first regresses X on x and y, and the second regresses Y on x and y. The results show

$$A = 2.032, B = -0.004, C = 517909.198,$$
$$D = 0.004, E = 2.032, F = 5250353.802$$

The other method is to estimate the transformation coefficients using the following equation in matrix form:

$$\begin{bmatrix} C & F \\ A & D \\ B & E \end{bmatrix} = \begin{bmatrix} n & \Sigma x & \Sigma y \\ \Sigma x & \Sigma x^2 & \Sigma xy \\ \Sigma y & \Sigma xy & \Sigma y^2 \end{bmatrix}^{-1} \cdot \begin{bmatrix} \Sigma X & \Sigma Y \\ \Sigma xX & \Sigma xY \\ \Sigma yX & \Sigma yY \end{bmatrix}$$

where n is the number of control points and all other notations are the same as previously defined. The transformation coefficients derived from the equation are the same as from the regression analysis.

$A = Sx \cos(t)$ (4.3)
$B = Sy [k \cos(t) - \sin(t)]$ (4.4)
$D = Sx \sin(t)$ (4.5)
$E = Sy [k \sin(t) + \cos(t)]$ (4.6)

where Sx is the change of scale in x, Sy is the change of scale in y, t is the rotation angle, and k is the shear factor. The skew angle can be derived from arctan (k). The equations for A and D can be used to solve for t and, in turn, Sx. The equations for B and E can be used to solve for k and, in turn, Sy and the skew angle. ARC/INFO lists the transformation coefficients and the geometric properties of transformation in the output from the TRANSFORM command (Box 4.3).

4.6.3 Root Mean Square Error

Control points play a crucial role in geometric transformation. The quality of the control points with respect to their estimated and actual locations determines the accuracy of geometric transformation and the positional accuracy of digitized map features (Bolstad et al. 1990). One measure of the goodness of the control points is the **root mean square** (**RMS**) error, which measures the deviation between the actual location of the control points, as projected from their longitude and latitude readings into the output map, and the esti-

mated location, as digitized on the input map. Typically, RMS errors are listed for each control point, and for the average from all control points. The equations for computing the RMS errors are as follows:

$$\text{RMS for a tic} = sqrt\,[(\text{act}X - \text{est}X)^2 + (\text{act}Y - \text{est}Y)^2] \quad (4.7)$$

$$\begin{aligned}\text{Average} \\ \text{RMS}\end{aligned} = sqrt\,[(\text{sum of squares of deviation in } X \text{ and } Y) / (\text{number of control points})] \quad (4.8)$$

The output from the TRANSFORM command in ARC/INFO lists the average RMS and x-error and y-error for each control point (Box 4.4). To ensure the accuracy of geometric transformation, the RMS error should be within a tolerance value. The tolerance value, which is not included in the published metadata standards, should vary depending on the accuracy and the map scale of the source map. If the RMS error exceeds the tolerance, either the control points need to be re-digitized or a new set of control points has to be entered. If the RMS is within the acceptable range, then the assumption is that this same level of accuracy can also be applied to map features, which are usually bounded within the control points.

Box 4.3   **Output from ARC/INFO's TRANSFORM Command**

Using the data from Box 4.2, the following shows the output from the TRANSFORM command that is related to the transformation coefficients and the geometric transformation:

Scale (X, Y) = (2.032, 2.032); Skew (degrees) = (-0.014)

Rotation (degrees) = (0.102); Translation = (517909.198, 5250353.802)

Affine $X = Ax + By + C$
$\qquad Y = Dx + Ey + F$
$A = 2.032; B = -0.004; C = 517909.198;$
$D = 0.004; E = 2.032; F = 5250353.802$

The positive rotation angle means a rotation counterclockwise from the x-axis, and the negative skew angle means a shift clockwise from the y-axis. Both angles are very small, meaning that the change from the original rectangle to a parallelogram through the affine transformation is very slight.

The above assumption relating RMS to the location accuracy of digitized map features can be wrong if gross errors are made in digitizing the control points or in inputting the longitude and latitude readings of the control points. Suppose we shift the locations of control points 2 and 3 (the two control points to the east) on the previous third quad by increasing their x values by a constant. The RMS error would remain about the same because the object formed by the four control points retains the shape of a parallelogram. The soil lines, however, would deviate from their locations on the source map. The same problem occurs, if we increase the x values of control points 1 and 2 (the upper two control points) by a constant and decrease the x values of control points 3 and 4 (the lower two control points) by a constant (Figure 4.19). In fact, the RMS error would be well within the tolerance value as long as the object formed by the shifted control points remains as a parallelogram.

Longitude and latitude readings printed on paper maps are sometimes erroneous. This would lead to acceptable RMS errors but significant location errors of digitized map features. Suppose the latitude readings of control points 1 and 2 (the up-per two control points) are off by 10" (e.g., 47°27'20" instead of 47°27'30"). The RMS error from transformation would be acceptable but the soil lines would deviate from their locations on the source map (Figure 4.20). The same problem occurs if the longitude readings of control points 2 and 3 (the two control points to the east) are off by 30" (e.g., −116°37'00" instead of −116°37'30"). Again, this happens because the affine transformation works with parallelograms. Although we tend to take for granted the accuracy of published maps, erroneous longitude and latitude readings are quite common, especially with inset maps (maps that are smaller than the regular size) and oversized maps (maps that are larger than the regular size).

GIS users typically use the four corner points of the source map as control points. This practice makes sense because the exact readings of longitude and latitude are usually shown at those points. Moreover, using the corner points as control points, it helps the process of joining the map with its surrounding maps. The practice of using four corner control points, however, does not preclude the use of more control points if additional points with known locations are available. As mentioned

---

Box 4.4   RMS Report from ARC/INFO's TRANSFORM Command

Output from the TRANSFORM command includes RMS statistics. The following shows the RMS report using the data from Box 4.3.

The output shows that the average deviation between the input and output locations of the control points is 0.280 meter based on the UTM coordinate system, or 0.00046 inch (0.138 divided by 300) based on the digitizer unit. This RMS error is well within the acceptable range. The individual x and y errors suggest that the error is slightly lower in the y direction than the x direction and the average RMS error is equally distributed among the four control points.

RMS Error (input, output) = (0.138, 0.280)

Tic ID	Input x / Output x	Input y / Output y	x error	y error
1	465.403	2733.558		
	518843.844	5255910.5	−0.205	−0.192
2	5102.342	2744.195		
	528265.750	5255948.5	0.205	0.192
3	5108.498	465.302		
	528288.063	5251318.0	−0.205	−0.192
4	468.303	455.048		
	518858.719	5251280.0	0.205	0.192

Figure 4.19

Inaccurate location of soil lines due to input or estimated tic location errors. The thin lines represent correct soil lines and the thick lines incorrect soil lines. In this case, the x values of the upper two tics were increased by 0.2" while the x values of the lower two tics were decreased by 0.2" on a third quad (15.4" x 7.6").

Figure 4.20

Incorrect location of soil lines due to output or actual tic location errors. The thin lines represent correct soil lines and the thick lines incorrect soil lines. In this case, the latitude readings of the upper two tics were off by 10" (e.g., 47°27'20" instead of 47°27'30") on a third quad.

earlier, a least squares solution is used in an affine transformation when more than three points are used as control points. Therefore, the use of more control points means a better coverage of the entire map in transformation. In other situations, control points that are closer to the map features should be used instead of the corner points. This ensures the location accuracy of the map features.

KEY CONCEPTS AND TERMS

Affine transformation: A geometric transformation method commonly used in GIS, which allows rotation, translation, skew, and differential scaling on the rectangular object, while preserving line parallelism.

Data conversion: Conversion of spatial data to the format compatible with the GIS package.

Differential correction: A method that uses data from a base station to correct noise errors in GPS data.

Digital Line Graphs (DLGs): Digital representations of point, line, and area features from USGS quads including contour lines, spot elevations, hydrography, boundaries, transportation, and the U.S. Public Land Survey System.

Digitizing: The process of converting data from analog to digital format.

Digitizing table: A table with a built-in electronic mesh that can sense the position of the cursor and can transmit its x-, y-coordinates to the connected computer.

Direct translation: Use of a translator or algorithm in a GIS package to directly convert spatial data from one format to another.

Federal Geographic Data Committee (FGDC): A U.S. multi-agency committee that coordinates the development of standards for spatial data.

Framework data: Data that most organizations regularly use for GIS activities.

Geometric transformation: The process of converting a map or an image from one coordinate system to another by using a set of control points and transformation equations.

Global Positioning System (GPS) data: Longitude, latitude, and elevation data for point locations made available through a navigational satellite system and a receiver.

Metadata: Data that provide information about spatial data.

Neutral format: A public format such as SDTS that can be used for data exchange.

On-screen digitizing: Manual digitizing on the computer monitor by using a data source such as a DOQ as the background.

Remotely sensed data: Data such as digital orthophotos and satellite images that are acquired by a sensor from a distance.

Root mean square (RMS) error: In geometric transformation, the RMS measures the deviation between the actual location and the estimated location of the tics.

Scanning: A digitizing method that converts an analog map into a scanned file in raster format, which can then be converted back to vector format through tracing.

Spatial Data Transfer Standard (SDTS): Public data formats for transferring spatial data such as DLGs, DEMs, and digital orthophotos from the USGS.

TIGER (Topologically Integrated Geographic Encoding and Referencing): A database prepared by the U.S. Bureau of the Census that contains legal and statistical area boundaries, which can be linked to the census data.

Vectorization: The process of converting raster lines into vector lines through tracing.

APPLICATIONS: VECTOR DATA INPUT

The applications section covers methods for vector data input. Task 1 uses existing data on the Internet. Task 2 lets you digitize several polygons directly on the computer screen. Tasks 3 and 4 show you how you can generate a theme, from an ASCII file or a .dgn file so that the theme can be used in ArcView. The last two tasks require use of ARC/INFO. Task 5 is an exercise on raster to vector conversion using a scanned file. Task 6 is an exercise on geometric transformation.

Task 1: Download a digital map from the Internet

Like many other states, Idaho has a website that lets GIS users download ARC/INFO coverages in either export (e00) format or shapefile format. Maintained by the Idaho Department of Water Resources, the website also provides metadata for the coverages. Task 1 shows you how to import an export file in ArcView, and convert the imported coverage to a shapefile.

1. Go to http://www.idwr.state.id.us/gisdata/, and click GIS Data. Then click Statewide Data.

2. The Statewide Data table lists coverages that can be downloaded and the choices of Preview gif, e00, Shapefile, and Metadata. Click on Metadata for County Boundaries. The metadata gives you information about the coverage. Now RIGHT click on the dot for e00 and select "Save Link As" (or "Save Target As" using Internet Explorer) from the menu. Name the file *idcounty.e00* and provide the path to save the file.

3. The first part of this exercise is to import *idcounty.e00* in ArcView. Select Import71 from the ArcView program group. In the import dialog, enter the path to *idcounty.e00* for the Export File Name and the path to *idcounty* (name of the imported coverage) for the Output Data Source. Import71 converts the import file into an ARC/INFO coverage.

4. Start ArcView, and open a new view. Navigate the path to *idcounty* and add the coverage to view. Check the box next to *idcounty* to view the coverage.

5. Next you want to convert *idcounty* to a shapefile. Make *idcounty* active, and select Convert to Shapefile from the Theme menu. Name the shapefile *idcounty.shp* and specify its path. Add *idcounty.shp* to view. *Idcounty.shp* looks exactly the same as *idcounty*.

Task 2: On-Screen digitizing using ArcView

What you need: *landuse.shp*, a background map for digitizing.

On-screen digitizing is technically similar to manual digitizing. The differences are (1) you use the mouse rather than the digitizer's cursor for digitizing, (2) the resolution of the computer monitor is much coarser than a digitizer, and (3) you need a coverage, a shapefile, or an image (e.g., a digital orthophoto) as the background for digitizing. This task lets you digitize several polygons off *landuse.shp* and make a new shapefile.

1. Start ArcView, open a new view, and add *landuse.shp* to view. Select Properties from the View menu and select meters for both the Map Units and Distance Units. *Landuse.shp* is measured in meters and has been projected.

2. Double click on *landuse.shp* in the Table of Contents to open its legend editor. Change the Legend Type to Single Symbol and the Symbol to a red, outline symbol. Then click Apply. The reason to symbolize *landuse.shp* in red is to distinguish it from the new shapefile you will digitize. You are not digitizing the entire *landuse.shp* for this task. Make *landuse.shp* active, and select Autolabel from the Theme menu. In the next dialog, select landuse_id for the Label Field.

The polygons in *landuse.shp* are labeled from 59 to 77. For this task, you will digitize polygons 72–76 in the lower left (Figure 4.21).

3. Select New Theme from the View menu. In the next dialog, select polygon for the Feature Type. Name the new shapefile *trial.shp* and specify its path. Notice the box next to *trial.shp* in the Table of Contents is in dashed lines, meaning that *trial.shp* is in edit mode. Double click on *trial.shp* to open its legend editor. Change the symbol for *trial.shp* to a black, outline symbol.

4. To make sure that features you digitize will meet and align properly, you need to use interactive snapping options. Make *trial.shp* active. Select Properties from the Theme menu. In the Theme Properties dialog, click on Editing in the table of contents. This opens two dialogs for Attribute Updating and Snapping. For this task, you need to only work with Snapping. Check the box for Interactive, enter 10 (meters) for the Tolerance, and click OK. The snapping tolerance of 10 meters means that if you digitize two points (vertices) within 10 meters, for example, the two points will be snapped together. As you will see in the next step, 10 meters is a large tolerance for digitizing.

5. Now you are ready to digitize *trial.shp*. Zoom in the area of polygons 72–76. Hold the Draw Rectangle tool down, and select Draw Polygon from the pull down menu. You have three digitizing options: left click on the mouse to digitize a point (vertex); right click on the mouse to open a popup menu, which contains such selections as Snap to Vertex and Snap to Intersection; and double click to finish digitizing. You can start with polygon 74: digitize vertices that make up its boundary. The last point you digitize for polygon 74 should be at the same location as the beginning point. Before you digitize the last point, right click on the mouse and select Snap to Vertex. The snapping tolerance (10 meters) that you have defined can now help you to snap the last point to the starting point. As soon as you double click the last point, handles appear around polygon 74. If you don't like what you have digitized, you can press the Delete key to remove polygon 74 and start over.

6. Polygon 74 shares a common border with polygon 75. So that you do not have to digitize the common border a second time, change the Draw Polygon tool to Draw Line to Append Polygon. To use this append tool, you must digitize the starting and end points of polygon 75 inside polygon 74 (Figure 4.22). Other than that, the procedure for digitizing is the same.

7. Polygons 72, 73, and 76 are island polygons. To digitize an island polygon, you can start at any vertex and end at the same point.

8. After you have digitized polygons 72–76, select Stop Editing from the Theme menu and save the edits.

9. Finally, make sure *trial.shp* is active and select Table from the Theme menu. The table should have five records, one for each polygon you have digitized.

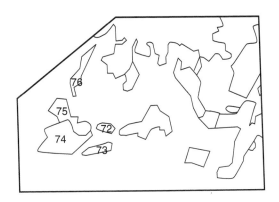

Figure 4.21
Polygons 72–76 are the ones to be digitized for Task 2.

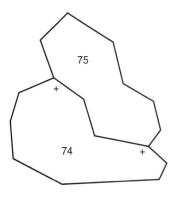

Figure 4.22
To use the Draw Line to Append Polygon tool, the starting point and the end point of the polygon to be appended (Polygon 75) must be within the polygon that has been digitized (Polygon 74), as indicated by the plus signs.

Task 3: Add Event Themes in ArcView

What you need: *events.txt*, a text file containing *x*-, *y*-coordinates of GPS readings.

Add Event Theme in ArcView adds a theme to view from an event table. An event table is essentially a text file that has *x*-, *y*-coordinates. The file *events.txt* is a text file that contains *x*-, *y*-coordinates of a series of points collected by GPS readings. Task 3 shows you how to use the file to create a new theme.

1. Start ArcView and open a new view.
2. Click on Tables and Add. In the Add Table dialog, first change the file type to Delimited Text and then select *events.txt*. The field Easting in *events.txt* is the *x*-coordinate and the field Northing the *y*-coordinate.
3. Select Add Event Theme from the View menu. In the Add Event Theme dialog, specify Easting as the *x* field and Northing as the *y* field. Click OK. A new theme called *events.txt* is added to the Table of Contents. Check the box next to it and you will see the points collected from the GPS readings.

Task 4: Read a .dgn File in ArcView

What you need: *boundary.dgn*, a Microstation design file showing the state and county boundary of Idaho in two layers.

ArcView can read CAD (Computer Aided Design) drawing files and create new themes from them. The supported drawing file formats are Microstation design (.dgn) files and AutoCAD's DXF and DWG files. Task 4 lets you read a simple .dgn file.

1. Start ArcView and load the Cad Reader extension.
2. Open a new view and select Add Theme from the View menu. In the next dialog, select Feature Data Source for the Data Source Type, navigate to *boundary.dgn*, and click OK.
3. Check the box next to *boundary.dgn* to view it. *Boundary.dgn* consists of two layers: the state border and the county boundary. To view the two layers, do the following: double click *boundary.dgn* to open its legend editor, select Unique Value for the Legend Type and Layer for the Value Field, and click Apply. Layer 48 contains the state border, and Layer 47 contains the county boundary.

Task 5: Use Scanning for Spatial Data Input

What you need: *hoytmtn_gd*, an ARC/INFO grid converted from a scanned file.

As explained in the text, scanning for digitizing involves two steps: scanning and raster-to-vector conversion. Task 5 performs the second step. A scanned file is an image file, commonly in TIFF (Tagged Image File Format) or RLC (Run Length Compressed) format. Working with a grid, which uses ESRI's proprietary format for raster data, is easier than working with an image file in ARC/INFO. A scanned file of a soil map has been converted to a grid called *hoytmtn_gd* for Task 5. In the following, you will convert *hoytmtn_gd* to an ARC/INFO coverage.

1. Before going to ArcTools, you need to create an empty coverage *hoytmtntrace* that will eventually be filled with arcs from the tracing of *hoytmtn_gd*:
   [ARC] create hoytmtntrace hoytmtn_gd
   By specifying *hoytmtn_gd* as an argument in the CREATE command ensures that *hoytmtntrace* has the same map extent (BND) as *hoytmtn_gd*.

2. Start ArcTools:
   [ARC] arctools

3. Select Edit Tools in the ArcTools menu and click OK.

4. Select File in the Edit Tools menu, and select Grid Open in the File pulldown.

5. Choose *hoytmtn_gd* in the Select an Edit Grid menu. Click OK. This action results in the display of *hoytmtn_gd* in the ArcEdit window, and the opening of the Grid Editing menu. The grid *hoytmtn_gd* contains text on data source. What you do in Steps 6–9 is to remove the text using the grid editing tools in ArcScan.

6. Click the Sketch button in the Grid Editing menu to open the Sketch menu.

7. Type 0 as the Fill value in the Sketch menu.

Note. Hoytmtn_gd is a bi-level, or black-and-white, grid file. The fill value of 0 is the same as the background.

8. Go to the ArcEdit window and use Extent in the Pan/Zoom pulldown to find the data source. Use Fullview to get the full view of *hoytmtn_gd*.

9. After you see the data source clearly in the ArcEdit window, click on the fill tool (the Solid Square icon) in the Sketch menu. Press the mouse and drag a box around the data source. The data source disappears as you release the mouse.

10. Select File in the Edit Tools menu, and select Coverage Open from the File pulldown.

11. In the Select an Edit Coverage menu, click on *hoytmtntrace* in the Coverages list. Click the Create New Feature button to open the Create Feature menu.

12. In the Create Feature menu, select Arc in the Feature Class list. Click Apply. This action adds Arc to the Available Feature list. Make sure that Arc remains highlighted.

13. Zoom into *hoytmtn_gd* so that you can see lines in the ArcEdit window.

14. Click on the Trace button in the Edit Arcs & Nodes menu. Press the cursor at the start of a line along the border of *hoytmtn_gd* that is well connected with other lines. A green line appears at the starting point; it is called the tracer.

Note. As in manual digitizing, you do not want to trace the same line twice. That could happen if you select a starting point in the middle of an arc.

15. Press 8 on the keyboard for semi-automatic tracing. Lines that have been traced turned into green. The green lines then turn into yellow after all the connected lines to the starting point have been traced and the built-in algorithms for line straightening and generalization have been applied to the traced lines.

16. Lines that remain untraced in *hoytmtn_gd* are those of island polygons, polygons along the border, and tic marks. Click the Trace button in the Edit Arcs & Nodes menu, press the cursor along an untraced line, and press 8 on the keyboard.

17. Repeat the above step until all lines are traced. Each time you press the cursor at a point, the tracer appears; you can change the tracer direction by middle clicking on the mouse. Left click on the mouse switches to the option of manual tracing. Manual tracing allows you to trace one line segment at a time.

Note. You need to change the tracer direction often in manual tracing because you trace one arc at a time.

18. Select File in the Edit Tools menu, and select Coverage Save in the File pulldown. This saves *hoytmtntrace* with its traced lines.

Task 6: Use TRANSFORM on a newly digitized map in ARC/INFO

What you need: a digitized soils map called *hoytmtn*, and a text file called *tic.geo*.

The digitized map *hoytmtn* is measured in inches. It has four control points that correspond to the four corner points, with the following longitude and latitude values in degree-minute-second (DMS):

TIC-ID	LONGITUDE	LATITUDE
1	− 116 00 00	47 15 00
2	− 115 52 30	47 15 00
3	− 115 52 30	47 07 30
4	− 116 00 00	47 07 30

The text file *tic.geo* contains the longitude and latitude values of the four tics. (Open *tic.geo* to see how the file is prepared.) The first step is to use the PROJECT command with the file option to convert the longitude and latitude values in *tic.geo* into UTM coordinates and save the new coordinates in *tic.utm*.

```
[ARC] project file tic.geo tic.utm
Project: input
Project: projection geographic
Project: units dms
Project: parameters
Project: output
Project: projection utm
Project: units meters
Project: zone 11
Project: parameters
Project: end
```

The choice of a real-world coordinate system is made through the PROJECT command. PRO-JECT is a dialog command, meaning that after you enter the command, you must provide projection parameters such as UTM zone by using PRO-JECT's subcommands. The dialog, as shown in the preceding paragraphs, is organized into two sections. The first section describes the input data, and the second the output data. Each section ends with the subcommand of parameters to see if any parameters are required for the specified projection. In this case, no parameters are required for the geographic grid (i.e., longitude and latitude values) or the UTM coordinate system.

The output file, *tic.utm*, has the UTM coordinates of the four tics as follows:

TIC-ID	x	y
1	575672.2771	5233212.6163
2	585131.2232	5233341.4371
3	585331.3327	5219450.4360
4	575850.1480	5219321.5730

The next step is to create an empty coverage, *hoytmtn2*, by copying the tics and the bnd from *hoytmtn*:

[ARC] create hoytmtn2 hoytmtn

The tics of *hoytmtn2* are measured in inches, similar to those of *hoytmtn*. But with the output from the PROJECT command you can now update the tics of *hoytmtn2* to their UTM coordinates in Tables. The final step is to use TRANSFORM to convert the rest of the coverage into UTM coordinates by using the tics as control points:

[ARC] transform hoytmtn hoytmtn2

REFERENCES

Anderson, J.R., et al. 1976. A Land Use and Land Cover Classification System for Use with Remote Sensor Data. *U. S. Geological Survey Professional Paper 964*. Washington, D.C.: U.S. Government Printing Office.

Bolstad, P.V., P. Gessler, and T.M. Lillesand. 1990. Positional Uncertainty in Manually Digitized Map Data. *International Journal of Geographical Information Systems* 4: 399–412.

Clarke, K.C. 1995. *Analytical and Computer Cartography.* 2d ed. Englewood Cliffs, NJ: Prentice Hall.

Guptill, S.C. 1999. Metadata and Data Catalogues. In P.A. Longley, M.F. Goodchild, D.J. MaGuire, and D.W. Rhind (eds.) *Geographical Information Systems.* 2d ed. New York: John Wiley & Sons, pp. 677–92.

Kennedy, M. 1996. *The Global Positioning System and GIS.* Ann Arbor, MI: Ann Arbor Press, Inc.

Lange, A. F., and C. Gilbert. 1999. Using GPS for GIS Data Capture. In P.A. Longley, M.F. Goodchild, D.J. MaGuire, and D.W. Rhind (eds.) *Geographical Information Systems.* 2d ed. New York: John Wiley & Sons, pp. 467–79.

Moffitt, F.H., and J.D. Bossler. 1998. *Surveying.* 10th ed. Menlo Park, CA: Addison-Wesley.

Onsrud, H. J., and G. Rushton (eds.). 1995. *Sharing Geographic Information.* New Brunswick, NJ: Center for Urban Policy Research.

Pettofrezzo, A.J. 1978. *Matrices and Transformations.* New York: Dover Publications, Inc.

Robinson, A.H., J.L. Morrison, P.C. Muehrcke, A.J. Kimerling, and S.C. Guptill. 1995. *Elements of Cartography.* 6th ed. New York: John Wiley & Sons.

Sperling, J. 1995. Development and Maintenance of the TIGER Database: Experiences in Spatial Data Sharing at the U.S. Bureau of the Census. In H. J. Onsrud and G. Rushton (eds.). *Sharing Geographic Information.* New Brunswick, NJ: Center for Urban Policy Research, pp. 377–96.

Taylor, P.J. 1977. Quantitative Methods in Geography: An Introduction to Spatial Analysis. Boston: Houghton Mifflin Company.

Verbyla, D.L., and K. Chang. 1997. *Processing Digital Images in GIS.* Santa Fe, NM: OnWord Press.

5

SPATIAL DATA EDITING

5.1 INTRODUCTION

Spatial data editing refers to the removal of errors from, and updating of, digital maps. Newly digitized maps, no matter how carefully prepared, always have some errors. Digital maps downloaded from the Internet may contain errors from initial digitizing or from outdated data sources. Digital maps showing roads, land parcels, forest inventory, and other data require regular revision and updating. Updating digital maps is included in this chapter because the updating process is basically the same as correcting errors on a newly digitized map.

Spatial data editing covers two types of errors. Location errors such as missing polygons or distorted lines relate to inaccuracies of map features, while others such as dangling arcs and unclosed polygons relate to logical inconsistencies among map features. To correct location errors, one often has to reshape individual arcs and digitize new arcs. To correct topological errors, one must be knowledgeable about the topological relationships required and use a topology-based GIS package to help make corrections.

Spatial data editing can go beyond individual digital maps. When a study area covers more than one source map, editing must be expanded to cover errors in matching lines across the map border. Spatial data editing may also include line simplification, line smoothing, and transferring of map features between maps. These feature manipulations are not directly related to location or topological errors. Because they do involve manipulations of polygons, lines, and points, these operations are also included in this chapter.

This chapter is organized into six sections. Section 1 discusses location and topological errors. Topological editing and non-topological editing are considered in section 2. Section 3 provides an overview of topological editing and use of the global and local methods for correcting topological errors. Section 4 covers edgematching, an editing operation to make sure that lines from adjacent maps meet perfectly along the border. Section 5 discusses editing operations for non-topological digital maps. Section 6 concludes the chapter with a discussion of other feature manipulations.

5.2 TYPE OF DIGITIZING ERRORS

5.2.1 Location Errors

Location errors relate to the location of map features on a digitized map. Assuming that the source map is correct, the ultimate goal of digitizing is to duplicate the source map in digital format. A **check plot**, which is a plot of the digitized map at the same scale as the source map, can be used to determine if the goal has been achieved. GIS users can superimpose the check plot on the source map to see how well they match. Typically a producer of digital maps decides on the tolerance of location error. Natural Resources Conservation Service (NRCS, formerly the Soil Conservation Service), for example, stipulates that each digitized soil boundary in the Soil Survey Geographic (SSURGO) database shall be within 0.01inch (0.254 millimeter) line width of the source map. At the scale of 1:24,000, this tolerance represents 20 feet (6 to 7 meters) on the ground.

Several scenarios may explain the discrepancies between digitized lines and lines on the source map. The first is human errors in manual digitizing. Human error is not difficult to understand: when a source map has hundreds of polygons and thousands of lines, one can easily miss some lines or connect the wrong points.

The second scenario consists of errors in scanning and tracing. As described in Chapter 4, a tracing algorithm usually has problems when raster lines meet or intersect, are too close together, are too wide, or are too thin and broken. Digitizing errors from tracing include collapsed lines, misshapen lines, and extra lines (Figure 5.1).

The third scenario consists of errors in converting the digitized map to real-world coordinates (Chapter 4). To make a check plot at the same scale as the source map, one must convert the digitized map to real-world coordinates by using a set of control points. With erroneous control points, this conversion can cause discrepancies between digitized lines and source lines. Unlike seemingly random errors from the first two scenarios, discrepancies from the conversion often exhibit regular patterns, such as the degree of discrepancy increasing in one direction. To correct this type of location errors, one must re-digitize or re-select control points and re-run geometric transformation.

Location errors also include duplicate lines from manual digitizing or tracing. Each arc on a

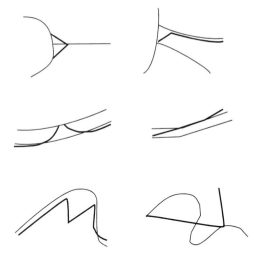

Figure 5.1
Common types of digitizing errors from tracing. The thin lines are lines on the source map, and the thick lines are lines from tracing.

Figure 5.2
Digitizing errors from duplicate lines include sliver polygons and missing labels for the sliver polygons.

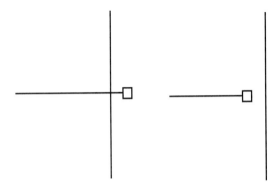

Figure 5.3
Digitizing errors of overshoot (left) and undershoot (right).

line or polygon map is supposed to be digitized only once. But a complex map may have lines that were digitized twice or even more times due to human errors. Duplicate lines occur frequently in tracing because semi-automatic tracing follows continuous lines even if some of the lines have already been traced. Duplicate lines from either manual digitizing or tracing are not going to be on top of one another because of the high resolution of a digitizing table in manual digitizing and the multiple-pixel raster line in tracing. If the topology of the map is built, these duplicate lines intersect one another to form a series of tiny polygons, and each polygon has a missing label (Figure 5.2). Duplicate lines are sometimes difficult to detect without zooming in.

5.2.2 Topological Errors

Topological errors violate the topological relationships used in a GIS package. A common digitizing error occurs when two arcs that are supposed to meet at a node do not meet perfectly. This type of error is called **undershoot** if a gap exists between the arcs, and **overshoot** if one arc is overextended (Figure 5.3). The result of both cases is a **dangling arc**, which has the same polygon on its left and right sides, and a **dangling node** at the end of the arc. Dangling nodes also occur when a polygon is not closed perfectly (Figure 5.4).

Dangling nodes are acceptable in special cases. Manual digitizing using a topology-based GIS package requires that the starting and end points of an arc be nodes. Therefore, small tributaries on a stream map and dead-end streets on a street map have dangling nodes attached to the arcs. These are acceptable dangling nodes.

Another type of node error is called **pseudo node**. A pseudo node appears along a continuous arc and divides the arc unnecessarily into separate arcs (Figure 5.5). Some pseudo nodes are acceptable or even necessary. For example, to show different segments of a road having different attributes such as speed limit, pseudo nodes may be placed at points where attribute values change. Another example applies to isolated polygons not

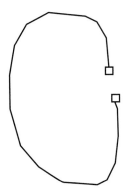

Figure 5.4
Digitizing errors of an unclosed polygon.

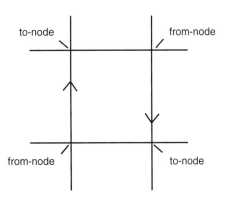

Figure 5.6
The from-node and to-node of an arc determine the arc's direction.

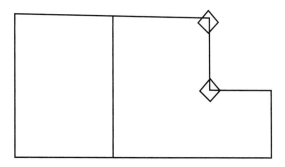

Figure 5.5
Pseudo nodes, shown by the diamond symbol, are nodes that are not located at line intersections.

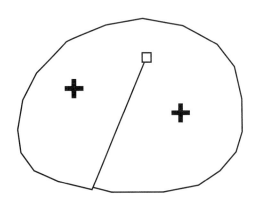

Figure 5.7
Digitizing error of multiple labels due to unclosed polygons.

connected to other polygons. An arbitrary starting point (and end point) must be chosen to digitize the polygon. That point then becomes a pseudo node.

The arc direction may become a source of topological errors. For example, in hydrologic applications, all streams must follow the downstream direction. In other words, the starting point (also called the *from-node*) of a stream must be at a higher elevation than the end point (also called the *to-node*). Another example is a street coverage. The direction of an arc or street is not an issue for two-way streets but is an issue for one-way streets (Figure 5.6).

A georelational data model uses the label to link a polygon to its attribute data. A polygon therefore should have one, and only one, label. Unclosed polygons are often the cause for multiple label errors (Figure 5.7). When a gap exists between polygons, the two polygons are treated as a single polygon topologically. Consequently, the two labels, one for each polygon, become a case of multiple labels. The multiple label error disappears when the gap is filled.

5.3 TOPOLOGICAL AND NON-TOPOLOGICAL EDITING

The nature of spatial data editing varies among GIS users: it depends on whether they use topological or non-topological GIS data, whether they are GIS data producers or users, and what GIS package they use. A topology-based GIS package can detect and display topological errors, and has functionalities to ease the removal of topological errors. Examples of topology-based GIS packages include ARC/INFO, AutoCAD Map, MGE, ILWIS, and SPANS (Box 5.1). They have similar capabilities in making sure that lines meet correctly, polygons are closed, and polygons are properly labeled. Some GIS packages are not topology-based, but have the capability of performing topological editing. For example, IDRISI has a separate software product for defining the spatial relationship between map features in vector format.

A non-topological GIS package cannot detect topological errors or build topology, although it can be used for digitizing and editing map features. Examples of non-topological GIS packages include ArcView and MapInfo. ArcMap in ArcInfo 8 is designed for data display and editing, but its current version does not have the functionality of displaying topological errors.

5.4 TOPOLOGICAL EDITING

Assuming that a map has just been digitized, this section describes the process of using a topology-based GIS package to edit the map. When appropriate, ARC/INFO commands are referenced.

5.4.1 An Overview

The editing process begins with constructing the topology of the map to be edited. This step ensures that the computer can recognize individual nodes, arcs, and polygons on the map. Also, the process of building the topology between map features can remove some of the digitizing errors.

The next step is to tell what types of digitizing errors exist on the map. A topology-based GIS package can easily detect topological errors and denote these errors with special symbols. Location errors such as missing lines or polygons require manual checking by comparing a check plot with its source map.

The third step is to remove digitizing errors. Digitizing errors may be removed individually or as a group. A specific action is usually needed to remove a specific type of digitizing error. After digitizing errors have been removed, the map's topology must be re-built to accommodate changes made during editing. Only rarely can one remove all digitizing errors in a single editing session.

Box 5.1   **GIS Packages for Topological Editing**

ARC/INFO, AutoCAD Map, MGE, and SPANS are all capable of performing topological editing. They are similar in editing operations. Only ARC/INFO commands are referenced in this chapter, but the following list highlights the editing capabilities of the other three packages:

- AutoCAD Map: build topology, repair undershoots, snap clustered nodes, remove duplicate lines, simplify linear objects, and correct other errors.

- MGE (Base Mapper): build topology, remove redundant linear data segments, fix undershoots and overshoots, and create an intersection at a line crossing.
- SPANS: identify and repair knots, overshoots, and dangles; build topology for line and area data; and use edgematching to join adjacent line data.

More typically, one must repeat the editing process before the map is free of digitizing errors.

5.4.2 Correcting Digitizing Errors

5.4.2.1 Global Method

Topology-based GIS packages typically include functionalities that will apply some specified tolerances to an entire map to remove digitizing errors. These are called global methods, and one example is the CLEAN command in ARC/INFO. CLEAN uses two tolerances on an input coverage: **dangle length** and **fuzzy tolerance**. The dangle length specifies the minimum length for dangling arcs on the output coverage. A dangling arc is removed if it is shorter than the specified dangle length. Therefore, the dangle length is useful in removing minor overshoot errors (Figure 5.8). The fuzzy tolerance specifies the minimum distance between arc vertices on the output coverage. The tolerance applies to vertices along an arc and vertices along two nearby arcs. The latter case can be useful in removing duplicate lines if the duplicate lines are within the specified fuzzy tolerance (Figure 5.9).

Besides the use of dangle length and fuzzy tolerance, CLEAN will automatically insert a node at a line intersection if the intersection has not been digitized as a node. Thus, CLEAN enforces the arc-node relationship in topology. This functionality is especially important if the manual digitizing followed spaghetti digitizing and ignored the arc-node relationship.

Because it applies to an entire coverage, the global method must be used cautiously (Box 5.2). If the dangle length is set too large, it will remove undershoots as well as overshoots (Figure 5.10). Undershoots should be extended to fill the gap and should not be removed. The fuzzy tolerance, if it is

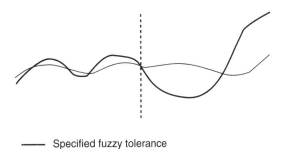

——— Specified fuzzy tolerance

Figure 5.9
The fuzzy tolerance specified by the CLEAN command can snap duplicate lines if the gap between the duplicate lines is smaller than the specified fuzzy tolerance. In this diagram, the duplicate lines to the left of the dashed line will be snapped but not those to the right.

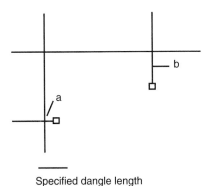

——— Specified dangle length

Figure 5.8
The dangle length specified by the CLEAN command can remove an overshoot if the overextension is smaller than the specified length. In this diagram, the overshoot (a) is removed and the overshoot (b) remains.

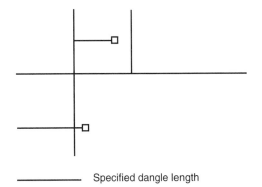

——— Specified dangle length

Figure 5.10
If the dangle length is set too large, it can remove overshoots, which should be removed, and undershoots, which should not be removed. In this diagram, the dangling arc at the top represents an undershoot. It will be removed using the specified dangle length.

U nless specified, the parameters of dangle length and fuzzy tolerance are given the default values in ARC/INFO. The default value of dangle length is 0. The default value of fuzzy tolerance depends on whether the coverage is measured in digitizer units (inches) or in real-world coordinates (meters or feet). It is 0.002 inch in digitizer units, or, in real-world co-ordinates, 1/10,000 of the width or height of the map extent, whichever is greater. The default fuzzy toler-ance value therefore increases as the size of the cov-erage increases in area. For example, the default value is two meters if the larger dimension of a coverage measures 20,000 meters, and 20 meters if the dimen-sion measures 200,000 meters. This change in the de-fault value presents a problem to an organization, which may have to piece together its map coverage from smaller coverages. A large fuzzy tolerance can easily distort map features. The solution is to use a constant fuzzy tolerance for a project by entering its value with the CLEAN command.

set too large, can snap arcs that are not duplicate arcs and thus distort map features on a digitized coverage (Figure 5.11).

5.4.2.2 Local Method

The local method deals with selected digitizing errors from a digitized map. To facilitate local edit-ing, GIS users can use tolerances. The tolerance for snapping nodes, called **nodesnap** in ARC/ INFO, is as useful in editing as in digitizing. The nodesnap tolerance should not be set too large as to snap wrong nodes. Also, when a node is snapped to a new location within the nodesnap tolerance, the node moves the arc segment that is attached to it. Therefore, the positional accuracy of arcs may suf-fer due to the snapping of nodes. This is another reason why the nodesnap tolerance must be set carefully.

The tolerance that specifies a search radius for selecting features for editing is called **editdistance** in ARC/INFO (Figure 5.12). If the editdistance is too small, the computer cannot select a feature like an arc. If it is too large, the computer may select a feature within the search radius, which is not the target feature for editing.

Both nodesnap and editdistance may be set with a specific value or interactively. Because nodesnap affects the geometry of spatial data, an agency may use a threshold value in editing. If set

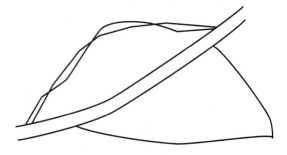

——— Specified fuzzy tolerance

Figure 5.11
If the fuzzy tolerance is set too large, it can remove duplicate lines as well as map features such as a small stream channel as shown in the diagram.

interactively, nodesnap and editdistance must be adjusted frequently during an editing session as the GIS user zooms in and out constantly. Zooming in provides a clearer picture of features to be edited, whereas zooming out allows the GIS user to move to other parts of the coverage for editing. Each zooming in or zooming out changes the map scale and the applicability of nodesnap and editdistance.

Most GIS packages allow GIS users to undo edits that are made inadvertently. Sometimes a GIS user may have deleted a set of arcs and then realizes some of the deleted arcs are needed.

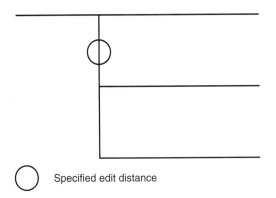

Specified edit distance

Figure 5.12
Editdistance should be set large enough for the computer to select the arc to be edited. But, if it is set too large, the computer may select the wrong arc.

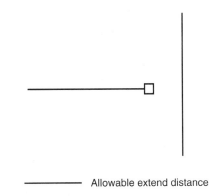

Allowable extend distance

Figure 5.13
The EXTEND command uses a specified allowable distance to extend a dangling arc to meet with an existing arc.

ARC/INFO's UNDO command, also known as OOPS, can undo the last edit made. Repeated use of UNDO can remove all edits made in a single edit session.

The following summarizes the use of local methods in removing digitizing errors (Box 5.3).

- Undershoot: The dangling arc must be extended to meet with the target arc at a new node by either digitizing or using a functionality such as the EXTEND command in ARC/INFO (Figure 5.13). The SPLIT command in ARC/INFO can split an arc by inserting a new node at the specified location.
- Overshoot: An overshoot can be removed by selecting the extension, which is a separate arc after the coverage has been "cleaned," and deleting it (Figure 5.14).
- Duplicate Arcs: One solution is to carefully select extra arcs and delete them, and the other is to re-digitize arcs after deleting all duplicate arcs in a box area.
- Wrong Arc Directions: The arc direction can be altered by using a command, such as ARC/INFO's FLIP command, to change the relative position of the from-node and to-node.
- Pseudo Nodes: By first setting the two arcs on each side of a pseudo node to have the

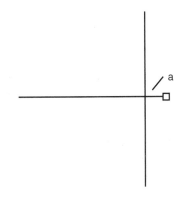

Figure 5.14
To remove an overshoot, simply select the overextension and delete it.

same ID value, the node can be removed by using ARC/INFO's UNSPLIT command (Figure 5.15). Occasionally, pseudo nodes may reveal missing arcs: the end points of the missing arc have been entered as nodes but not connected.

- Label Errors: For missing labels on a polygon coverage, new labels with proper ID values must be added. If a polygon has multiple labels, the extra labels can be selected and deleted. As discussed earlier, however, multiple labels may indicate

---

Box 5.3 **Errors That Cannot be Categorized**

All experienced GIS users have encountered digitized errors that cannot be easily categorized. Often these errors represent errors made on top of other errors. Checking the source map is probably the best way to take care of this kind of error.

---

unclosed polygons, and the problem can be resolved by closing the polygons.

- Reshaping Arcs: To reshape an arc, vertices that make up the arc can be moved, added, or deleted. ARC/INFO's VERTEX command can be used to reshape an arc by adding, deleting, and moving points that make up the arc.

5.5 EDGEMATCHING

The discussion has so far concentrated on correcting digitizing errors on a single map. Many GIS projects, however, use multiple maps. For example, a forest fire coverage for a national forest may consist of a number of maps, each representing a unit area or a USGS 1:24,000 scale quad. Each map is digitized and edited separately. After all maps are finished, they must be joined to make the final coverage (Box 5.4). **Edgematching** is a necessary operation before joining maps because lines from two maps rarely meet perfectly along the border (Figure 5.16).

ARC/INFO uses menus to guide GIS users through the process of edgematching. One coverage in edgematching is called the snap coverage and the other the edit coverage. Arcs on the edit coverage are adjusted to match those on the snap coverage. Before adjusting arcs on the edit coverage, one must snap nodes at ends of arcs from the two coverages using a snapping distance as the tolerance. The edit coverage can be adjusted either with the entire coverage or with one section of the coverage at a time.

After the lines are matched across the map border and the edit coverage has been adjusted, the snap and edit coverages can be joined to form a

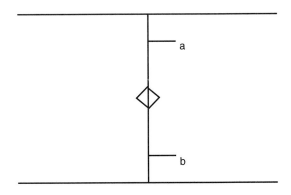

Figure 5.15
To remove a pseudo node, first select a and b, the arcs on both sides of the pseudo node. Make sure the arcs have the same ID value. Then use the UNSPLIT command to remove the pseudo node.

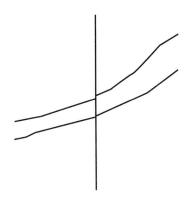

Figure 5.16
The diagram shows lines from two adjacent quads do not meet perfectly. The mismatch is only visible after zooming in.

The Geoprocessing extension to ArcView offers an operation called MERGE, which can be used to create one theme that contains the features of two or more themes. The merged theme is useful for data query or display from multiple themes. MERGE, however, differs from edgematching in ARC/INFO.

Themes that are merged are simply put side by side; the lines across the themes are not matched or joined. When zoomed in, gaps and mismatched lines may appear along the border between themes. Also, if a quad boundary frames each theme, the boundaries remain on the merged theme.

single coverage. The border separating the two coverages stays on the joined coverage, but it can be removed by the DISSOLVE command, an operation that can remove the boundary of adjacent polygons with the same label ID.

5.6 NON-TOPOLOGICAL EDITING

Non-topological editing works with location errors, primarily for updating or revision of spatial data. GIS users may update maps by using digital orthophotos as the backdrop, combine polygons, or split polygons. This section focuses on editing operations in ArcView.

5.6.1 Delete, Move, and Cut and Paste

The functions of delete, move, and cut and paste apply to one or more selected features, which may be points, lines, or polygons. Each polygon in a shapefile is a unit, separate from other polygons. Therefore, moving a polygon means placing the polygon on top of an existing polygon while creating a void area in its original location—a maneuvering that is very different from topology-based editing (Figure 5.17).

5.6.2 Reshape

Moving, deleting, or adding points (called vertices in ArcView) on a line can alter the shape of the line. ArcView has a vertex edit tool, which allows GIS users to select a vertex and drag it to a new location (Figure 5.18) or delete it (Figure 5.19). The vertex edit tool can also add a vertex by

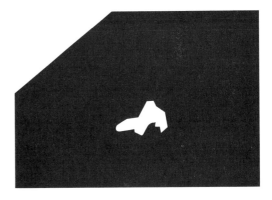

Figure 5.17
After a polygon is moved in ArcView, a void area appears in its location. GIS users, who are used to topology-based operations, may find this ArcView operation strange.

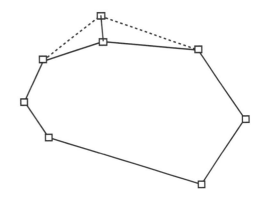

Figure 5.18
Reshape a line in ArcView by moving a selected vertex.

simply clicking its intended location on the line (Figure 5.20).

The vertex edit tool can also be used to reshape polygons. When the tool is applied to a selected polygon, it can change the shape of the polygon without affecting the polygon's neighbors. When the tool is applied to a common boundary or a common vertex between polygons, it will change the shapes of those adjacent polygons simultaneously.

5.6.3 Split and Merge

A new line that crosses the existing lines can split the lines, including the new line itself, into separate line segments. Line merging, on the other hand, groups the selected lines into one line. Polygons can also be split or merged. Splitting polygons means subdividing a polygon into two or more polygons by drawing one or more split lines through the polygon (Figure 5.21). But the split line must extend beyond the polygon boundary (e.g., overshoot) for ArcView to recognize the split. Merging groups the selected polygons, if they are adjacent to each other, into a single polygon (Figure 5.22). If the selected polygons are not

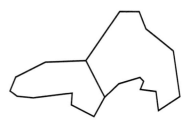

Figure 5.21
A polygon can be split into two in ArcView by drawing a split line across the polygon boundary.

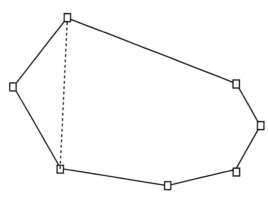

Figure 5.19
Reshape a line in ArcView by deleting a selected vertex.

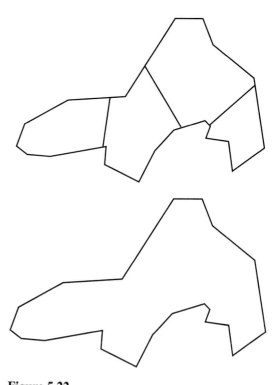

Figure 5.22
Two or more selected polygons can be grouped into one in ArcView.

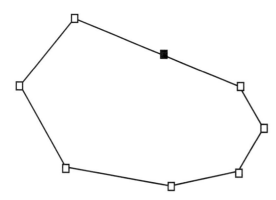

Figure 5.20
Add a vertex to a line in ArcView.

adjacent, the result is a single polygon with multiple parts.

5.6.4 Manipulation of Polygon Features

Because polygons in a shapefile are separate and can overlap, ArcView allows GIS users to create new polygons representing the intersection of polygons, create doughnut polygons (i.e., polygons with holes inside them), and remove area of overlap between polygons. These manipulations are not possible with topology-based polygon coverges.

5.7 OTHER TYPES OF MAP FEATURE MANIPULATION

5.7.1 Line Simplification, Densification, and Smoothing

Line simplification refers to the process of simplifying or generalizing a line by removing some of its points. Line simplification is a common practice in map display (Robinson et al. 1995) and a common topic under generalization in digital cartography and GIS (McMaster and Shea 1992; Weibel and Dutton 1999). When a map digitized from the 1:100,000 scale source map is to be used at the scale of 1:1,000,000, lines become congested and illegible because of the reduced map space. The lines must be simplified by removing some of their points.

The **Douglas-Peucker algorithm**, perhaps the best-known algorithm for line simplification, is used in ARC/INFO and other GIS packages (Douglas and Peucker 1973). The algorithm works line-by-line and with a specified tolerance. For a specific line, the process starts with a trend line connecting the end points of the line (Figure 5.23). Deviations of the intermediate points from the trend line are then calculated. In the case of Figure 5.23A, the points with the largest deviation for each curve are connected to the end points of the original line to form new trend lines, followed by calculation of deviations of the points from the new trend lines. This process continues until no deviations exceed the tolerance, and the result is a simplified line (often with sharp angles) that connects the trend lines. In the case of Figure 5.23B,

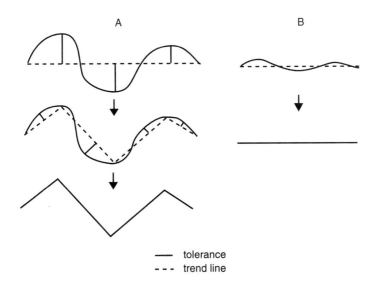

— tolerance
- - - trend line

Figure 5.23
The Douglas-Peucker line simplification algorithm is an iterative process, which requires use of a tolerance, trend lines, and calculation of deviations of vertices to the trend line. See text for explanation.

because deviations of the points are all smaller than the tolerance, the simplified line is the straight line connecting the end points. Figure 5.24 shows the result of line simplification using ARC/INFO's GENERALIZE command.

Line densification is the process of adding new points to select line or lines in a map at a specified interval. The operation is relatively simple. Starting at the starting point of a line, the operation adds a new point at every specified interval (Figure 5.25). Existing points are retained, and the shape of the line is not altered.

Line smoothing also adds new points to lines, but the location of new points is generated by mathematical functions. Figure 5.26 shows the result of smoothing using ARC/INFO's SPLINE command. The most common method for spline consists of cubic polynomials, which have the general form (Davis 1986)

$$y = \beta_1 + \beta_2 x + \beta_3 x^2 + \beta_4 x^3 \qquad (5.1)$$

A cubic polynomial can pass exactly through four points. For an arc with more than four points, one must use a succession of polynomial segments and smooth out abrupt changes in curvature between segments.

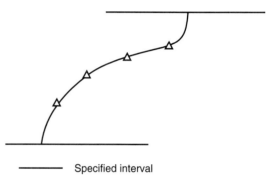

———— Specified interval

Figure 5.25
Line densification adds new vertices to an arc at a specified interval.

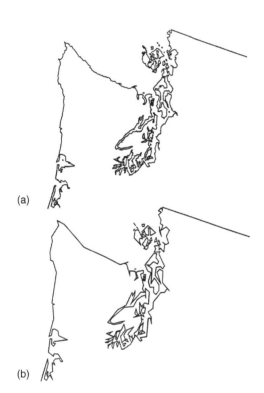

(a)

(b)

Figure 5.24
Two views of the Puget Sound area in Washington. Created from ARC/INFO's GENERALIZE command with a weed tolerance of 2 km, Map (b) is a generalized version of Map (a).

Figure 5.26
The ARC/INFO SPLINE command smoothes the polygon boundary by adding new vertices generated mathematically. The smoothed boundary is shown in thicker line.

5.7.2 Transferring Map Features from One Map to Another

The function of transferring map features is useful if a map has lines or polygons that are needed in another map. ARC/INFO has the commands of GET and PUT for transferring map features. GET copies all features of the specified type (i.e., point, line, or polygon) from one coverage to another. PUT copies only selected features from one coverage to another.

GET and PUT are particularly useful when different map coverages share common boundaries. For example, a soil map may share some common boundaries with a land use map. After the land use map is digitized, those boundaries it shares with the soil map can be transferred to the soil map. Not only does this transfer save time in digitizing, but it also ensures the maps match spatially for data analysis.

KEY CONCEPTS AND TERMS

Check plot: A plot of a digitized map for error checking.

Dangle length: A tolerance used in ARC/INFO that specifies the minimum length for dangling arcs on the output coverage.

Dangling arc: An arc that has the same polygon on both its left and right sides and a dangling node at the end of the arc.

Dangling node: A node at the end of an arc that is not connected to other arcs.

Douglas-Peucker algorithm: A computer algorithm for line simplification.

Edgematching: An operation in joining adjacent coverages.

Editdistance: A tolerance used in ARC/INFO that specifies a search radius for selecting features for editing.

Fuzzy tolerance: A tolerance used in ARC/INFO that specifies the minimum distance between arc vertices on the output coverage.

Line densification: The process of adding new vertices to a line.

Line simplification: The process of simplifying or generalizing a line by removing some of the line's vertices.

Line smoothing: The process of smoothing a line by adding new vertices, which are typically generated by a mathematical function such as spline, to the line.

Location errors: Errors related to the location of map features such as missing lines or missing polygons.

Nodesnap: A tolerance used in ARC/INFO for snapping nodes.

Non-topological editing: Spatial data editing using a non-topological GIS package such as ArcView.

Overshoot: One type of digitizing error that results in an overextended arc.

Pseudo node: A node appearing along a continuous arc.

Spatial data editing: The process of removing errors or updating on digital maps.

Topological errors: Errors related to the topology of map features such as dangling arcs and missing or multiple labels.

Undershoot: One type of digitizing error that results in a gap between arcs.

APPLICATIONS: SPATIAL DATA EDITING

This application section covers three tasks. Task 1 covers correction of topological errors in ARC/INFO. Task 2 uses ArcTools, ARC/INFO's menu-driven utility program, to perform edge-match and mapjoin. Task 3 reviews editing capabilities of ArcView.

Task 1: Correct Topological Errors in ARC/INFO

What you need: *editmap1* (Figure 5.27), a coverage with topological errors.

The coverage *editmap1* has several types of digitizing errors: overshoot, undershoot, unclosed polygon, missing label, and multiple labels. Task 1 shows you how you can correct these digitizing

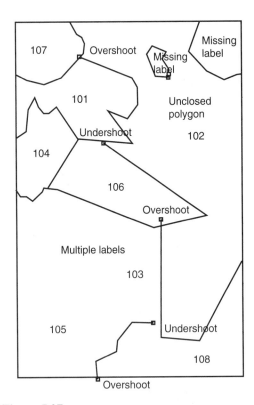

Figure 5.27
Editmap1 has topological errors.

errors. While working on this task, you need to zoom in to get a better look at map features for editing and zoom out to move from one part of the coverage to another part. You also need to define nodesnap, editdistance, and other tolerances and re-define them as you zoom in and out.

1. Go to ArcEdit and set the environment for *editmap1*:
   [ARC]arcedit
   :display 9999
   :mapex editmap1
   :editcov editmap1
   :drawenv arc node errors label
   :nodecolor dangle 2 /* color dangling nodes in red
   :nodecolor pseudo 3 /* color pseudo nodes in green
   :draw
2. To remove an overshoot:
   :editfeature arc /* specify arc as the feature to be edited
   :select /* zoom in an overshoot and select the over-extended arc
   :delete
3. To remove an undershoot:
   :editfeature arc
   :select /* zoom in an undershoot and select the arc
   :extend /* specify a distance so that the dangling arc can meet its target
4. To close an unclosed polygon:
   :editfeature node
   :nodesnap closest * /* specify a circle for nodes to be snapped
   :move /* select the node to move, press 4 to move, then select the node to move to
   Note: *An alternative to moving a node to close an unclosed polygon is to digitize an arc connecting the two dangling nodes.*
5. To add a label:
   :editfeature label
   :add /* click the mouse within the polygon that has a missing label

Note: If the label value is not what you want, press 8 for digitizing options. This will bring up a menu, which allows you to define the new user-id or label value.

6. To remove a label:
   :editfeature label
   :select /* select the label to remove
   :delete

Note: An alternative to removing a label is to move it to a polygon that has a missing label. To move a label, you will first select the label and then type "move."

7. After you have removed digitizing errors in *editmap1*, you need to save the changes before quitting ArcEdit. Then, you must rebuild the topology of *editmap1* by using CLEAN or BUILD.

Task 2: Edgematch/Mapjoin/Dissolve

What you need: *qhoytmtn* and *qmrblemtn*, two soil coverages that need to be matched and joined (Figure 5.28). Both coverages are based on the Universal Transverse Mercator (UTM) coordinate system and measured in meters.

The soil coverages have been digitized from two separate quads. They must be edgematched and joined, if the study area of a project covers both quads. EDGEMATCH is a menu-driven operation in ArcEdit. MAPJOIN in Arc is a dialog command, which can join up to 50 coverages.

1. Go to ArcEdit and start the EDGEMATCH command:
   Arc: arcedit
   Arcedit: display 9999
   Arcedit: edgematch

2. In the Coverage menu, specify *qhoytmtn* as the Edit Coverage, *qmrblemtn* as the Snap Coverage, and Node as the snap coverage feature class. Click Apply. This action initiates the opening of the Edgematching menu and the display of *qhoytmtn* (in white) and *qmrblemtn* (in red) in the ArcEdit window.

3. The Edgematching menu includes several parameter settings, of which the Snap Environment is probably the most important. Click the Snap Environment button. The Snapping Distance specifies the tolerance

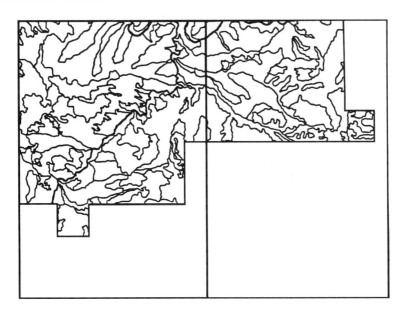

Figure 5.28
Qhoytmtn (right) and qmrblemtn (left) are two coverages to be edgematched and joined.

within which the features from *qhoytmtn* and *qmrblemtn* are linked. A default value of 14.02 meters is given to the Snapping Distance, but it can be changed interactively or by typing in a new value. You can use 7 meters as the snapping distance for this task. For the node to be snapped to within the search area, you can choose the closest one.

4. Click the Add Automatically button in the Edgematching menu. Use Zoom/Pan to check each link. All features from *qhoytmtn* and *qmrblemtn* are linked correctly except for the top two lines, which need to be linked interactively. Click the Add Interactively button in the menu. Zoom in the area of the first line to be linked so that you can see the end node of the line in each quad. Click on the end node in *qhoytmtn* and then click on its corresponding end node in *qmrblemtn*. An arrow linking the two nodes should appear. Do the same for the other line. Press 9 to exit.

5. Click the Adjust button. The arrow symbols now become square symbols. Again,

examine the result of Adjust by using Zoom/Pan. If everything looks right, click the Save button. Exit ArcEdit and save edits.

6. Edgematching has altered the topology of the edit coverage. Therefore, you need to re-build the topology for *qhoytmtn*:
Arc: build qhoytmtn poly

7. Now the two coverages have been edgematched. The next step is to join *qhoytmtn* and *qmrblemtn* to create *qandq*:
Arc: mapjoin qandq poly all /* enter *qhoytmtn* and *qmrblemtn* as the coverages to join

8. You can display *qandq* in either ArcPlot or ArcEdit. Although the two quads are joined, the quad boundary still remains. To remove the artificial quad boundary, you can use the DISSOLVE command in Arc:
Arc: dissolve qandq qandq2 #all /* use all attribute items to dissolve *qandq*

9. Figure 5.29 shows *qandq2*.

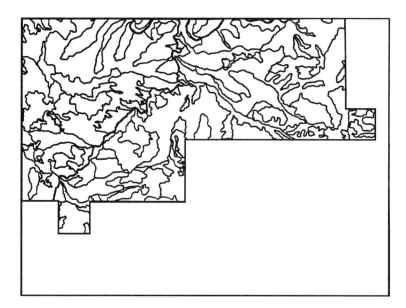

Figure 5.29
Qandq2 is a seamless coverage created from qhoytmtn and qmrblemtn.

Task 3: Spatial Data Editing Using ArcView

What you need: *editmap2.shp* (Figure 5.30), *editmap3.shp* (Figure 5.31).

Task 3 covers three common edit functions in ArcView: merging polygons, splitting a polygon, and reshaping the polygon boundary. You will work with *editmap2.shp*, while *editmap3.shp* shows how *editmap2.shp* looks like after editing.

1. Start ArcView, open a new view, and add *editmap2.shp* and *editmap3.shp* to view. Spatial data editing in ArcView only applies to shapefiles.
2. Activate *editmap2.shp* and select Start Editing from the Theme menu. A dashed line around the check box for *editmap2.shp* indicates that the theme is in edit mode. The first part of Task 3 is to merge Polygons 74 and 75. Click on the Pointer tool. Click inside Polygon 74, and then click inside Polygon 75 while holding down the shift key. The handles now appear around the two polygons. Select Union Features from the Edit menu.
3. The second part of Task 3 is to split Polygon 71. Click the Drawing tool and select the Draw Line to Split Polygon tool. To split a polygon, the split line must cross over the polygon boundary; in other words, you want to have "overshoots" at both ends of the split line. Click the left mouse button where you want the split line to start, click each vertex that makes up the split line, and double-click the end vertex.

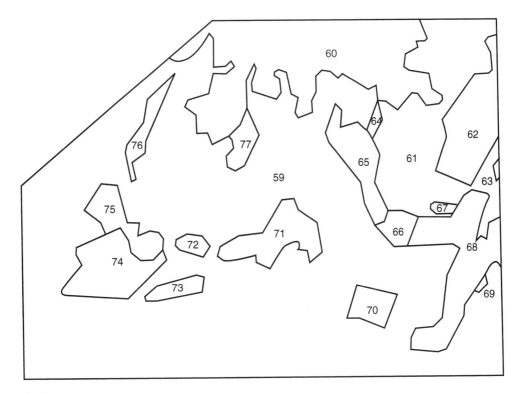

Figure 5.30
Editmap2.shp is a polygon shapefile to be edited in ArcView for polygon merging, polygon splitting, and reshaping of the polygon boundary.

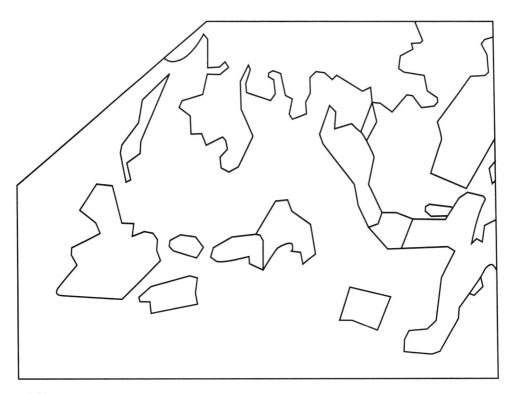

Figure 5.31
Editmap2.shp after editing.

4. The third part of Task 3 is to reshape Polygon 73 by extending its southern border in the form of a rectangle. The strategy in reshaping the polygon is to add three new vertices and to drag the vertices to form the new shape. Zoom in the area around Polygon 73. Click on the Vertex Edit tool. Do the following to add a new vertex: click inside Polygon 73 to see the existing vertices in the form of small squares; move the cursor to where a new vertex is to be added; and, when you see a cross-hair symbol, left click on the mouse. Create a new vertex (vertex 1) anywhere along the southern border of Polygon 73. To drag a vertex, click the vertex a couple of times until the vertex changes to a square,

which is linked to two circles (i.e., anchor vertices). Now, drag vertex 1 to where the new border is going to be (use *editmap3.shp* as a guide), and release the mouse button. Click inside Polygon 73 again (the square symbols should appear around the polygon again). Next, add another vertex (vertex 2) along the line connecting vertex 1 and the original SE corner of Polygon 73. Click vertex 2 a couple of times and drag vertex 2 to the SE corner of the new boundary. Do the same to form the SW corner of the new boundary.

REFERENCES

Davis, J. C. 1986. *Statistics and Date Analysis in Geology,* 2^d ed. New York: John Wiley & Sons.

Douglas, D.H., and T. K. Peucker. 1973. Algorithms for The Reduction of The Number of Points Required to Represent a Digitized Line Or Its Caricature. *The Canadian Cartographer* 10: 110–22.

McMaster, R. B., and K. S. Shea. 1992. *Generalization in Digital Cartography*. Washington, D.C.: Association of American Geographers.

Robinson, A.H., J.L. Morrison, P.C. Muehrcke, A.J. Kimerling, and S.C. Guptill. 1995. *Elements of Cartography*, 6th ed. New York: John Wiley & Sons.

Weibel, R., and G. Dutton. 1999. Generalizing Spatial Data and Dealing With Multiple Representations. In P.A. Longley, M.F. Goodchild, D.J. MaGuire, and D.W. Rhind (eds.). *Geographical Information Systems*, 2^d ed. New York: John Wiley & Sons, pp. 125–55.

6

ATTRIBUTE DATA INPUT AND MANAGEMENT

6.1 INTRODUCTION

GIS involves both spatial and attribute data: spatial data relate to the geometry of map features, whereas attribute data describe the characteristics of the map features. Attribute data are stored in tables. Each row of a table represents a map feature, and each column represents a characteristic. The intersection of a column and a row shows the value of a particular characteristic for a particular map feature.

The difference between spatial and attribute data is well defined with vector-based map features. The **georelational data model** uses a split data system by storing spatial data and attribute data in separate files. Linked by the feature ID, the two sets of data files are synchronized so that both data can be queried, analyzed, and displayed. The **object-oriented data model** also distinguishes spatial data from attribute data but stores both data in a single database, which saves the processing overhead.

The previous two chapters discussed spatial data input and editing. To complete the discussion of vector data, this chapter covers attribute data input and management. Because the object-oriented data model is still under development in GIS (Chapter 3), most materials covered in this chapter apply to the georelational data model. It should be noted, however, that both data models operate in a relational database environment.

This chapter is divided into five sections. Section 1 provides an overview of attribute data in GIS. Section 2 discusses the relational database model, data normalization, and types of data relationship. Sections 3 and 4 cover attribute data input and verification respectively. Section 5 discusses

creation of new attribute data by data classification and computation.

6.2 ATTRIBUTE DATA IN GIS

6.2.1 Linking Attribute Data and Spatial Data

The georelational data model links spatial data and attribute data by the feature ID (Figure 6.1). Each map feature has a unique label ID. Attribute data are stored in a table, called the **feature attribute table**, which contains the label ID and a default set of attributes, such as area and perimeter for the polygon features. Each row of the feature attribute table represents a map feature, and each column describes an attribute of the map feature (Figure 6.2). A row is also called a **record** or a tuple, and a column is also called a **field** or an item.

The feature attribute table may be the only table needed if a map has only several attributes besides the default fields. But this is not the case with most GIS projects. For example, a soil map unit can have over 80 estimated soil physical and chemical properties, interpretations, and performance

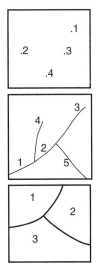

Point-ID	Item 1	Item 2	. . .
1			
2			
3			
4			

Line-ID	Item 1	Item 2	. . .
1			
2			
3			
4			
5			

Polygon-ID	Item 1	Item 2	. . .
1			
2			
3			

Figure 6.1
Attribute data on the right are linked to spatial data by the label ID of map features.

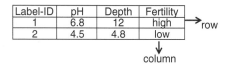

Figure 6.2
A feature attribute table consists of rows and columns. Each row represents a map feature, and each column represents a property or characteristic of the map feature.

data. Some of these data apply directly to the soil unit or spatial data. Others, such as interpretation and performance data, are derived from soil properties. To store all these attributes in a single table will require many repetitive entries, a process that wastes both time and computer space. Moreover, the table will be extremely difficult to use and update.

The alternative to storing all attribute data in a single table is to manage attribute data in separate tables and to use a database management system (DBMS). A **DBMS** is a set of computer programs for managing an integrated and shared database (Laurini and Thompson 1992). A DBMS provides tools for data input, search, retrieval, manipulation, and output.

Most commercial GIS packages include database management tools for accessing local and external databases. ARC/INFO uses INFO, and ArcView uses dBASE, to handle local data files. External databases may be managed using Oracle, Informix, SYBASE, Access, FoxPro, SQL Server, or other DBMS. MapInfo uses a setup similar to ArcView. MGE offers database management tools from Oracle and Informix and provides an interface to other DBMS. AutoCAD Map uses VISION to manage its database.

The use of a DBMS has other advantages beyond its GIS applications. Often a GIS is part of an enterprise-wide information system, and attribute data needed for the GIS may reside in various departments of the same organization. Therefore, the GIS must function within the overall information system and interact with other information technologies. One effect of the object-oriented data model is the total immersion of GIS in a DBMS.

6.2.2 Type of Attribute Data

One method for defining attribute data is to use the data types allowed in a GIS package. Data types used in a GIS and in computer programming include character strings, integers, floating points or real numbers, dates, and time intervals. Each field in an attribute table is defined with a data type, which applies to the domain of the field.

Another method is to define attribute data by measurement scale. The measurement scale concept groups attribute data into nominal, ordinal, interval, and ratio data (Stevens 1946; Chang 1978). **Nominal** data describe different kinds or different categories of data such as land use types or soil types. **Ordinal** data differentiate data by a ranking relationship. For example, cities may be grouped into large, medium, and small cities by population size. **Interval** data have known intervals between values such as temperature readings. For example, a temperature reading of 70°F is warmer than 60°F by 10°F. **Ratio** data are the same as interval data except that ratio data are based on a meaningful, or absolute, zero value. Population densities are an example of ratio data, because a density of 0 is an absolute zero. In GIS applications the four measurement scales may be grouped into two higher-level categories: **categorical** including nominal and ordinal scales, and **numerical** including interval and ratio scales.

Measurement scales are important to data display and data analysis in GIS. The choice of map symbols depends on the data to be displayed. For example, different sized symbols are not appropriate for displaying nominal data. Chapter 8 discusses in more detail the data-symbol relationship.

GIS analysis often involves computation, which is limited to numerical data. For example, descriptive statistics such as mean and standard deviation can only be derived from numeric data. But GIS projects, especially those dealing with suitability analysis, commonly assign scores to nominal or ordinal data and to use these scores in computation (Chrisman 1997). For example, a suitability analysis for a housing development may assign different scores to different soil types and then combine these scores with scores from other variables in computing the total suitability score. Assigning scores to categorical data requires more information that is not in the base data. Scores in this case represent interpreted data.

Data types and measurement scales are obviously related. Character strings are appropriate for nominal and ordinal data. Integers and real numbers are appropriate for interval and ratio data, depending on whether decimal digits are included or not. But there are exceptions. For example, a study may classify the potential for groundwater contamination as high, medium, and low, but enter the information as numeric data using a look-up table. The look-up table may show 1 for low, 2 for medium, and 3 for high. It would be wrong to say that the medium potential is twice as high as the low potential because numbers in this case are just numeric codes. GIS packages also have functionalities to convert between data types (e.g., from character strings to numbers, or from numbers to character strings). GIS users must pay attention to the nature of attribute data before using them in analysis.

6.3 THE RELATIONAL DATABASE MODEL

A database is a collection of interrelated tables in digital format. There are four types of database design: flat file, hierarchical, network, and relational (Figure 6.3) (Laurini and Thompson 1992; Worboys 1995). A **flat file** contains all data in a large table. An extended feature attribute table is like a flat file. Another common example of a flat file is a spreadsheet. A **hierarchical database** organizes its data at different levels and uses only the one-to-many association between levels. The simple example in Figure 6.3 shows the hierarchical levels of zoning, parcel, and owner. Each level is divided into different branches. A **network database** builds connections across tables, as shown by the linkages between the tables in Figure 6.3. One problem with both the hierarchical and the network database designs is that the linkages between tables must be known in advance and built into the computer code. This requirement tends to make a complicated and inflexible database.

Most GIS vendors use the relational database model for data management (Codd 1970; Codd 1990). A **relational database** is a collection of tables, also called relations, which can be connected to each other by keys. A **key** represents one or more attributes whose values can uniquely identify a record in a table. Therefore, a key common to two tables can establish connection between corresponding records in the tables. In Figure 6.3, the key connecting zoning and parcel is the zone code and the key connecting parcel and owner is the PIN (parcel ID number). When used together, the keys can relate zoning and owner.

Compared to other database designs, a relational database is simple and flexible. It has two distinctive advantages. First, each table in the database can be prepared, maintained, and edited separately from other tables. This feature is important because, with the increased popularity of GIS technology, more data are being recorded and managed in spatial units. Second, the tables can remain separate until a query or an analysis requires attribute data from different tables be linked together. Because the need for linking tables is often temporary, a relational database is efficient for both data management and data processing.

6.3.1 MUIR: An Example of Relational Database

The Natural Resources Conservation Service (NRCS), formerly the Soil Conservation Service, produces the **Soil Survey Geographic** (**SSURGO**) database nationwide. NRCS collects SSURGO data from field mapping and archives the data in 7.5-minute quadrangle units for a soil survey area. A soil survey area may consist of a county, multiple counties, or parts of multiple counties. The SSURGO database represents the most detailed level of soil mapping in the United States. Linked to SSURGO is a **Map Unit Interpretations Record** (**MUIR**) attribute database, which contains about 88 estimated soil physical and chemical properties, interpretations, and performance data

(a) Flat file

PIN	Owner	Zoning
P101	Wang	Residential (1)
P101	Chang	Residential (1)
P102	Smith	Commercial (2)
P102	Jones	Commercial (2)
P103	Costello	Commercial (2)
P104	Smith	Residential (1)

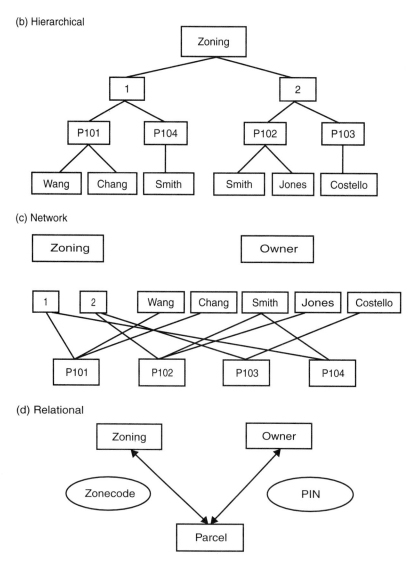

(b) Hierarchical

(c) Network

(d) Relational

Figure 6.3
Four types of database design: (a) flat file, (b) hierarchical, (c) network, and (d) relational.

in a series of tables. This comprehensive attribute database is on the Internet (http://www.statlab .iastate.edu/soils/muir/).

NRCS structures the MUIR database by soil survey area, map units within the survey area, and map unit components within the map unit (Figure 6.4). Four keys are essential in relating tables in the database: state code, soil survey area ID, map unit symbol, and sequence number. The first two keys define the state and the area of the soil survey. The map unit symbol is a numeric code for each soil polygon in the survey. The sequence number key indicates the components of a soil map unit, because a unit may consist of a single dominant soil or two or more soil components. Additionally, various other keys are used to relate soil tables to look-up tables. Examples of look-up tables include soil classification and plants' scientific and common names.

GIS users tend to be overwhelmed by the sheer size of the MUIR database. Actually, the database is not difficult to use if one has a proper understanding of the relational database model and the keys used in relating MUIR tables. This chapter uses the MUIR database to illustrate the type of relationship between tables. Chapter 9 uses the database as an example in data exploration.

6.3.2 Normalization

Preparing a relational database such as MUIR must follow certain rules. One of the rules is called normalization. **Normalization** is a process of decomposition, taking a table with all the attribute data and breaking it down to small tables while maintaining the necessary linkages between them (Vetter 1987). Normalization is designed to achieve several objectives:

- To avoid redundant data in tables that waste space in the database and may cause data integrity problems.
- To ensure that attribute data in separate tables can be maintained and updated separately and can be linked whenever necessary.
- To facilitate a distributed database.

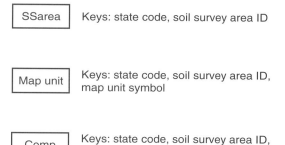

Figure 6.4

The three most important tables in the MUIR database. Ssarea is the soil survey area table, map unit is the soil survey map unit table, and comp is the map unit components table. Each table has keys that can relate to other tables in the database.

Table 6.1 shows attribute data for a parcel map. Table 6.1 is unnormalized because it contains redundant data. Owner addresses are repeated for Smith and residential and commercial zoning are entered twice. An unnormalized table can be difficult to prepare and manage. For example, the number of values for the fields of owner and owner address in Table 6.1 varies from record to record, making it difficult to define the fields and to store field values. Also, if ownership changes, the table must be updated with all the attribute data. The same difficulty applies to operations in which values are added and deleted.

Table 6.2 represents the first step in normalization. Often called the first normal form, Table 6.2 no longer has multiple values in its cells but the problem of data redundancy has increased. The parcels P101 and P102 are repeated twice except for their owners and owner addresses. Smith's address is repeated twice. And the zoning descriptions of residential and commercial are repeated three times each. Also, identification of the owner address is not possible with PIN alone but requires a compound key of PIN and Owner.

Figure 6.5 represents the second step in normalization. In place of Table 6.2 are three small tables for parcel, owner, and address tables. PIN is the key relating the parcel and owner tables.

TABLE 6.1	An unnormalized table					

PIN	Owner	Owner address	Sale date	Acres	Zone code	Zoning
P101	Wang	101 Oak St	1-10-98	1.0	1	residential
	Chang	200 Maple St				
P102	Smith	300 Spruce Rd	10-6-68	3.0	2	commercial
	Jones	105 Ash St				
P103	Costello	206 Elm St	3-7-97	2.5	2	commercial
P104	Smith	300 Spruce Rd	7-30-78	1.0	1	residential

TABLE 6.2	First step in normalization					

PIN	Owner	Owner address	Sale date	Acres	Zone code	Zoning
P101	Wang	101 Oak St	1-10-98	1.0	1	residential
P101	Chang	200 Maple St	1-10-98	1.0	1	residential
P102	Smith	300 Spruce Rd	10-6-68	3.0	2	commercial
P102	Jones	105 Ash St	10-6-68	3.0	2	commercial
P103	Costello	206 Elm St	3-7-97	2.5	2	commercial
P104	Smith	300 Spruce Rd	7-30-78	1.0	1	residential

Owner name is the key relating the address and owner tables. The relationship between the parcel and address tables can be established through the keys of PIN and owner name. The only problem with the second normal form is data redundancy with the fields of zone code and zoning.

The final step in normalization with the above example is summarized in Figure 6.6. A new table, zone, is created to take care of the remaining data redundancy problem with zoning. Zone code is the key relating the parcel and zone tables. Unnormalized data in Table 6.1 are now fully normalized.

Although it can achieve objectives consistent with the relational database model, normalization does have a major drawback of slowing down data access. To find the addresses of parcel owners, for example, one must link three tables (parcel, owner, and address) and employ two keys (PIN and owner name). One way to increase the performance in

data access is to reduce the level of normalization by, for example, removing the address table and including the addresses in the owner table. Database design must therefore consider other factors besides normalization.

6.3.3 Type of Relationship

A relational database usually contains three types of relationships between tables: one-to-one, one-to-many, and many-to-one (Figure 6.7). To explain these relationships, we need to define the **source** (from) table and the **destination** (to) table. For example, if the purpose is to add attribute data from a table to the feature attribute table, then the feature attribute table is the destination table and the other table is the source table.

The simplest of the three, the **one-to-one** relationship means that one and only one record in the

	PIN	Sale date	Acres	Zone code	Zoning
Parcel table	P101	1-10-98	1.0	1	residential
	P102	10-6-68	3.0	2	commercial
	P103	3-7-97	2.5	2	commercial
	P104	7-30-78	1.0	1	residential

	PIN	Owner name
Owner table	P101	Wang
	P101	Chang
	P102	Smith
	P102	Jones
	P103	Costello
	P104	Smith

	Owner name	Owner address
Address table	Wang	101 Oak St
	Chang	200 Maple St
	Jones	105 Ash St
	Smith	300 Spruce Rd
	Costello	206 Elm St

Figure 6.5
Separate tables from the second step in normalization.

	PIN	Sale date	Acres	Zone code
Parcel table	P101	1-10-98	1.0	1
	P102	10-6-68	3.0	2
	P103	3-7-97	2.5	2
	P104	7-30-78	1.0	1

	Owner name	Owner address
Address table	Wang	101 Oak St
	Chang	200 Maple St
	Jones	105 Ash St
	Smith	300 Spruce Rd
	Costello	206 Elm St

	PIN	Owner name
Owner table	P101	Wang
	P101	Chang
	P102	Smith
	P102	Jones
	P103	Costello
	P104	Smith

	Zone code	Zoning
Zone table	1	residential
	2	commercial

Figure 6.6
Separate tables after normalization.

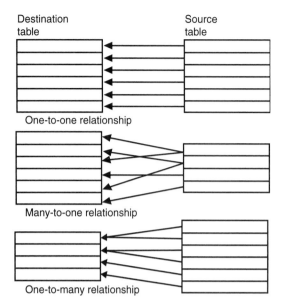

Destination table

Source table

One-to-one relationship

Many-to-one relationship

One-to-many relationship

Figure 6.7
Three types of data relationship between tables: one-to-one, many-to-one, and one-to-many.

destination table is related to one and only one record in the source table. The **one-to-many** relationship means that one record in the destination table may be related to more than one record in the source table. The **many-to-one** relationship is just the opposite: two or more records in the destination table may be related to one record in the source table. The designation of the destination and source tables can therefore determine if the relationship is one-to-many or many-to-one.

How to designate the destination and source tables depends on the data stored in the tables and the information sought. This is illustrated in the following two examples.

6.3.3.1 Example 1: Type of Relationship between Normalized Tables

Figure 6.6 shows four tables from normalization: parcel, owner, address, and zone. Suppose the question is to find who owns a selected parcel. To answer the question, one can set up the parcel table as the destination table and the owner table as the source table. The relationship between the tables is

one-to-many: one record in the parcel table may correspond to more than one record in the owner table.

Suppose the question is now changed to find land parcels owned by a selected owner. The proper setup is to have the owner table as the destination table and the parcel table as the source table. The relationship is many-to-one: more than one record in the owner table may correspond to one record in the parcel table. The same is true between the parcel table and the zone table. If the question is to find the zoning code for a selected parcel, it is a many-to-one relationship. If the question is to find land parcels that are zoned commercial, it is a one-to-many relationship.

6.3.3.2 Example 2: Type of Relationship in the MUIR Database

One approach to better understanding the MUIR database is to sort out the relationship between tables. Most relationships are either one-to-many or many-to-one. For example, the many-to-one relationship exists between the map unit components table (destination table) and the soil classification table (source table) because different soil components may be grouped into the same soil class (Figure 6.8). On the other hand, the one-to-many relationship applies to the map unit components table (destination table) and the soil profile layers table (source table) because a soil component may have multiple profile layers (Figure 6.9).

6.3.3.3 Type of Relationship and Data Display

The type of relationship can influence how data are displayed. Suppose the source table is an owner table, and the destination table is the feature attribute table of a parcel map. To display the parcel ownership would be simple if the relationship between the tables is one-to-one or many-to-one. The one-to-one relationship would have one unique symbol for each parcel, and the many-to-one relationship would have one symbol for either one or more than one parcel.

The one-to-many relationship would create a problem in symbolization. Because a parcel may

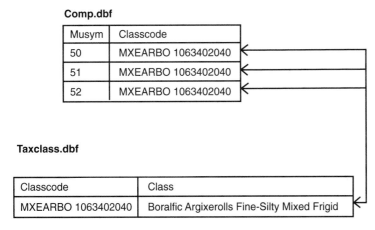

Comp.dbf

Musym	Classcode
50	MXEARBO 1063402040
51	MXEARBO 1063402040
52	MXEARBO 1063402040

Taxclass.dbf

Classcode	Class
MXEARBO 1063402040	Boralfic Argixerolls Fine-Silty Mixed Frigid

Figure 6.8
An example of the many-to-one relationship in the MUIR database for Latah County, Idaho. Three map unit components in comp.dbf are related to the same soil class in the taxclass.dbf.

Comp.dbf

Muid	Seqnum	Musym
610001	1	1

Layer.dbf

Muid	Seqnum	Layernum	Laydepl	Laydeph
610001	1	1	0	7
610001	1	2	7	18
610001	1	3	18	58
610001	1	4	58	62

Figure 6.9
An example of the one-to-many relationship in the MUIR database for Latah County, Idaho. One map unit component in comp.dbf is related to four layers in layer.dbf.

have more than one owner, it would not be right to show a parcel with multiple owners with an owner who happens to be the first on the list of owners. One solution to the problem is to use a different symbol design for multiple ownership.

6.3.4 Attribute Data Management Using ArcView

A major strength of Windows-based GIS packages is their flexibility in working with tables and the different relationships between tables. ArcView is used here as an example to illustrate the flexibility. ArcView has two functions for managing tables: JOIN and LINK.

JOIN in ArcView brings together the destination table and the source table, given that each table has a key. The keys, called the relate field, do not have to have the same name or definition. JOIN is usually recommended for the one-to-one or many-to-one relationship. Given the one-to-one relationship, two tables are joined by record. Given the many-to-one relationship, a record value in the source table is assigned to more than one matching record in the destination table. JOIN does not work well with the one-to-many relationship, because only the first matching record value in the source table is assigned to a record in the destination table.

LINK keeps the tables separate but highlights the associated records in the tables. LINK is therefore useful for the one-to-one, many-to-one, and one-to-many relationships. It can also work with two or more tables simultaneously. One disadvantage of LINK is that attributes from the linked but separate tables cannot be used together in a data query statement (Chapter 9).

6.3.5 Attribute Data Management Using ARC/INFO

ARC/INFO is a command-driven GIS package. Although it has the same attribute data management capabilities as ArcView, the operations are cumbersome. The command JOINITEM joins two tables permanently with a shared item. The shared item must be defined exactly the same in both tables. JOINITEM can work with the one-to-one or many-to-one relationship. The RELATE operation, which may have one or more relates, can join feature attribute tables with internal or external attribute tables. After it is initiated, a relate connects a record in one table to a corresponding record in another table. A relate is temporary and is dropped at the end of an ARC/INFO session. Similar to JOINITEM, RELATE is best used when the relationship between tables is either one-to-one or many-to-one.

The CURSOR command is recommended if the relationship is one-to-many. CURSOR functions in the same way as LINK in ArcView: it highlights source table records that correspond to a single record in the destination table. CURSOR then allows the user to step through the highlighted (selected) records one record at a time.

6.4 ATTRIBUTE DATA ENTRY

6.4.1 Field Definition

The first step in attribute data entry is to define each field in the table. A field definition usually includes data width, data type, and number of decimal places. The width refers to the number of spaces to be reserved for a field. The width should be large enough for the largest number or the longest string in the data. Spaces for the negative sign and the decimal point should also be included in the width. The data type must follow data types

allowed in the GIS package and may be numeric or character. The number of decimal places is part of the definition for the real numeric data type.

Character and numeric type definitions can be confusing at times. For example, the soil map unit in the MUIR database is defined as character, although the map unit is coded as numbers such as 610001, 610002, and so on. These are ID numbers to identify soil map units; therefore, these unit numbers are nominal data rather than interval or ratio data. In other words, computations such as subtracting 610001 from 610002 are meaningless with soil map unit numbers. But treating soil map unit numbers as numeric may be useful in certain types of data query, such as use of a numeric range from 610001 to 610010 in a query.

6.4.2 Methods of Data Entry

Attribute data entry is like digitizing for spatial data entry. Attribute data need to be entered by typing. A map with 4000 polygons, each with 50 fields of attribute data, requires entering 200,000 values. How to reduce time and effort in attribute data entry is of interest to any GIS user.

As in finding existing data for GIS projects, it is best to determine if a government agency or an organization has already entered attribute data in digital format. If the answer is yes, then one need simply import the digital data file into a GIS. The data format is important in importing. Most GIS packages can import dBASE and ASCII files as well as data from database servers such as Oracle, Access, Sybase, and Informix.

If attribute data files do not exist, then typing is the option. But the amount of typing can vary depending on which method or command is used. The ARC/INFO command UPDATE, for example, can be used for attribute data entry but requires the typing of values one record at a time.

One way to save time in typing is to take advantage of the relational database model by having a look-up table. For example, instead of entering attribute data for each polygon in a soil map, one can use a key and a look-up table. First, each soil polygon is assigned a map unit symbol. Next, a look-up table is prepared to list soil attributes for each map unit symbol. By using the map unit symbol as the key, one can then join or link data from the look-up table to the soil map. The same look-up table can be used for other maps in the same soil survey as well.

Attribute data entry using a GIS package is usually more cumbersome than using a word processing or spreadsheet package. A word processing package offers cut-and-paste, find-and-replace, and other functions that are helpful to a typist, especially a poor typist. A GIS package typically does not offer these word-processing options. Therefore, it is wise to enter attribute data using a "typist friendly" package, export the data file in a format compatible with the GIS package, and import the file.

6.5 ATTRIBUTE DATA VERIFICATION

Attribute data verification consists of two parts. The first is to make sure that attribute data are properly linked to spatial data: the label ID should be unique and should not contain null values. The second is to verify the accuracy of attribute data. Data verification is difficult because inaccuracies may be attributed to a large number of factors including observation errors, out-of-date data, and data entry errors.

There are at least two methods for checking data entry errors. First, attribute data can be printed for manual verification. This is like using check plots to check the accuracy of spatial data. Second, computer programs can be written to verify data accuracy. For example, NRCS has developed automated data validation procedures to catch obvious errors in soil attribute data such as data type, data length, data range, and key data field. The procedures also test attribute data against a master set of valid attributes to catch logical errors such as having incompatible components in the same soil map unit. Automated data validation procedures require expert knowledge about attribute data and a well-designed computer program that can catch all possible errors.

One advantage of using the geodatabase model in ArcInfo 8 is its validation rules (Zeiler 1999). Directly related to attribute data verification is the rule about attribute domains, which groups objects into classes, or subtypes, by a valid range of values or a valid set of values for an attribute. Suppose the feature class of land parcels has the subtypes of residential, commercial, and industrial and the field zoning has the value of 1 for residential, 2 for commercial, and 3 for industrial. This set of zoning values can be enforced whenever the field zoning is edited. Therefore, if a zoning value of 9 is entered, the value will be flagged or rejected because it is outside the valid set of values. Similar constraint using a valid numeric range instead of a valid set of values can be applied to lot size or height of building.

6.6 CREATING NEW ATTRIBUTE DATA FROM EXISTING DATA

New attribute data can be created from existing data through data manipulation. New data are often specific to a project. One common scenario is to simplify existing data by grouping the data values into a small number of classes. The other is to compute new data from existing data.

6.6.1 Attribute Data Classification

Data classification reduces a data set to a small number of classes based on the values of an attribute or attributes. For example, elevations may be grouped into lower than 500 meters, 500 to 1000 meters, and so on. Another example is to classify by elevation and slope: class 1 has lower than 500 meters in elevation and 0 to 10% slope, class 2 has lower than 500 meters in elevation but 10 to 20% slope, and so on.

Operationally, creating new attribute data by classification involves three steps: defining a new field for saving the classification result, selecting a data subset through query, and assigning a value to the selected data subset. The second and third steps are repeated until all records are classified and assigned new field values. The main advantage of data classification is that it reduces or simplifies a data set so that the new data set can be more easily used in GIS analysis or modeling.

6.6.2 Attribute Data Computation

One method for creating new data is to perform computation with the existing data. Suppose a map has the length measure in meters. A new field called feet can be computed by length · 3.28.

Another method is to create interpreted data by incorporating expert information in computation. Suppose the quality of wildlife habitat can be numerically scored by the evaluation of slope, aspect, and elevation. The first step in accomplishing the task is to develop a scoring system for each variable. Then, the total score indicating the quality of wildlife habitat can be computed by summing the scores for slope, aspect, and elevation.

In the previous example, the computation of the total score may also involve weights for the variables. Suppose a wildlife specialist believes that elevation is three times as important as slope and aspect. In that case, the total score will be computed by

slope score + aspect score + 3 · elevation score

Weighted or not weighted, the total score becomes a new attribute, which can be further classified and analyzed.

KEY CONCEPTS AND TERMS

Categorical data: Data that are measured at a nominal or an ordinal scale.

Database management system (DBMS): A set of computer programs for managing an integrated and shared database for such tasks as data input, search, retrieval, manipulation, and output.

Destination table: The table to assign data to.

Feature attribute table: A table that stores attribute data of a digital map.

Field: A column in a table that describes an attribute of the map feature. Also called *column* or *item*.

Flat file: A database that contains all data in a large table.

Georelational data model: A common model used in a GIS package that stores spatial data in binary files and attribute data in tables, and links the two data components by the IDs of map features.

Hierarchical database: A database that is organized at different levels and uses the one-to-many association between levels.

Interval data: Data with known intervals between values, such as temperature readings.

Key: A common field used in joining or linking tables.

Many-to-one relationship: One type of data relationship, in which two or more records in the destination table are related to one record in the source table.

Map Unit Interpretations Record (MUIR): An attribute database linked to SSURGO.

Network database: A database that is based on the built-in connections across tables.

Nominal data: Data that show different kinds or different categories, such as land use types or soil types.

Normalization: The process of taking a table with all the attribute data and breaking it down to small tables while maintaining the necessary linkages between them in a relational database.

Numerical data: Data that are measured at an interval or a ratio scale.

Object-oriented data model: A data model that stores spatial data and attribute data in a single database. An example is ArcInfo 8's geodatabase data model.

One-to-many relationship: One type of data relationship, in which one record in the destination table is related to more than one record in the source table.

One-to-one relationship: One type of data relationship, in which one record in the destination table is related to one and only one record in the source table.

Ordinal data: Data that are ranked, such as large, medium, and small cities.

Ratio data: Data with known intervals between values and are based on a meaningful zero value, such as population densities.

Record: A row in a table that represents a map feature. Also called *row* or *tuple*.

Relational database: A database that consists of a collection of tables and uses keys to connect the tables.

Soil Survey Geographic (SSURGO) database: A database maintained by the Natural Resources Conservation Service (NRCS), formerly the Soil Conservation Service, which archives soil survey data in 7.5-minute quadrangle units.

Source table: The table to assign data from.

APPLICATIONS: ATTRIBUTE DATA ENTRY AND MANAGEMENT

This applications section starts with a simple method for attribute data entry in Task 1. Tasks 2 and 3 cover linking tables and joining tables respectively. Tasks 4 and 5 show how new attributes can be created through data classification and data computation respectively. All five tasks use ArcView. Although both packages have similar functionalities for attribute data entry and management, ArcView is more user-friendly and flexible than ARC/INFO.

Task 1: Enter attribute data in ArcView

What you need: *landat.shp,* a feature theme with 19 records.

For Task 1, you will add a new field to a table and enter the field values. Because it works with one field and one record at a time, this data entry method is good for updating rather than entering a large amount of data.

1. Start ArcView, open a new view, and add *landat.shp* to view. Select Table from the Theme menu to open the *landat.shp* theme table.
2. First you need to add a field to the theme table before you can enter the field values. Select Starting Editing from the Table menu. Notice that the field names in the theme table become non-italic signaling editing is enabled. To add a field, select Add Field from

the Edit menu. In the Field Definition dialog, change the name to lucode, select Number as the type, change the width to 4, and leave the decimal places as 0.

3. Now you are ready to enter lucode values. Click the Edit tool. Then click the first cell under lucode, type in its lucode value according to the table below, and enter it. Type and enter the remaining lucode values.
4. Select Stop Editing from the Table menu. Answer Yes to "Save Edits?"
5. After lucode has been added as an attribute to *landat.shp*, you can display lucode values. Activate the view. Double click *landat.shp* in the Table of Contents to open the Legend Editor. Select Unique Values as the legend type and lucode as the value field. Click Apply.

Landat-ID	Lucode	Landat-ID	Lucode
59	400	69	300
60	200	70	200
61	400	71	300
62	200	72	300
63	200	73	300
64	300	74	300
65	200	75	200
66	300	76	300
67	300	77	300
68	200		

Task 2: Link tables in ArcView

What you need: *wp.shp,* a forest stand theme; *wpdata.dbf* and *wpact.dbf,* two attribute data files that can be linked to *wp.shp. Wpdata.dbf* includes vegetation and land type data, and *wpact.dbf* includes activity records.

ArcView offers LINK and JOIN for attribute data management. Task 2 uses LINK to link a feature theme table to two separate dBASE files. The data files are part of a relational database.

1. Start ArcView, open a new view, and add *wp.shp* to view. Select Table from the Theme menu to open the *wp.shp* theme table.
2. Make the Project window active. Click Tables and Add to open the Add Table dialog. Make sure that the File Type is dBASE in the dialog. Click on *wpdata.dbf* and *wpact.dbf* to add them as new tables.
3. Arrange the three tables of attributes of *wp.shp*, *wpdata.dbf*, and *wpact.dbf* so that they are all visible on the monitor. *Wpdata.dbf* and *wpact.dbf* are dBASE files containing additional attributes of *wp.shp*. What you want to do next is to link *wpdata.dbf* and *wpact.dbf* to the *wp.shp* theme table. In linking, *wpdata.dbf* and *wpact.dbf* are called the source tables and the *wp.shp* theme table is called the destination table.
4. First, link *wpdata.dbf* to the *wp.shp* theme table by using the field of Id in both tables as the relate item. Click on Id in *wpdata.dbf* and Id in the *wp.shp* theme table. Then select Link from the Table menu. Repeat the same procedure to link *wpact.dbf* to the *wp.shp* theme table.
5. Click a record in the *wp.shp* theme table. The record is highlighted, so are its related records in *wpdata.dbf* and *wpact.dbf* and the selected map feature in *wp.shp*.

Task 3: Join tables in ArcView

What you need: *wp.shp* and *wpdata.dbf,* same as Task 2.

Task 3 asks you to join a dBASE file to a feature theme table. The choice between JOIN and LINK in ArcView depends on the task. JOIN combines attribute data from different tables into a single table, making it possible to use all attribute data in query, classification, or computation. Tables that are linked remain separate, thus limiting attribute data manipulation to individual tables.

1. Start ArcView, open a new view, and add *wp.shp* to view. Select Table from the Theme menu to open the *wp.shp* theme table.
2. Make the Project window active. Click Tables and Add to open the Add Table dialog. Click on *wpdata.dbf* to add the table.
3. At this point, you have opened two tables: the theme table (Attributes of *wp.shp*) and *wpdata.dbf*. Next, you want to join the data from *wpdata.dbf*, the source table, to the theme table, the destination table, by using Id in both tables as the key.
4. Click on the Id field in *wpdata.dbf* to make it active. Click on Id in the theme table to make it active.
5. Click the Join button to join *wpdata.dbf* to the *wp.shp* theme table.

Task 4: Attribute Data Classification in ArcView

What you need: *wp.shp* and *wpdata.dbf,* same as Task 3.

You have joined a dBASE file to the *wp.shp* theme table in Task 3. Task 4 demonstrates how this expanded theme table and its attribute data can be used for data classification and creation of a new attribute.

1. Make sure that *wp.shp* is still in view, and attribute data from *wpdata.dbf* are still joined to the *wp.shp* theme table.

2. You want to classify values of the field Elev into 4 classes. Elev represents average elevation in a vegetation stand and is measured in hundreds of feet. Elev in *wp.shp* ranges from 24 to 52. One record has an Elev value of 0, because the polygon is not under the jurisdiction of the national forest. The four classes of Elev are <= 40, 41–45, 46–50, and > 50.

3. First, you need to add a new field, called Elevzone, to save the results of classification. Follow the same procedure as in Task 1: select Starting Editing from the Table menu, select Add Field from the Edit menu, and define the new field as having the name of Elevzone, Type of Number, Width of 2, and Decimal Places of 0.

4. Click the Query Builder button to open the Query Builder Box. Prepare the logical expression as: ([Elev] > 0) and ([Elev] <= 40). Click on New Set in the dialog. Those records that meet the logical expressions are now highlighted in the *wp.shp* theme table. Click the Promote button so that the highlighted records are at the top of the table. Close the Query Builder dialog.

5. Select Calculate from the Field menu to open the Field Calculator dialog. The lower left corner of the dialog is the display area for the computation expression. Notice that the field, Elevzone, is shown above the display area with the equal sign. Enter 1 in the display area and click OK. Now the highlighted records in the table all have the value of 1 under Elevzone, or are classified into Elevzone 1.

6. Click the Query Builder button again and prepare the logical expressions as: ([Elev] > 40) and ([Elev] <= 45). Click on New Set in the dialog. Open the Field Calculator dialog, and enter 2 in the display area. Now the classification of Elevzone 2 is done.

7. Repeat the same procedure to complete the classification of Elevzone 3 and 4. To save results of the classification, select Stop

Editing from the Table menu and answer Yes to save edits.

Task 5: Attribute Data Computation

What you need: *wp.shp* and *wpdata.dbf,* same as Task 3.

You have created a new field from data classification in Task 4. Another common method for creating new fields is computation. Task 5 shows how a new field can be created and computed from existing attribute data.

1. If *wp.shp* is still in view, make it active; otherwise, add *wp.shp* to view and make it active. Open its theme table.

2. The field area in the *wp.shp* theme table is measured in square meters. You want to convert the area measurement to acres in this task. Select Start Editing from the Table menu. Notice that the field names in the theme table become non-italic when editing is enabled. Select Add Field from the Edit menu. In the Field Definition dialog, change the name to acres, select Number as the type, change the width to 8, and change the decimal places to 2. After you click OK and dismiss the dialog, the new field, acres, is added to the *wp.shp* theme table.

3. Select Calculate from the Field menu to open the Field Calculator dialog. The lower left corner of the dialog is the display area for the computation expression. Notice that the field, acres, is shown above the display area with the equal sign. Double click the field area, double click the / request, type 1000000, double click the * request, and then type 247.11. The completed calculation expression should read: [Area] / 1000000 * 247.11. Click OK in the Field Calculator dialog and dismiss it. The acres field is now populated with the calculated values in acres.

4. To save results of the calculation, select Stop Editing from the Table menu and answer Yes to save edits.

REFERENCES

Chang, K. 1978. Measurement Scales in Cartography. *The American Cartographer* 5: 57–64.

Chrisman, N. 1997. *Exploring Geographic Information Systems*. New York: John Wiley & Sons.

Codd, E. F. 1970. A Relational Model for Large Shared Data Banks. *Communications of the Association for Computing Machinery* 13: 377–87.

Codd, E. F. 1990. *The Relational Model For Database Management, Version 2*. Reading, MA: Addison-Wesley Publishing Company.

Laurini, R., and D. Thompson. 1992. *Fundamentals of Spatial Information Systems*, London: Academic Press.

Stevens, S.S. 1946. On the Theory of Scales of Measurement. *Science* 103: 677–80.

Vetter, M. 1987. *Strategy for Data Modelling*. New York: John Wiley & Sons.

Worboys, M.F. 1995. *GIS: A Computing Perspective*. London: Taylor & Francis.

Zeiler, M. 1999. *Modeling Our World: The ESRI Guide to Geodatabase Design*. Redlands, CA: ESRI Press.

RASTER DATA

7.1 INTRODUCTION

The vector data model uses the geometric objects of point, line, and area to construct spatial features. Although ideal for discrete features with well-defined locations and shapes, the vector data model does not work well with spatial phenomena that vary continuously over the space such as precipitation, elevation, and soil erosion. A better option for representing continuous phenomena is the raster data model.

The raster data model uses a regular grid to cover the space and the value in each grid cell to correspond to the characteristic of a spatial phenomenon at the cell location. Conceptually, the variation of the spatial phenomenon is reflected by the changes in the cell value. Raster data have been described as field-based, as opposed to object-based vector data (Burrough and McDonnell 1998).

A wide variety of data used in GIS are encoded in raster format. They include digital elevation data, satellite images, digital orthophotos, scanned maps, and graphic files. Most GIS packages can display raster and vector data simultaneously, and can convert from raster to vector data or from vector to raster data. Raster data also introduce a large set of data analysis functions to GIS. In many ways, raster data complement vector data in GIS applications. Integration of both types of data has therefore become a common feature in a GIS project.

Because the data model determines how GIS data are structured, stored, and processed, raster data are covered in this chapter, separate from vector data. This chapter is divided into six sections. Section 1 discusses the basic elements of raster data. Section 2 describes different types of raster data. Section 3 provides an overview of data structure, data compression, and data files. Section 4 explains projection and geometric transformation of raster data. Section 5 discusses conversion between raster and vector data. Section 6 offers examples of integration of raster data and vector data in GIS.

7.2 ELEMENTS OF THE RASTER DATA MODEL

A raster data model is variously called a grid, a raster map, a surface cover, or an image in GIS. Grid is adopted in this section. A grid consists of rows, columns, and cells. The origin of rows and columns is at the upper left corner of the grid.

Rows function as *y*-coordinates and columns as *x*-coordinates in a two-dimensional coordinate system. A cell is defined by its location in terms of row and column.

Raster data represent points by single cells, lines by sequences of neighboring cells, and areas by collections of contiguous cells (Figure 7.1). Each cell in a grid carries a value, either an **integer** or a **floating-point** value (a value with decimal digits). Integer cell values typically represent categorical data. For example, a land cover model may use 1 for urban land use, 2 for forested land, 3 for water bodies, and so on. Floating-point cell values represent continuous data. For example, a precipitation model may have precipitation values of 20.15, 12.23, and so forth. A floating-point grid requires more computer memory than an integer grid—an important factor for a GIS project that covers a large area. Also, data query and display of a floating-point grid should be based on value ranges such as 12.0–19.9, instead of individual values.

Each cell value in a grid represents the characteristic of a spatial phenomenon at the location

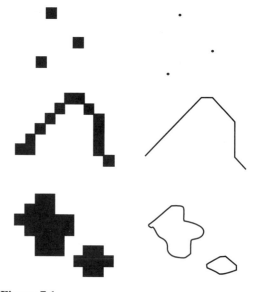

Figure 7.1

Representation of point, line, and area features using the raster format (left) and the vector format (right).

denoted by its row and column. The raster data model does not clearly separate spatial and attribute data and, unlike vector data, raster data have little use for database management. Although an integer grid can have a value attribute table, the table is used primarily for summarizing the cell values and their frequencies.

One option to store multiple attributes in a grid is to use some kind of cell ID as the cell value. For example, a grid representing counties in the Pacific Northwest may use the county FIPS (U.S. Federal Information Processing Standards) code as the cell value and store county population and income data in separate fields. The primary function of such a value attribute table is data storage. The grid can be used for displaying population or income data, but any query or analysis with the grid is limited to the FIPS codes. The strong association between a grid and its cell values is an important difference between the raster data model and the vector data model.

The cell size determines the resolution of the raster data model. A cell size of 30 meters means that each cell measures 900 square meters (30 × 30 meters). A large cell size cannot represent the precise location of spatial features, thus increasing the chance of having mixed features such as forest, pasture, and water in a cell (Box 7.1). These problems lessen when a grid uses a smaller cell size. But a small cell size increases the data volume and the data processing time.

A grid is normally projected onto a coordinate system such as the UTM coordinate system. Therefore, raster data can be displayed with vector data if they are based on the same coordinate system. The area extent of a grid can also be determined by reading the real-world coordinates of its lower left corner and upper right corner points. It should be noted that a real-world coordinate system has its origin at the lower left corner, different from the origin of a grid.

Although the raster data model has its weakness in representing the precise location of spatial features, it has the distinctive advantage in having fixed cell locations (Tomlin 1990). In computing algorithms, a grid can be treated as a matrix with rows and columns, and its cell values can be stored into a two-dimensional array. All commonly used programming languages can easily handle arrayed variables. Raster data are therefore much easier for data manipulation, aggregation, and analysis than vector data.

7.3 TYPES OF RASTER DATA

7.3.1 Satellite Imagery

Remotely sensed satellite data are recorded in raster format. The U.S. **Landsat** satellites have produced the most widely used imagery worldwide since 1972. Landsat 1, 2, and 3 acquired images by the Multispectral Scanner (MSS) with a

---

Box 7.1   Rules in Determining a Categorical Cell Value

If a large cell does cover forest, pasture, and water on the ground, which category should be entered for the cell? Probably the most common method is to enter the category that occupies the largest area percentage of the cell. But in some situations the majority rule may not be the best option. For example, studies of endangered species would be more inclined to use the presence/absence rule than the majority rule

(Chrisman 1997). As long as there is presence of an endangered species in a cell, no matter how much of the cell it occupies, the cell would be coded with a value for presence. Similarly, the determination of cell values may be based on a ranking of the importance of spatial features to the study. If pasture is deemed to be more important than forest and water, a mixed cell of all three features may be coded as pasture.

spatial resolution of about 79 meters. The spatial resolution relates to the ground pixel size. A spatial resolution of 79 meters, for example, means that each pixel on the satellite image corresponds to a ground pixel of 79 meters by 79 meters. MSS images were terminated in 1992.

When launched in 1982, Landsat 4 carried a new sensor, the Thematic Mapper (TM) scanner. TM was a significant improvement over MSS, providing a spatial resolution of 30 meters. A second TM was launched aboard Landsat 5 in 1984. Landsat 6 failed to reach its orbit after launch in 1993. Landsat 7 was launched successfully in April 1999, carrying an Enhanced Thematic Mapper-Plus (ETM+) sensor (http://landsat7.usgs.gov/). The enhanced sensor is designed to seasonally monitor small-scale processes on a global scale such as cycles of vegetation growth, deforestation, agricultural land use, erosion and other forms of land degradation, snow accumulation and melt, and urbanization. The spatial resolution of Landsat 7 imagery is 30 meters.

The U.S. National Oceanic and Atmospheric Administration (NOAA) uses weather satellites as an aid to weather prediction and monitoring. NOAA's Polar Orbiting Environmental Satellites (POES) carry the **AVHRR** (Advanced Very High Resolution Radiometer) scanner, which provide data useful for large-area land cover and vegetation mapping (http://edcdaac.usgs.gov/1KM/1kmhome-page.html/). AVHRR data have a spatial resolution of 1.1 kilometers, which may be too coarse for some GIS projects. But the coarse spatial resolution is offset by the daily coverage and reduced volume of AVHRR data.

The French **SPOT** satellite series began in 1986 (http://www.spot.com/). Each SPOT satellite carries two sensors. The panchromatic sensor acquires single-band imagery with 10-meter spatial resolution, while the multi-spectral sensor captures 20-meter imagery in three wavelengths. High spatial resolution makes SPOT images good spatial data sources for GIS projects. Other important satellite programs have also been established since the late 1980s in India (http://www.isro.org/) and Japan (http://www.nasda.go.jp/).

The privatization of the LANDSAT program in the U.S. in 1985 has opened the door for private companies to gather and market remotely sensed data using various platforms and sensors. One such company is Space Imaging (http://www.spaceimag-ing.com/). Space Imaging's Ikonos satellites are designed to acquire panchromatic images with a 1-meter spatial resolution and multi-spectral images with a 4-meter resolution. The Ikonos satellite succeeded to reach its orbit in September 1999, thus becoming the first commercial 1-meter resolution satellite. A resolution of 1 meter is high enough to detect ground objects such as cars, small houses, fires, and troop deployments. Other companies such as EarthWatch Incorporated (http://www.digitalglobe.com/) also plan to market high-resolution imagery.

The pixel value in a satellite image represents light energy reflected or emitted from the Earth's surface (Lillesand and Kiefer 2000). The measurement of light energy is based on spectral bands from a continuum of wavelengths known as the electromagnetic spectrum. Panchromatic images are comprised of a single spectral band, whereas multi-spectral images are comprised of multiple bands. Landsat TM images, for example, have seven spectral bands: blue, green, red, near infrared, mid-infrared I, thermal-infrared, and mid-infrared II. Unlike a GIS package, an image processing system must be able to access and process multi-band imagery. This difference is usually the basis for separating GIS from image processing packages, although image processing packages such as ERDAS (http://www.erdas.com/) and ER Mapper (http://www.ermapper.com/) have data processing functions that can easily be used in GIS.

By analyzing the pixel values, an image processing system can extract a variety of themes from satellite images, such as land use and land cover, hydrography, water quality, and areas of eroded soils. Satellite images can be displayed in black and white or in color. Satellite images can also simulate color photographs if they have pixel values from the blue, green, and red spectral bands. The image looks like a color photograph if bands 3, 2, and 1 are assigned to red, green, and blue respectively, and a color infrared photograph if bands 4, 3, and 2 are assigned to red, green, and blue respectively.

7.3.2 Digital Elevation Models

A **digital elevation model** (**DEM**) consists of an array of uniformly spaced elevation data. A DEM is point-based, but it can easily be converted to raster data by placing each elevation point at the center of a cell. Most GIS users in the United States use DEMs from the U.S. Geological Survey (USGS). USGS offers four types of DEM data: 7.5-minute DEM, 30-minute DEM, 1-degree DEM, and Alaska DEM. Each DEM file contains, in addition to elevation data, the header information including units of measurement, minimum and maximum elevations, and projection parameters as well as statistics on the accuracy of the data. The USGS has a website that updates the status of DEM data by state (http://mcmcweb.er.usgs.gov/status/dem_stat.html/).

7.3.2.1 The 7.5-minute DEM

The 7.5-minute DEMs provide elevation data at a spacing of 30 meters or 10 meters on a grid measured in UTM (Universal Transverse Mercator) coordinates. Each DEM covers a 7.5-by-7.5-minute block that corresponds to the USGS 1:24,000 scale quad (Box 7.2). The 7.5-minute DEMs are grouped into four levels by production method, with level 1 having the least accurate data and level 4 the most accurate data. Only level 1 and level 2 DEMs are currently available for most states. Level 1 DEMs have the vertical accuracy of 15 meters or better, whereas level 2 DEMs have the vertical accuracy of 7 meters. Level 2 DEMs, which are derived from contour lines (Kumler 1994), can have resolutions at 30 meters or 10 meters.

7.3.2.2 The 30-minute DEM

The 30-minute DEMs provide elevation data at a spacing of two arc seconds on the geographic grid (about 60 meters in the mid-latitudes). Each DEM covers a 30-by-30-minute block, corresponding to the east or west half of a USGS 30-by-60-minute, 1:100,000 quad. The vertical accuracy of the 30-minute DEMs is equal to or better than 25 meters (80 feet), one-half of a contour interval of the 1:100,000 quad.

7.3.2.3 The 1-degree DEM

The one-degree DEMs provide elevation data at a spacing of three arc-seconds on the geographic grid (about 100 meters in the mid-latitudes). Each DEM covers a 1-by-1-degree block, corresponding to the east or west half of a USGS 1-by-2-degree, 1:250,000 quad. The Defense Mapping Agency (DMA, now the National Imagery and Mapping Agency or NIMA) originally produced the one-degree DEMs by interpolation from digitized

---

Box **7.2**   **No-Data Slivers in 7.5-minute DEM**

Although a 7.5-minute DEM corresponds to a USGS 1:24,000 scale quad, a sliver of no data often shows up along the border of the DEM. This is because the bounding coordinates of the DEM do not form a rectangle and thus cannot match the USGS quad perfectly. The 7.5-minute DEM data are stored as a series of profiles, with a constant spacing of 30 meters along and between each profile. When DEMs are cast on the UTM coordinate system, the bounding coordinates of a 7.5-minute DEM form a quadrilateral, rather than a rectangle. The no-data value is used to fill the uneven rows and columns.

contour lines. The vertical accuracy of elevation data is about 30 meters (100 feet).

7.3.2.4 Alaska DEMs

USGS provides 7.5-minute and 15-minute Alaska DEMs, with spacing referenced to latitude and longitude. The spacing for the 7.5-minute Alaska DEM is 1 arc second of latitude by 2 arc seconds of longitude, and the spacing for the 15-minute Alaska DEM is 2 arc seconds of latitude by 3 arc seconds of longitude.

7.3.2.5 Non-USGS DEMs

A basic method for producing DEMs is to use a stereoplotter and aerial photographs with overlapped areas. The stereoplotter creates a three-dimensional model, which allows the operator to compile elevation data as well as orthophotos. Although this method can produce highly accurate DEM data at a finer resolution than USGS DEMs, it is expensive for coverage of large areas.

There are alternatives to the use of a stereoplotter. One alternative that has become popular with GIS users is to generate DEMs from satellite imagery such as the SPOT stereo model. Software packages for extracting elevation data from SPOT images on the personal computer are commercially available. Besides imagery data, the data extraction process requires ground control points, which can be measured in the field by GPS (Global Positioning System) with differential correction. The quality of such DEMs depends on the software package and the quality of the inputs.

Intermap Technologies is a Canadian company that specializes in producing commercial DEM data (http://www.intermaptechnologies.com/). The company offers two types of DEMs. Produced from stereo radar data, the first type provides elevation data at a spacing of 5 to 15 meters and with a vertical resolution of 2 to 5 meters. Radar represents an active remote sensor: it emits microwave pulses and measures returned energy from ground objects. Radar can therefore penetrate cloud cover, which usually presents a major problem with aerial photography. The second type of DEMs from Intermap Technologies is derived from satellite images; their vertical resolutions range from 15 to 30 meters.

7.3.2.6 Global DEMs

DEMs are now available on the global scale. GTOPO30 is a global DEM with a horizontal grid spacing of 30 arc seconds (approximately 1 kilometer) (http://edcdaac.usgs.gov/gtopo30/gtopo30.html/). ETOPO5 (Earth Topography-5 Minute) data cover both the land surface and ocean floor of the Earth, with a grid spacing of 5 minutes of latitude by 5 minutes of longitude (http://edcwww.cr.usgs.gov/glis/hyper/guide/etopo5/).

7.3.3 Digital Orthophotos

A **digital orthophoto quad** (**DOQ**) is a digitized image prepared from an aerial photograph or other remotely sensed data, in which the displacement caused by camera tilt and terrain relief has been removed. A digital orthophoto is geo-referenced and can be registered with topographic and other maps. The USGS began to produce DOQs in 1991 from 1:40,000 scale National Aerial Photography Program aerial photographs and expects to have complete coverage of the conterminous United States by 2004.

The standard USGS DOQ format is either a 3.75-minute quarter quad or 7.5-minute quad in the form of a black and white, color infrared, or natural color image. Most GIS users use 3.75-minute quarter quads in black and white. These quarter quads have a one-meter ground resolution (i.e., each pixel in the image measures 1-by-1 meter on the ground) and have pixel values representing 256 gray levels, similar to a single-band satellite image.

For decades, mapmakers have produced photomaps by superimposing map symbols on aerial photographs, either rectified or unrectified. Continuing with this tradition, DOQs combine the image characteristics of a photograph with the geometric qualities of a map and can be easily integrated with GIS. DOQs are the ideal background for data display or for updating of digital maps.

7.3.4. Binary Scanned Files

A **binary scanned file** is a scanned image containing values of 1 or 0. Scanned files for tracing are examples of binary scanned files (Chapter 4). Maps to be digitized are typically scanned at 300 or 400 dpi (dots per inch).

7.3.5 Digital Raster Graphics (DRG)

A **digital raster graphic** (**DRG**) is a scanned image of a USGS topographic map. The USGS scans the 7.5-minute topographic map at 250 dpi, thus producing a DRG with a ground resolution of 2.4 meters (8 feet). The USGS uses up to 13 colors on each 7.5-minute DRG. With 2.5-minute grid ticks as control points, the DRG is geo-referenced to the UTM coordinate system.

7.3.6 Graphic Files

Maps, photographs, and images can be stored as digital graphic files. Many of the popular graphic files are in raster format, such as TIFF (Tag(ged) Image File Format), GIF (Graphics Interchange Format), and JPEG (Joint Photographic Experts Group). The USGS distributes the DOQs in TIFF or GeoTIFF formats. GeoTIFF is a geo-referenced version of the TIFF format. By having the geographic reference information of the image, DOQs can be readily used with other GIS data.

7.3.7 GIS Software-Specific Raster Data

GIS packages use raster data that are imported from DEMs, satellite images, scanned images, graphic files, and ASCII files or are converted from vector data. These raster data are named differently. For example, ARC/INFO, ArcView, and MGE name raster data grids, GRASS raster map layers, IDRISI images, and PCI surface covers.

Based on a proprietary format, **ARC/INFO grids** are stored as either integer or floating-point grids. An integer grid has a table that stores cell values, and a floating-point grid does not. Both ARC/INFO and ArcView can convert a floating-point grid to an integer grid, and vice versa. They can also convert vector-based coverages and shapefiles to grids, and vice versa.

7.4 RASTER DATA STRUCTURE, COMPRESSION, AND FILES

7.4.1 Data Structure

Raster data structure refers to the storage of raster data so that they can be used and processed by the computer. The simplest data structure is called the **cell-by-cell encoding** method: a raster model is stored as a matrix, and its cell values are written into a file by row and column (Figure 7.2). Functioning at the cell level, this method is an ideal choice if the cell values of a raster model change continuously.

Digital elevation models use the cell-by-cell data structure because the neighboring elevation values are rarely the same. Satellite images also use the method for data storage. With multiple spectral bands, however, each cell or pixel in a

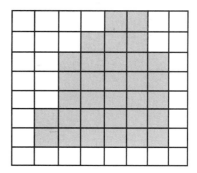

Row 1: 0 0 0 0 1 1 0 0
Row 2: 0 0 0 1 1 1 0 0
Row 3: 0 0 1 1 1 1 1 0
Row 4: 0 0 1 1 1 1 1 0
Row 5: 0 0 1 1 1 1 1 0
Row 6: 0 1 1 1 1 1 1 0
Row 7: 0 1 1 1 1 1 1 0
Row 8: 0 0 0 0 0 0 0 0

Figure 7.2
The cell-by-cell data structure records each cell value by row and column.

satellite image has more than one value. Multi-band imagery is typically stored in the following three formats. The band interleaved by line (.bil) method stores the first value of every row sequentially, followed by the second value of every row, and so on in one image file. The band sequential (.bsq) method stores the values for each band of satellite data sequentially in one image file. Using the band interleaved by pixel (.bip) method, each row of an image is stored sequentially, row 1 all bands, row 2 all bands, and so on.

The cell-by-cell encoding method becomes inefficient if a raster model contains many redundant cell values. For example, a binary scanned file from a soil map would have many 0s representing non-inked areas and only occasional 1s representing the inked soil lines. Raster models such as binary scanned files can be more efficiently stored using the **run-length encoding** (RLE) method, which records the cell values by row and by group (Figure 7.3). Each group includes a cell value and the number of cells with that value. If all cells in a row contain the same value, only one group is recorded, thus saving the computer memory.

Working along one row at a time, the RLE method is not efficient for recording two-dimensional features, often called regions in the literature. Three methods are designed for two-dimensional regions: chain codes, block codes, and quad trees (Worboys 1995; Burrough and McDonnell 1998).

The **chain code** method represents the boundary of a region by using a series of cardinal directions and cells. For example, N1 means moving north by 1 cell and S4 means moving south by 4 cells. Figure 7.4 illustrates the chain coding of a region. A variation of the method is to code the cardinal directions numerically: for example, 0 for east, 1 for north, 2 for west, and 3 for south.

The **block code** method uses square blocks to represent a region. A unit square represents a cell, a 4-square block represents 2-by-2 cells, a 9-square block represents 3-by-3 cells, and so on. Using the medial axis transform (Rosenfeld 1980), each square block is coded only with the location of a cell (e.g., the lower left of the block), and the

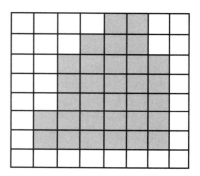

Row 1: 5 6
Row 2: 4 6
Row 3: 3 7
Row 4: 3 7
Row 5: 3 7
Row 6: 2 7
Row 7: 2 7

Figure 7.3
The run length encoding method records the cell values in runs. Row 1, for example, has two adjacent cells in columns 5 and 6 that are gray or have the value of 1. Row 1 is therefore encoded with one run, beginning in column 5 and ending in column 6. The same method is used to record other rows.

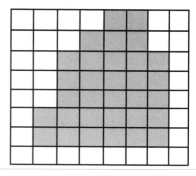

N1 E1 N3 E1 N1 E1 N1 E1 S2 E1 S4 W5

Figure 7.4
Starting at the lower left cell of the region, the chain codes method records the region's boundary by using the principal direction and the number of cells. In this example, the recording follows a clockwise direction.

side length of the block. Figure 7.5 shows the block encoding of a region.

Regional **quad tree** uses recursive decomposition to divide a grid into a hierarchy of quadrants (Samet 1990) (Figure 7.6). A quadrant having cells with the same value will not be subdivided, and it is stored as a leaf node. Leaf nodes are coded with the value of the homogeneous quadrant. A quadrant having different cell values will be subdivided until a quadrant at the finer level contains only one value. Recursive decomposition refers to this process of continuous subdivision. Regional quad tree is an efficient method for storing area data, especially if the data represent a small number of categories. The method is also efficient for data processing (Samet 1990).

GIS packages use different raster data structures. Both GRASS and IDRISI use raster data that are stored using either the cell-by-cell or RLE method. SPANS uses a quad-tree data structure. ARC/INFO grids use a hierarchical tile-block-cell data structure. Depending on its size, a grid is divided into a number of tiles or treated as a single tile. Each tile is subdivided into a series of rectangular blocks, and each block is made up of square cells. Cell values are stored by block. Each block is stored as one variable-length record using either the cell-by-cell or RLE method.

7.4.2 Data Compression

Data compression refers to the reduction of raster data volumes, a topic closely related to data structure. A raster data file stored using the RLE method is called the run-length compressed (RLC) file, because the storage method by group saves the computer memory. A binary scanned file of a 7.5-minute soil quad map, scanned at 300 dpi, can be over 8 megabytes (MB). When encoded using the RLE method, for example, the same file is reduced to about 800,000 bytes at a 10:1 compression ratio. Some raster data such as DEMs and satellite images are difficult to compress because of their continuously changing values.

Graphic files such as TIFF, GIF, and JPEG files are regularly compressed using a variety of image compression algorithms. TIFF and GIF files use **lossless compression**, which allows the original image to be precisely reconstructed, whereas JPEG files use **lossy compression**, which can achieve high compression ratios but cannot reconstruct fully the original image. Image degradation through lossy compression can result in "blocky" appearances. DOQs from the USGS are often distributed in quarters, because even a quarter DOQ is about 45–50 MB in size. Data compression is therefore useful for the sharing of image files.

MrSID (Multi-resolution Seamless Image Database) is a compression technology originally developed at the Los Alamos National Laboratory and later awarded to LizardTech, Inc. (http://www.lizardtech.com/). Multi-resolution means that MrSID has the capability of recalling the image data at different resolutions or scales. Seamless means that MrSID can compress a large image such as a DOQ with sub-blocks and eliminates the artificial block boundaries during the compression process. MrSID assumes that the level of detail in

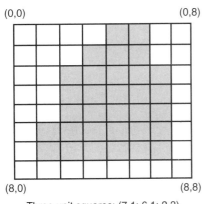

Three unit squares: (7,1; 6,1; 2,3)

One 4-squares : (2,4)

One 25-squares : (7,2)

Figure 7.5

The block codes method works with square blocks. The region in the diagram is divided into three unit square blocks, one 4-square block, and one 25-square block. Each block is encoded with the coordinates of its lower left corner. The origin of the coordinates is at the upper left corner of the grid.

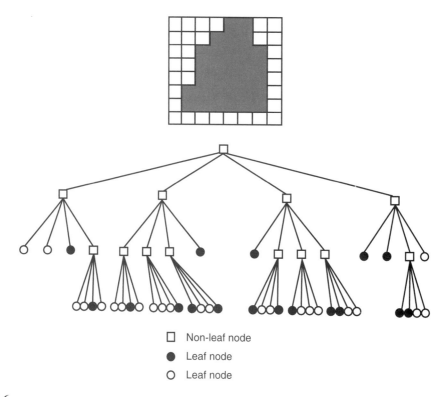

Figure 7.6
The regional quadtree method divides a grid into a hierarchy of quadrants. The division stops when a quadrant is made of cells of the same value. A quadrant that cannot be subdivided is called a leaf node.

an image is not constant. Therefore, MrSID encodes the detailed parts of the image at high resolution and the less detailed parts at low resolution.

Data compression programs, such as PKZip and WinZip for PCs and gunzip for the UNIX environment, can work with any type of data files. Given simple binary scanned files, PKZip and WinZip can achieve extremely high compression ratios. For example, over 100 MB of binary scanned files can be put on a 1.44 MB floppy disk after compression.

7.4.3 Raster Data Files

To import raster data for use, a GIS package must have information about the data structure and, if applicable, the compression method. This information is usually contained in the header file, which is like metadata in function. For example, the header file, often denoted by the extension .hdr, for a satellite image contains information about the image data, such as the data structure method, numbers of rows and columns, number of spectral bands, and number of bits per pixel per band. Two optional files may accompany satellite images. The statistics file, often with the extension .stx, describes statistics such as minimum, maximum, mean, and standard deviation for each spectral band in an image. The color file, often with the extension .clr, associates colors with different pixel values in an image.

Raster data used in a GIS package are organized into several different files. For example, each ARC/INFO grid has its own separate directory, which contains at least six files. The first three files describe the area extent of the grid, the statistical data about the cell values, and, in the case of an integer grid, the cell value and the number of cells that have the same value. Of the next three files, the header file stores information about the grid's cell size, blocking factor, and compression technique. The other two files store information on the data and spatial indexing for the first tile in the grid.

7.5 PROJECTION AND GEOMETRIC TRANSFORMATION OF RASTER DATA

Map projection and geometric transformation are just as important to raster data as to vector data. Unless they are projected onto the same coordinate system, raster data such as satellite imagery and DEMs cannot be registered spatially with vector data such as roads, streams, and other map features in a GIS database.

Projected raster data are still based on rows and columns but the rows and columns are measured in real-world coordinates. For example, an elevation grid may be defined as follows:

- Rows: 463, Columns: 318, Cell size: 30 m
- UTM coordinates at the lower left corner: 499995, 5177175
- UTM coordinates at the upper right corner: 509535, 5191065

The numbers of rows and columns can be derived from the bounding UTM coordinates of the grid and the cell size (i.e., 463 = (5191065 − 5177175) / 30, and 318 = (509535 − 499995) / 30). One can also derive the UTM coordinates defining each grid cell. For example, the cell in Row 1, Column 1 has the UTM coordinates of 499995, 5191035 at the lower left corner and 500025, 5191065 at the upper right corner.

Geometric transformation of satellite imagery is often called **geo-referencing** in image processing (Verbyla and Chang 1997; Lillesand and Kiefer 2000). Two common methods for geo-referencing are affine transformation and polynomial equations. Affine transformation can geo-reference an image through rotation, translation, and scaling. For example, a rotation adjustment can correct a satellite image with the northeast orientation, rather than the north orientation, because the satellite orbit is from northeast to southwest. An affine transformation of image data uses the same transformation equations as for vector data:

$$x' = Ax + By + C \qquad\qquad (7.1)$$
$$y' = Dx + Ey + F \qquad\qquad (7.2)$$

The differences are that (1) x and y represent column number and row number and (2) the coefficient E is negative because of the different origins of an image and a coordinate system. Transformation of image data, however, requires the additional work of selecting control points. Because control points are not part of an image, control points, also called **ground control points**, must come from a different source. Ideally, control points are those features that show up clearly as single distinct cells on the satellite image. For example, control points may be selected at road intersections, rock outcrops, small ponds, or distinctive features along shorelines. After control points are identified, their real-world coordinates can then be obtained from digital maps or GPS readings.

Because ground control points are selected by the user and are subject to errors in interpretation, the affine transformation of satellite images is usually an iterative process. The affine transformation is run with an initial set of control points. The root mean square (RMS) error is examined. If the RMS error exceeds the required tolerance value, for example, one pixel, then the control points that contribute most to the RMS error are removed, and a different set of control points is entered for the next affine transformation. This process continues until a satisfactory RMS error is obtained.

A file must be created after an affine transformation to store the values of the six coefficients in the order of A, D, B, E, C, F. ARC/INFO and ArcView, for example, use a world file, which is an ASCII file and has the letter w in the file extension for easy recognition. Using a world file (e.g., tmrect.blw), an image file (e.g., tmrect.bil) can be converted to real-world coordinates and registered with vector-based map features.

The use of polynomial equations for geo-referencing is also called "**warping**" or "rubber sheeting." A polynomial equation provides a mathematical model for differential scaling and rotating across an image. The degree of complexity of the model is expressed by the order of the polynomial, which may range from 2 to 5. The 2nd order polynomial, for example, uses the following equations for transformation:

$$x' = a_0 + a_1x + a_2y + a_3xy + a_4x^2 + a_5y^2 \quad (7.3)$$
$$y' = b_0 + b_1x + b_2y + b_3xy + b_4x^2 + b_5y^2 \quad (7.4)$$

GIS packages that allow integration of image data with vector and raster data are usually capable of performing geometric transformation. For example, ArcView has the Warp Environment extension that lets the user select control points from a feature theme and use them in geometric transformation.

Map projection, re-projection, or geometric transformation creates a new grid. Data **re-sampling** is therefore required to fill each cell of the new grid with the value of the corresponding cell or cells in the original grid. Three common re-sampling methods for filling the new raster grid are nearest neighbor, bilinear interpolation, and cubic convolution (Figure 7.7).

The **nearest neighbor** re-sampling method fills each cell of the new grid with the nearest cell value from the original grid. The **bilinear interpolation** method fills each cell of the new grid with the weighted average of the four nearest cell values from the original grid. The **cubic convolution** method fills each cell of the new grid with the weighted average of the 16 nearest cell values from the original grid. Nearest neighbor is primarily used for categorical data such as land cover types because the method will not change the cell values. All three methods can be used for continuous data such as elevation or satellite imagery. The overall smoothness of the re-sampled output—at the expense of the processing time—increases from nearest neighbor to bilinear interpolation to cubic convolution.

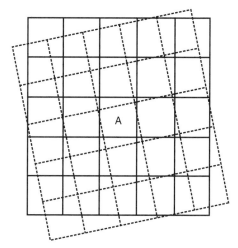

Figure 7.7
The grid in solid line is the new grid created by transforming the grid in dashed lines. Re-sampling is required to fill the value of each cell in the new grid such as cell A.

7.6 DATA CONVERSION

The conversion of vector data to raster data, for example, a polygon coverage to a grid, is called **rasterization**, and the conversion of raster data to vector data, for example, a grid to a polygon coverage, is called **vectorization** (Figure 7.8). Both types of conversion require use of computer algorithms (Piwowar et al. 1990).

Generally simpler than vectorization, rasterization involves three basic steps (Clarke 1995). The first step is to set up a grid with a specified cell size to cover the area extent of the coverage and to assign initially all cell values as zeros. The second step is to change the values of those cells that correspond to points, lines, or polygon boundaries. The cell value is set to one for a point, the line's value for a line, and the polygon's value for a polygon boundary. The third step is to fill the interior of the polygon outline with the polygon value. Errors from rasterization depend on the computer algorithm and the size of raster cell (Bregt et al. 1991).

Chapter 4 has already covered vectorization as part of the scanning method for spatial data entry.

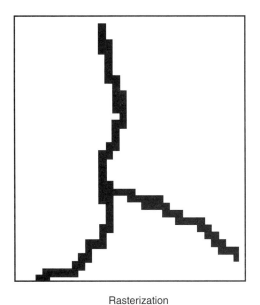

Rasterization   Vectorization

Figure 7.8
On the left is an example of conversion from vector to raster data, or rasterization. On the right is an example of conversion from raster to vector data, or vectorization.

In short, the process involves the following steps: raster line thinning, vector line extraction, topological reconstruction, and line smoothing.

Most GIS packages have functionalities that can convert between vector data and raster data. ARC/INFO, for example, has the command POLYGRID to convert a polygon coverage to a grid and the command GRIDPOLY to convert a grid to a polygon coverage. Similar commands are available for points and lines. ArcView has menu selections for conversion to shapefile and conversion to grid.

7.7 INTEGRATION OF RASTER AND VECTOR DATA

Depending on the type of raster data, integration of raster data and vector data in GIS can take place in data display, data processing, data conversion, or data analysis. DEMs are input data, which must be processed to extract topographic features such as contour, slope, aspect, drainage network, and watersheds (Jenson and Domingue 1988; Band 1993; Dymond et al. 1995). A DEM may be processed in its raster format, or in the form of a TIN (Triangulated Irregular Network) (Chapter 12).

DOQs, DRGs, and graphic files are essentially pictures. They are useful as the background for data display, or as the source for spatial data input or revision (Box 7.3). But they cannot be used for data processing in the same way as DEMs or binary scanned files. Binary scanned files can be used for tracing and digitizing (Chapter 4).

Satellite images present a different case from other raster data types. Satellite images are pictures like DOQs, but they provide timely data with multiple bands. When used as pictures, georeferenced satellite images can be displayed with other spatial features. When used in image processing, satellite images can create maps such as land cover, vegetation, urbanization, snow accumulation, and environmental degradation (Mesev et al. 1995; Hinton 1996). Maps created from the

---

Box 7.3   **Linking Vector Data with Images**

Hot Link in ArcView can link a point, line, or polygon feature in vector format with an image. A field called "link to" in the feature attribute table can instruct the computer to find the image that is linked to the feature. For example, a map showing houses for sale can be set up so that, when a house (point) is clicked, it opens a picture of the house in a graphic file.

---

processing of satellite images may remain in raster format, or they may be converted to vector format.

Integration between vector data and remotely sensed data has a long history and is well documented in the literature (Ehlers et al., 1989; Ehlers et al. 1991; Wilkinson 1996; Hinton 1996). Remotely sensed data can benefit GIS users in the creation and revision of maps of land cover/land use and environmental conditions. Likewise, vector data can provide ancillary information useful for image processing. Image stratification is a good example. This method uses GIS data to divide the landscape into major areas of different characteristics or different elevation zones and then treats these areas or zones separately in image processing and classification. Another example is to use vector data in selecting control points for the geo-referencing of remotely sensed data and in selecting sample sites for ground truth.

GIS and remote sensing data can be integrated at different levels (Ehlers et al. 1989). At the simple level, GIS and remote sensing reside in two separate software packages and the linkage between them is through data exchange. At the next level GIS and remote sensing software packages still exist separately but share a common user interface, and are capable of simultaneous display. The highest level is total integration: GIS and remote sensing reside in one software unit with combined processing. The current status of data integration is near the second level. Most GIS packages allow simultaneous displays of raster data and vector data, and use a common user interface to access remote sensing and GIS data. But data conversion must be performed first if analysis of both raster and vector data is required.

KEY CONCEPTS AND TERMS

ARC/INFO grid: A proprietary ESRI format for raster data.

AVHRR (Advanced Very High Resolution Radiometer) scanner: A weather satellite that provides large-area land cover data at a 1 km spatial resolution.

Bilinear interpolation: A re-sampling method that fills each cell of the new raster grid with the weighted average of the four nearest cell values from the original grid.

Binary Scanned File: A scanned file containing values of 1 or 0.

Block code: A raster data structure that uses square blocks to represent a region.

Cell-by-cell encoding: A raster data structure that stores cell values in a matrix by row and column.

Chain code: A raster data structure that represents the boundary of a region by using a series of cardinal directions and cells.

Cubic convolution: A re-sampling method that fills each cell of the new raster grid with the weighted average of the 16 nearest cell values from the original grid.

Data Compression: Reduction of data volumes, especially for raster data.

Digital elevation model (DEM): A digital model with an array of uniformly spaced elevation data in raster format.

Digital orthophoto quad (DOQ): A digitized image prepared from an aerial photograph or other remotely sensed data, in which the displacement caused by camera tilt and terrain relief has been removed.

Digital raster graphic (DRG): A scanned image of a USGS topographic map.

Floating-point grid: A grid that contains cells of continuous values.

Geo-referencing: The process of using a set of control points to convert images from image coordinates to real-world coordinates.

Ground control points: Control points used in geo-referencing an image.

Integer grid: A grid that contains cell values of integers.

Landsat: An orbiting satellite that provides repeat images of the Earth's surface. Landsat 7 was launched in April 1999.

Lossless compression: One type of data compression that allows the original image to be precisely reconstructed.

Lossy compression: One type of data compression that can achieve high compression ratios but cannot reconstruct fully the original image.

Nearest neighbor: A re-sampling method that fills each cell of the new raster grid with the nearest cell value from the original grid.

Quad tree: A raster data structure that divides a raster model into a hierarchy of quadrants.

Rasterization: Conversion of vector data to raster data.

Re-sampling: The process of filling each cell of a new grid with the value of the corresponding cell or cells in an original grid.

Run-length encoding (RLE): A raster data structure that records the cell values by row and by group. A run-length encoded file is also called run-length compressed (RLC) file.

SPOT: A French satellite that provides a 10 m spatial resolution for panchromatic images and 20 m spatial resolution for multi-spectral images.

Vectorization: Conversion of raster data to vector data.

Warping: The process of using polynomial equations for geo-referencing an image. Also called *rubber sheeting*.

APPLICATIONS: RASTER DATA

This applications section lets you use ArcView to view two types of raster data: DEM and Landsat TM imagery. ArcView can work with feature data, grid data, and imagery data. Additionally, ArcView can import data files such as USGS DEMs and convert them to grids. Tasks 1 and 3 require use of the Spatial Analyst extension.

Task 1: View USGS DEM Data in ArcView

What you need: *filer.dem*, a USGS 7.5-minute DEM.

Task 1 lets you view a USGS 7.5-minute DEM and examine its properties.

1. Start ArcView, and load the Spatial Analyst extension.
2. Open a new view. Select Import Data Source from the File menu. In the next dialog,

choose USGS DEM as the import file type and click OK. Navigate to the workspace, where *filer.dem* resides. Double-click on *filer.dem*. Then, save the output grid as *filergrd* and add it to view.

3. Select Properties from the Theme menu. The Properties dialog shows you the cell size, numbers of rows and columns, the area extent (i.e., the lower left and upper right coordinates), and whether the grid is an integer or a floating-point grid.

4. ArcView displays *filergrd* using the default color scheme and data classification. You can change both by opening the *filergrd*'s legend editor.

Task 2: View a Satellite Image in ArcView

What you need: *tmrect.bil*, a Landsat TM image comprised of the first five bands.

Task 2 lets you view a Landsat TM image with five bands. By changing the color assignment to each of the bands, you can alter the view of the image.

1. Start ArcView. Open a new view and select Add Theme from the View menu. Select Image Data Source as the Data Source Type. Double-click on *tmrect.bil* to add it to view.

2. Select Properties from the Theme menu. The Theme Properties dialog shows the coordinates of the lower left and upper right of *tmrect.bil*.

3. Double-click on *tmrect.bil* to open its Image Legend Editor. The Image Legend Editor includes the choice of single band and multiband, controls of red, green, and blue colors, and buttons for image data processing.

4. First, set red, green, and blue for bands 3, 2, and 1, respectively. Click the Default button. You should see the image as a color photograph.

5. Next, set red, green, and blue for bands 4,3, and 2, respectively. Click the Default button. You should see the image as a color infrared photograph.

Task 3: Vector to Raster Conversion

What you need: *nwroads.shp* and *nwcounties.shp*, shapefiles showing major highways and counties in the Pacific Northwest.

As mentioned in the chapter, integration of vector and raster data in a GIS project often requires conversion of vector to raster data and vice versa. In Task 3, you will convert a line theme (*nwroads.shp*) and a polygon theme (*nwcounties.shp*) to grids. Covering Idaho, Washington, and Oregon, both themes are projected onto a Lambert conformal conic projection and are measured in meters.

1. Start ArcView, and load the Spatial Analyst extension. Open a new view and add *nwroads.shp* and *nwcounties.shp* to view. Select Properties from the View menu and specify meters for Map Units.

2. When a vector theme is converted to a grid, an attribute from the vector theme table must be chosen for the cell value of the grid. Activate *nwroads.shp* and open its theme table. The field Rte_num1 shows the highway number, which is a good choice for the cell value. Activate *nwcounties.shp* and open its theme table. You will use the field Fips, which is the county FIPS (Federal Information Processing Standards) code, for the cell value.

3. Activate *nwroads.shp*. Select Convert to Grid from the Theme menu. You need to work with several dialogs for the conversion. In the first dialog, specify *nwroads_gd* as the grid name and provide the path to save the grid. Next is the Conversion Extent dialog. Enter Same as *Nwroads.shp* for the Output Grid Extent, and 5000 m for the Output Grid Cell Size. In the Conversion Field dialog, pick Rte_num1. Click OK to join attribute data and to add the new grid to view.

4. Activate *nwcounties.shp*, and follow the same procedure as above to convert it to a grid. Name the grid *nwcounties_gd*, specify *nwcounties.shp* for the grid extent, specify

5000 m for the cell size, and pick Fips for the conversion field.

5. Compare *nwroads.shp* with *nwroads_gd*. The highways look blocky in *nwroads_gd*, because they are represented by a series of

5000 m x 5000 m cells. Compare *nwcounties.shp* with *nwcounties_gd*, and you will see the same effect of vector to raster conversion.

REFERENCES

Band, L.E. 1993. Extraction of Channel Networks and Topographic Parameters from Digital Elevation Data. In Kirkby M.J. and Beven K. (eds.). *Channel Network Hydrology*. Chichester, New York: John Wiley & Sons, pp. 13–42.

Bregt, A.K., J. Denneboom, H.J. Gesink, and Y. Van Randen. 1991. Determination of Rasterizing Error: A Case Study with the Soil Map of the Netherlands. *International Journal of Geographical Information Systems* 5: 361–67.

Burrough, P.A., and R. A. McDonnell. 1998. *Principles of Geographical Information Systems*. Oxford: Oxford University Press.

Chrisman, N. 1997. *Exploring Geographic Information Systems*. New York: John Wiley & Sons.

Clarke, K.C. 1995. *Analytical and Computer Cartography*. 2ᵈ ed. Englewood Cliffs, NJ: Prentice Hall.

Dymond, J.R., R.C. Derose, and G.R. Harmsworth. 1995. Automated Mapping of Land Components from Digital Elevation Data. *Earth Surface Processes and Landforms* 20: 131–37.

Ehlers, M., G. Edwards, and Y. Bedard. 1989. Integration of Remote Sensing with Geographical Information Systems: A Necessary Evolution. *Photogrammetric Engineering and Remote Sensing* 55: 1619–27.

Ehlers, M., D. Greenlee, T. Smith, and T. Star. 1991. Integration of Remote Sensing and GIS: Data and Data Access. *Photogrammetric Engineering and Remote Sensing* 57: 669–75.

Hinton, J.E. 1996. GIS and Remote Sensing Integration for Environmental Applications. *International Journal of Geographical Information Systems* 10: 877–90.

Jenson, S.K., and J.O. Domingue. 1988. Extracting Topographic Structure from Digital Elevation Data

for Geographical Information System Analysis. *Photogrammetric Engineering and Remote Sensing* 54: 1593–1600.

Kumler, M.P. 1994. An Intensive Comparison of Triangulated Irregular Networks (TINs) and Digital Elevation Models (DEMs). *Cartographica* 31 (2): 1–99.

Lillesand, T.M., and R.W. Kiefer. 2000. *Remote Sensing and Image Interpretation*. 4ᵗʰ ed. New York: John Wiley & Sons.

Mesev, T.V., P.A. Longley, M. Batty, and Y. Xie. 1995. Morphology from Imagery—Detecting and Measuring the Density of Urban Land-use. *Environment and Planning A* 27: 759–80.

Piwowar, J.M., E.F. LeDraw, and D.J. Dudycha. 1990. Integration of Spatial Data in Vector and Raster Formats in A Geographic Information System Environment. *International Journal of Geographical Information Systems* 4: 429–44.

Rosenfeld, A. 1980. Tree Structures for Region Representation. In H. Freeman and G.G. Pieroni (eds.). *Map Data Processing*. New York: Academic Press, pp. 137–50.

Samet, H. 1990. *The Design and Analysis of Spatial Data Structures*. Reading, Mass: Addison-Wesley.

Tomlin, C.D. 1990. *Geographic Information Systems and Cartographic Modeling*. Englewood Cliffs, N.J.: Prentice Hall.

Verbyla, D.L., and Chang, K. 1997. *Processing Digital Images in GIS*. Santa Fe, NM: OnWord Press.

Wilkinson, G.G. 1996. A Review of Current Issues in the Integration of GIS and Remote Sensing Data. *International Journal of Geographical Information Systems* 10: 85–101.

Worboys, M.F. 1995. *GIS: A Computing Perspective*. London: Taylor & Francis.

8

DATA DISPLAY AND CARTOGRAPHY

CHAPTER OUTLINE

8.1 INTRODUCTION

Maps are an interface to GIS (Kraak and Ormeling 1996). We view, query, and analyze maps. We plot maps showing results of query and analysis and use them in presentation and reports. As a visual tool maps are most effective in communicating spatial data, whether the emphasis is on the location or the distribution pattern of spatial data.

A map includes the title, body, legend, north arrow, scale, acknowledgement, and neat line (Figure 8.1). These elements work together to bring spatial information to the map reader. The map body is the most important part of a map because it contains the map information. Other elements of the map support the communication process. For example, the title suggests the subject matter, and the legend relates map symbols to spatial data. In practical terms, mapmaking may be described as the process of assembling elements for a map.

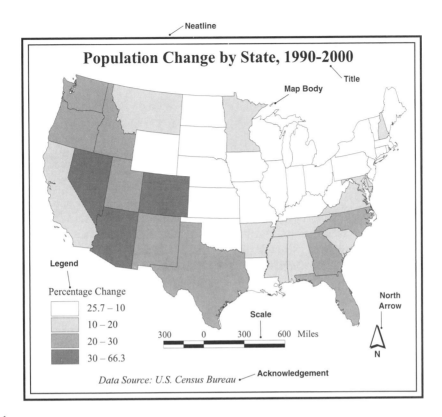

Figure 8.1
Common map elements.

Mapmaking should be guided by a clear idea of map design. A well-designed map can help the mapmaker communicate spatial information to the map reader, whereas a poorly designed map can confuse the map reader and even distort the information intended by the mapmaker.

Data display is one area that commercial GIS packages have greatly improved in recent years. Windows-based GIS packages are excellent for data display for two reasons. First, the mapmaker can simply point and click the graphic icons to construct a map. In comparison, a command driven GIS package requires the user to become familiar with many commands before making a map. Second, Windows-based packages have incorporated conventional design options into menu selections so that GIS users do not have to create their own options.

This chapter is divided into five sections. Section 1 discusses cartographic symbolization, including the data-symbol relationship and use of color. Section 2 considers different types of maps by map symbols. Section 3 provides an overview of typography, selection of type variations, and placement of text. Section 4 covers map design and the design elements of layout and visual hierarchy. Section 5 examines soft-copy versus hard-copy maps. Whenever appropriate, this chapter uses ArcView, a Windows-based GIS package, as an example.

8.2 CARTOGRAPHIC SYMBOLIZATION

8.2.1 Spatial Features and Map Symbols

Spatial features are characterized by their locations and attributes. To display a spatial feature on a map, we use a map symbol to indicate the feature's location and a visual variable, or visual variables, with the symbol to show the feature's attribute data. For example, a thick line in red may represent an interstate highway and a thin line in black may represent a state highway. The line symbol shows the location of the highway in both cases, but the line width and color—two visual variables with the line symbol—separate the interstate from the state highway.

Choosing the appropriate map symbol and visual variables is the main concern for data display (Robinson et al. 1995; Dent 1999; Slocum 1999). Choice of map symbol for raster data is not really an issue because map symbols consist of cells whether the spatial feature to be depicted is a point, line, or polygon. For vector data, however, conventional map symbols are grouped into point, line, and area symbols like the classification of feature types (Figure 8.2). The general rule in selecting map symbols for vector data is therefore the following: point symbols for point features, line symbols for line features, and area symbols for area features.

There are exceptions to the above data-symbol rule. A GIS project often uses volumetric or 3-D data such as elevation, temperature, and precipitation, but there are no volumetric symbols for them. Instead, point, line, and area symbols are used to represent volumetric data. Another common exception is the use of point symbols at the polygon

label locations to show aggregate data such as county or state populations.

Visual variables in cartographic symbolization include shape, size, texture, pattern, hue, value, and chroma (Figure 8.3). Hue, value, and chroma relate to color, which is covered in the next section. Size and texture (the spacing of symbol markings) suggest quantitative differentiation among the data mapped. For example, a map may use different sized circles to represent different sized cities. Pattern (the type of symbol markings), on the other hand, is more appropriate for displaying nominal or qualitative data. For example, a map may use different area patterns to show different land use types.

ArcView offers choices of pattern and texture through its fill, pen, marker, and color palettes. The fill palette has 50 fill patterns, the pen palette has 28 line patterns, and the marker palette has 50 icons or symbols. Choices from all three palettes can have colors assigned to them from the color palette, which includes 60 colors.

Display of raster data is again limited in the choice of visual variables. The visual variables of shape and size do not apply to raster data because of the use of cells. Using texture or pattern is

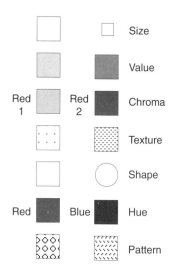

Figure 8.3
Visual variables in cartographic symbolization.

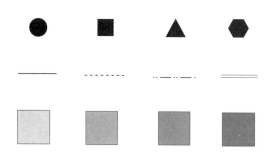

Figure 8.2
Type of map symbol. Top to bottom: point, line, and area symbols.

possible for low-resolution raster data but difficult with very small cells. Display of raster data is therefore limited to different colors or different shades of colors.

8.2.2 Use of Color

Because color adds a special appeal to a map, mapmakers will choose color maps over black and white maps whenever possible. But color is probably the most misused visual variable according to published critiques of computer-generated maps (Monmonier 1996). Use of color in mapmaking must begin with an understanding of the color dimensions of hue, value, and chroma.

Hue is the quality that distinguishes one color from another, such as red from blue. Hue can also be defined as the dominant wavelength of light making up a color. We tend to relate different hues with different kinds of data. **Value** is the lightness or darkness of a color, with black at the lower end and white at the higher end. We generally perceive darker symbols as being more important on a map (Robinson et al. 1995). Also called saturation or intensity, **chroma** refers to the richness, or brilliance, of a color. A fully saturated color is pure, whereas a low saturation approaches gray. We generally associate higher intensity symbols with greater visual importance.

The first rule of thumb in use of color is simple: hue is a visual variable better suited for qualitative data, whereas value and chroma are better suited for quantitative data. Qualitative data are nominal data, whereas quantitative data are data measured at an ordinal, interval, or ratio scale. It is not difficult to find 12 or 15 distinctive hues for a qualitative map. If a map requires more hues, then patterns, another qualitative visual variable, or text may be combined with hues to make up more map symbols.

Over the years cartographers have suggested several color schemes based on hue, value, and chroma for displaying quantitative data (Cuff 1972; Antes and Chang 1990; Mersey 1990; Brewer 1994). The common thread among these color schemes is that, through the selection of color symbols, the reader can easily perceive the progression from low to high values.

- The single hue scheme. This is the simplest and probably the most effective color scheme for displaying quantitative data. It uses a single hue but varies the combination of value and chroma to produce a sequential color scheme. ArcView offers "red monochromatic," "orange monochromatic," and so on for this color scheme.
- The double-ended, or diverging, scheme. This color scheme uses graduated colors between two dominant colors. For example, a double-ended scheme may progress from dark blue to light blue and then from light red to dark red. The double-ended color scheme is ideal for displaying data with positive and negative values, or increases and decreases. ArcView has "blues to reds dichromatic," "blues to oranges dichromatic," and so forth for this scheme.
- The part spectral scheme. This color scheme uses adjacent colors of the visible spectrum to show variations in magnitude. Under this color scheme, ArcView offers "yellow to orange to red," "yellow to green to dark blue," and others.
- The full spectral scheme. This color scheme uses all colors in the visible spectrum. It is not usually recommended for mapping quantitative data except for elevation and temperature. ArcView has separate color schemes for elevation, temperature, and precipitation.

8.3 TYPE OF MAP

Cartographers classify maps by function and by symbolization. By function, maps can be divided into the general reference and thematic maps. The **general reference map** is used for general purpose. An example would be a USGS quad map, which shows a variety of spatial features, including boundaries, hydrology, transportation, contour lines, settlements, and land covers. The **thematic map** is also called the special purpose map, because its main objective is to show the distribution pattern of a theme, such as the distribution of population densities by county in a state.

Cartographers also group maps by map symbols, either quantitative or qualitative. Several common types of quantitative maps are recognized in cartography (Figure 8.4):

- The **dot map** uses uniform point symbols to show spatial data, with each symbol representing a unit value (Box 8.1).
- The **graduated color map** uses a quantitative color scheme to show the

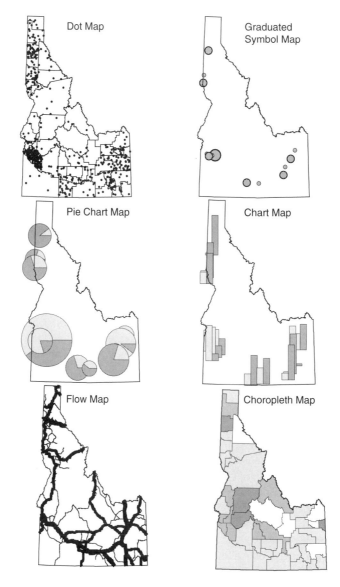

Figure 8.4
Six common types of quantitative maps.

The most difficult part of making a dot map is locating dots. Suppose a dot represents 500 persons and a county has a population of 5000. This means that the county should have 10 dots within its border. The next question is where to place the 10 dots. Ideally the dots should be placed at the locations of towns and villages, rather than in unsettled areas. Most computer mapping packages including ArcView, however, use a random method in placing dots. The random method can create an unrealistic dot map. One way to improve the accuracy of dot maps is to base them on the smallest administrative unit possible. Instead of placing dots at the county level, consult a census block group or a census block map with its population statistics to locate dots.

Cartographers distinguish between derived and absolute values (Chang 1978). Derived values are normalized values such as county population densities, whereas absolute values are magnitudes or raw data such as county population. County population densities, derived from dividing the county population by the area of the county, are independent of the county size. Two counties with equal populations but different sizes will have different population densities. Cartographers recommend the use of choropleth maps for derived values and graduated symbol maps for absolute values. If choropleth maps are used for mapping absolute values such as county populations, size differences among counties can severely distort map comparisons (Monmonier 1996).

variation in spatial data. Cartographers distinguish between the choropleth map and the dasymetric map. The **choropleth map** symbolizes derived data such as average income or population density based on administrative units such as counties or states (Box 8.2). The **dasymetric map** uses statistics and additional information to delineate areas of homogeneous values, rather than following administrative boundaries (Robinson et al. 1995).

- The **graduated symbol map** uses different sized symbols such as circles, squares, or triangles to represent different magnitudes. Two issues are important to this map type: the range of sizes, and the discernible difference between sizes. Both issues are obviously related to the number of classes, or graduated symbols, on the map.

- The **chart map** uses either pie charts or bar charts. A variation of the graduated circle, the pie chart can display two sets of quantitative data: its size can be made proportional to a value such as a county population, and its subdivisions can show the makeup of the value, such as the racial composition of the county population. Bar charts are useful for comparing data side by side.

- The **flow map** displays different quantities of flow data by varying the width of line symbols.

GIS has introduced a new classification of maps by data model. Maps prepared from vector data are the same as traditional maps, using point, line, and area symbols. Most of this chapter applies to the display of vector data. Maps prepared from raster data, although they may look like traditional maps, are based on cells. Raster data may be categorical such as vegetation types, or numeric such as elevation data. The color schemes for qualitative and quantitative maps described in this section can therefore be applied to mapping categorical and numeric raster data, respectively.

Times New Roman

Tahoma

Figure 8.5
Times New Roman is a serif typeface, and Tahoma is a sans serif typeface.

8.4 TYPOGRAPHY

A map cannot be understood without lettering or type on it. Lettering is needed for almost every map element. Mapmakers treat type as a map symbol because, like point, line, or area symbols, type has many variations. Using type variations to create a pleasing and coherent map is a major challenge to mapmakers.

8.4.1 Type Variations

Type varies in typeface, form, size, and color (Box 8.3). **Typeface** refers to the design character of the type, and type form refers to the variant of letterform. There are two main groups of typefaces: **serif** (with serif) and **sans serif** (without serif) (Figure 8.5). Serifs are small, finishing touches at the end of line strokes, which tend to make it easier to read running text in newspapers or books. Compared to serif types, sans serif types appear simpler and bolder. Although rarely used in books or other text intensive materials, sans serif type stands up well on maps with complex map symbols and remains legible even in small sizes. Sans serif types have an additional advantage in mapmaking because many of them come in a wide range of type variations.

Type form includes variations in **weight** (bold, regular, or light), **width** (condensed or extended), upright versus slanted (or roman versus italic), and uppercase versus lowercase (Figure 8.6). Type size measures the height of a letter in **points,** with 72 points to an inch. Printed letters look smaller than what their point sizes suggest. The point size is measured from a metal type block, which must accommodate the lowest point of the descender (such as p or g) to the highest part of the ascender (such as d or b). But no letters extend to the very edge of the block. The term **font** describes a complete set of all variants of a given typeface and size.

Box 8.3   **Terms for Type Variations**

Except for type size and type color, terms for type variations are not standardized. This lack of standardization can create confusion when GIS users turn to cartography textbooks for help in type design. The design character of the type is referred to as *type style* in Robinson et al. (1995), *typeface* in Dent (1999) and ArcView, and *font* in Slocum (1999). Type weight, type width, uppercase versus lowercase, and upright versus italic are collectively referred to as *type form* in Robinson et al. (1995), *letterforms* in Dent (1999), and *type style* in Slocum (1999) and ArcView. Finally, Dent (1999) and ArcView use the term *font* to mean a complete set of all characters of one size and design of a typeface.

Helvetica Normal

Helvetica Italic

Helvetica Bold

Helvetica Bold-Italic

Times Roman Normal

Times Roman Italic

Times Roman Bold

Times Roman Bold-Italic

Figure 8.6
Type variations in weight and roman/italic.

8.4.2 Selection of Type Variations

Type variations can function in the same way as visual variables for map symbols. Differences in typeface, type color, and roman versus italic are more appropriate for qualitative data, whereas differences in type size, weight, and uppercase versus lowercase are more appropriate for quantitative data. For example, a reference map showing different sizes of cities typically shows the largest cities in a large type size and in bold and capital letters, and the smallest cities in a small type size and in thin and lowercase letters.

Cartographers also recommend legibility, harmony, and conventions in type selection (Dent 1999). Legibility is difficult to control because it can be influenced by the choice of typeface, type size, placement of lettering, letter spacing, and contrast between type and the background symbol. GIS users have the additional problem of having to design a map on the computer monitor and to print it on a much larger plot. GIS users therefore need to experiment to ensure type legibility in all parts of the map.

Type legibility, however, should be balanced with harmony. The function of type is to communicate the map content. Thus type should be legible but should not draw too much attention. Mapmakers can generally achieve harmony by adopting only one or two typefaces on a map and uses other type variations for different elements or symbols (Figure 8.7). For example, many mapmakers use a sans serif type in the body of a map and a serif type for the map's title and legend (Robinson et al. 1995). Conventions in type selection include italic for names of water features, upper case and letter spacing for names of administrative units such as states or counties, and variations in type size and form for names of cities that are ranked in a hierarchical order.

Most GIS users have to work with fonts that have been loaded on the computer from the printer manufacturer or other software packages. These fonts are used for word-processing, graphics, and mapping. The only fonts that ArcView installs, such as ESRI Geometric Symbols and ESRI Oil, Gas, & Water, are actually point or marker symbols, not text fonts. Pre-loaded fonts such as script-cursive and display-decorative fonts are not suitable for mapmaking. ArcView users should be able to select from the remaining fonts enough type variations for their maps. If not, they can use other fonts by importing them into ArcView.

8.4.3 Placement of Text

The placing of lettering, or labeling, on a map is just as important as selecting type variations. As a general rule, lettering should be placed to show the location or the area extent of the named spatial feature. Cartographers recommend placing the name of a point feature to the upper right of its symbol, the name of a line feature in a block and parallel to the course of the feature, and the name of an area feature to indicate its area extent. Other general rules suggest aligning names with the map border or with the lines of latitude, and placing names entirely on land or on water.

Implementing labeling algorithms in a GIS package is no easy task (Mower 1993). Automated name placement presents several difficult problems to the computer programmer: names must be

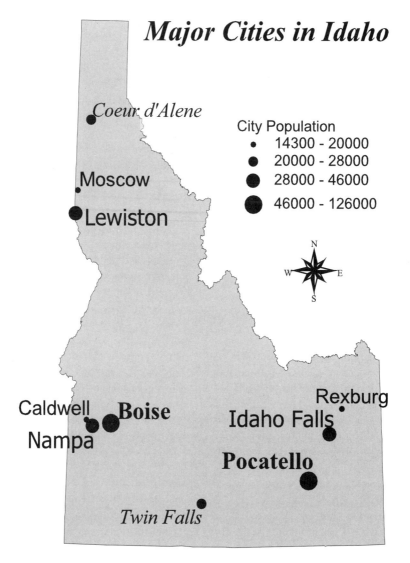

Figure 8.7
The look of the map is not harmonious because of the use of too many typefaces.

legible, names cannot overlap other names or symbols, names must be clear as to their intended referent symbols, and name placement must follow cartographic conventions. These problems worsen at smaller map scales, as competition for map space intensifies between names.

GIS users should not expect that labeling be completely automated. Usually, some interactive editing is needed to improve the final map's appearance. For this reason GIS packages offer more than one method of labeling. For example, ArcView provides three methods for labeling. The first method is to interactively enter the text by choosing a point location such as the beginning of a text string and then typing the text. This method is ideal for adding data source, acknowledgement, or explanatory notes on a map.

The second method is to interactively label a feature with a specified field value such as the name of a city. The location of the label can be pre-defined relative to the map feature, such as by placing the name of a city to its upper right. If the pre-defined location does not work out well, the label can later be moved.

The third method, which is probably the method of choice for most users, is to automatically label all, or selected, features using an attribute value. There are two options for this method. The first is to use a pre-defined position relative to the feature to be labeled and apply the same position to all features. ArcView offers a sec-

ond option called Find Best Placement, which uses an algorithm to evaluate the placement of a label in relation to other labels and adjusts labeling. Where labels are on top of one another because of congested map features, the overlapped label can be either left out or shown in a different color from other labels (Figure 8.8). One way to keep an overlapped label is to move it to a new location (Figure 8.9) and, if necessary, link it to the map feature with a leader line (Figure 8.10).

Perhaps the most difficult task in labeling is to label names of streams. The general rule states that the name should follow the course of the stream, either above or below it. The Find Best Placement option in ArcView can curve the name if it appears along a curvy part of the stream. But the appearance of the curved name depends on the smoothness of the corresponding stream segment, the length of the stream segment, and the length of the

Figure 8.8
Labeling major cities in the United States using the Find Best Placement option in ArcView. The result is generally not satisfactory. The names of Los Angeles and Phoenix are on top of one another. San Jose is north of San Francisco. Many names such as Seattle and Chicago are placed too far from their point symbols.

Figure 8.9
A revised version of Figure 8.8. City names are moved individually to be closer to their point symbols.

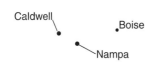

Figure 8.10
A leader line connects a point symbol to its label.

name. Placing every name in its correct position the first time is nearly impossible (Figure 8.11). Problem names must be removed and re-labeled. ArcView has a **spline text** tool, which can align a text string along a curved line that is digitized on screen (Figure 8.12).

8.5 MAP DESIGN

Like graphic design, **map design** is a visual plan to achieve a goal. The purpose of map design is to enhance map communication, which is particularly important for thematic maps. A well-designed map is balanced, coherent, ordered, and interesting to look at, whereas a poorly designed map is confusing and disoriented (Antes et al. 1985). Map design overlaps with the field of graphic arts, and many map design principles have their origin in visual perception. Cartographers usually study map design from the perspectives of layout or planar organization and of visual hierarchy.

8.5.1 Layout

Layout deals with the arrangement and composition of map elements such as the map body, the title, the legend, the scale, and the north arrow. Major concerns with layout are focus, order, and balance. A map, especially a thematic map, should have a clear focus, which is usually the map body or a part of the map body. To draw the map reader's attention, the focal element should be placed near the optical center of the map, which is just above the map's geometric center. The focal element should be differentiated from other map elements by contrast in line, texture, value, detail, and color.

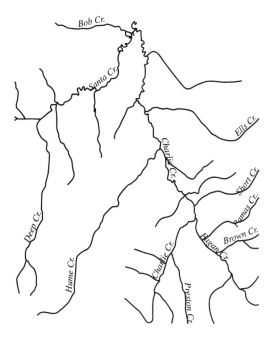

Figure 8.11
The Find Best Placement option in ArcView may not work with every label.

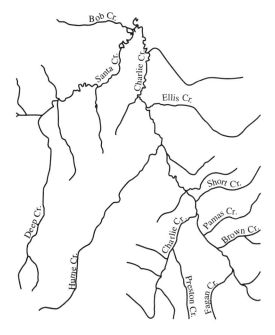

Figure 8.12
Problem labels in Figure 8.11 are replotted with the Spline Text tool in ArcView.

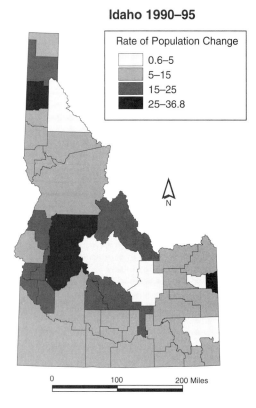

Idaho 1990–95

Rate of Population Change

	0.6–5
	5–15
	15–25
	25–36.8

Figure 8.13
Use a box around the legend to draw the map reader's
attention to it.

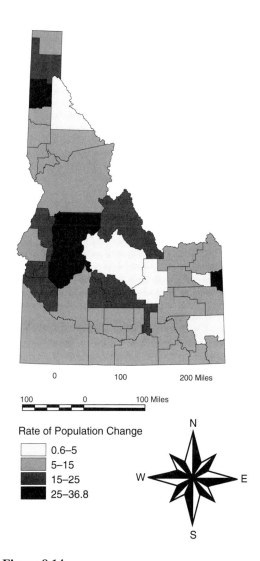

Rate of Population Change

	0.6–5
	5–15
	15–25
	25–36.8

Figure 8.14
A poorly balanced map.

After viewing the focal element, the reader
should be directed to the rest of the map in an ordered fashion. For example, the legend and the title are probably the next elements that the viewer
needs to look at after the map body. To smooth
the transition, the mapmaker should clearly place
the legend and the title on the map, with perhaps
even a box around them to draw attention to them
(Figure 8.13).

A finished map should look balanced. It
should not give the map reader an impression that
the map "looks" heavier on the top, bottom, or the
side (Figure 8.14). Balance, however, does not
suggest the breaking down of the map elements
and placing them, almost mechanically, in every
part of the map. Although in that case the elements

would be in balance, the map would be disorga-
nized and confusing. Mapmakers therefore should
deal with balance within the context of organization and map communication.

Cartographers used to use thumbnail sketches
to experiment with balance on a map. Now they
use computers to manipulate graphic elements on
a layout page. ArcView has two layout options.
The first is to open a new layout and to place in it

empty frames or containers for the map body, legend, title, and other map elements. This layout is like an electronic thumbnail sketch because each frame can be moved, reduced, or enlarged (Figure 8.15). After the layout design is completed, it can be linked to a view to fill in the contents of the frames. This option is particularly useful in producing a series of maps with the same look.

The second option is to link the layout directly with a view. ArcView offers five layout templates: landscape, portrait, landscape with inset, portrait with neat lines, and landscape with neat lines. After a template choice is made, ArcView displays the view with its title, legend, scale, and north arrow in the layout. As in the first option, ArcView lets the user graphically manipulate each map element in the initial layout.

The legend in a layout includes descriptions of all the themes from a view. As default, these descriptions are placed together as a single graphic element, which can become quite lengthy with multiple themes. A lengthy legend presents a problem in layout design, the solution to which is to use the built-in functions of Simplify and Group in ArcView. Simplify can separate a lengthy legend into graphic elements, and Group can group these graphic elements into several smaller legends. It is easier to manipulate and rearrange smaller legends in a layout (Figure 8.16). Because it will separate the map body into a set of points, lines, rectangles, and text graphics, Simplify cannot be applied to the map body. When simplified, the map body cannot be restored back to its original state, and its link to the legend is lost. If map symbols need to be changed within the map body, they should be changed in the view rather than in the layout.

8.5.2 Visual Hierarchy

Visual hierarchy is the process of developing a visual plan to introduce the 3-D effect or depth to maps. Mapmakers create the visual hierarchy by placing map elements at different visual levels according to their importance to the map's purpose. The most important element should be at the top of

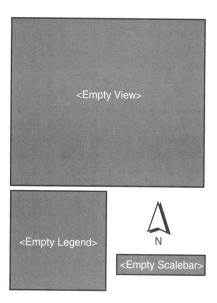

Figure 8.15
A new layout contains empty frames for the map body, legend, scale, and N-arrow.

the hierarchy and should appear closest to the map reader. The least important element should be at the bottom. A thematic map may consist of three or more levels in a visual hierarchy.

The concept of visual hierarchy is an extension of the **figure-ground relationship** in visual perception. The figure is more important visually, appears closer to the viewer, has form and shape, has more impressive color, and has meaning. The ground is the background. Cartographers have adopted the depth cues for developing the figure-ground relationship in map design (Robinson et al. 1995; Dent 1999).

Probably the simplest and yet most effective principle in creating a visual hierarchy is called interposition or superimposition. **Interposition** uses the incomplete outline of an object to make it appear behind another. Examples of interposition abound in maps, especially in newspapers and magazines. Continents on a map look more important or occupy a higher level in visual hierarchy if the lines of longitude and latitude stop at the coast. A map title, a legend, or an inset map looks more

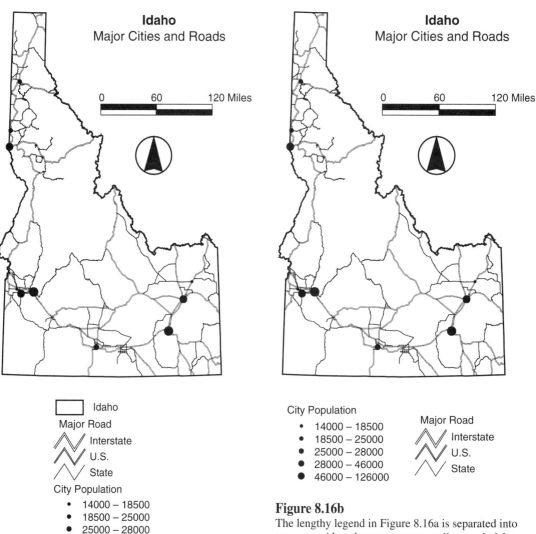

Figure 8.16a
A lengthy legend is confusing and can create a problem in layout design.

Figure 8.16b
The lengthy legend in Figure 8.16a is separated into two parts. Also, the unnecessary outline symbol for Idaho is removed from the legend.

prominent if it lies within a neat line, with or without the drop shadow. When the map body is deliberately placed on top of the neat line around a map, the map body will stand out more (Figure 8.17). Because interposition is so easy to use, it can be overused or misused. A map looks confusing if

several of its elements compete for the map reader's attention (Figure 8.18).

Sub-divisional organization is a map design principle that groups map symbols at the primary and secondary levels according to the intended visual hierarchy. Each primary symbol is given a distinctive hue, and the differentiation among the secondary symbols is based on pattern, texture, or orientation (Figure 8.19). For example, all tropical climates on a climatic map are shown in red, and

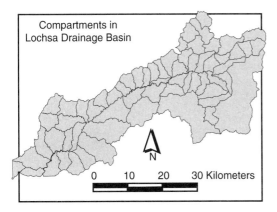

Figure 8.17
The interposition effect in map design.

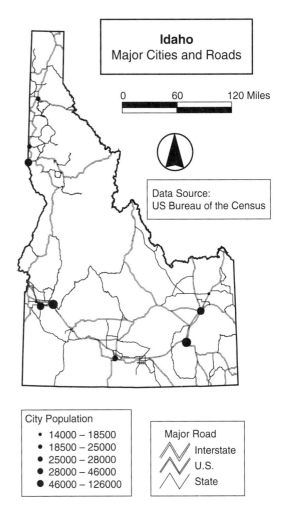

Figure 8.18
A map looks confusing if it has too many boxes to highlight individual map elements.

different tropical climates (e.g., wet equatorial climate, monsoon climate, and wet-dry tropical climate) are distinguished by line or dot patterns. Sub-divisional organization is most useful for maps with many map symbols, such as climate, soils, geology, and vegetation maps.

Contrast is a basic element in map design, important to layout as well as to visual hierarchy. Contrast in size or width can make a state outline look more important than county boundaries and larger cities look more important than smaller ones (Figure 8.20). Contrast in color can separate the figure from the ground. Cartographers often use a warm color (e.g., orange to red) for the figure and a cool color (e.g., blue) for the ground. Contrast in texture can also differentiate between the figure and the ground because the area containing more details or a greater amount of texture tends to stand out on a map. Like the use of interposition, too much contrast can create a confusing map appearance. For example, if bright red and green are used side-by-side as area symbols on a map, they appear to vibrate.

8.6 SOFT-COPY VERSUS HARD-COPY MAPS

GIS users design and make maps on the computer screen. These maps are in soft-copy form. Soft-

copy maps can be used in overhead computer projection systems, exported to other software packages, or converted to hard-copy maps. Like check plots for checking the accuracy of digitizing, hard-copy maps, or proofs, are useful for viewing the result of map design and production. A satisfactory proof can be used directly in presentation or further processed to make color separates for printing.

GIS users are often surprised that color symbols from their color printers do not exactly match those on the computer screen. This discrepancy results from the use of different media and color models. Most computer screens use **cathode ray tubes (CRTs)** for data display. A CRT screen has a

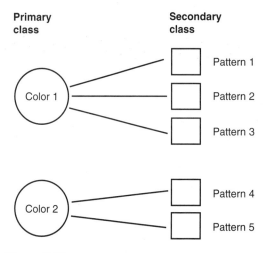

Primary class

Secondary class

Color 1
Color 2

Pattern 1
Pattern 2
Pattern 3
Pattern 4
Pattern 5

Figure 8.19
An example of sub-divisional organization.

built-in fine mesh of pixels, and each pixel has three phosphors, one red, one green, and one blue. These phosphors emit light when struck by electrons from three electron guns with the **RGB** (red, green, and blue) designation. Therefore, a color symbol we see on the computer screen is made of pixels, and the color of each pixel is actually a mixture of RGB.

The intensity of each primary color in a mixture determines its color. The number of intensity levels each primary color can have depends on the number of bit-planes assigned to the electron gun. Typically, a color display system has 8 bit-planes per gun, meaning that the system can generate 16.8 million colors ($2^8 \times 2^8 \times 2^8$, or $256 \times 256 \times 256$).

Many software packages offer the RGB color model for color specification. ArcView, for example, uses the RGB color model for multi-band image data. Color mixtures of RGB, however, are not intuitive (Figure 8.21). It is difficult to perceive that, for example, a mixture of red and green at full intensity is a yellow color. This is why other color

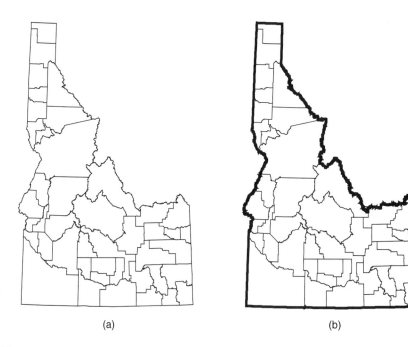

(a)   (b)

Figure 8.20
Contrast is missing in (a), whereas line contrast makes the state outline look more important than the county boundaries in (b).

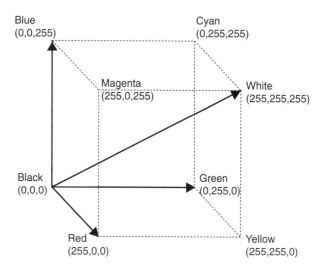

Figure 8.21
The RGB (red, green, and blue) color model.

models have been developed to specify colors to be used on the computer screen. ArcView, for example, has the **HSV** (Hue/Saturation/Value) color model for specification of custom colors. Like the RGB model, each color dimension in the HSV model ranges in value from 0 to 255, thus allowing for a theoretical total of 16.8 million colors.

Printed color maps differ from CRT color displays in two ways: color maps reflect rather than emit light; and the creation of colors on color maps is a subtractive rather than additive process. The three primary subtractive colors are cyan, magenta, and yellow. In printing, the three primary colors plus black form the four process colors of **CMYK.**

Color symbols are produced on a printed map in much the same way as on a computer screen. In place of phosphors and pixels are color dots, and in place of varying light intensities are percentages of area covered by color dots. A deep orange color on a printed map may represent a combination of 60% magenta and 80% yellow, whereas a light orange color may represent a combination of 30% magenta and 90% yellow. To match a color symbol on the computer screen with a color symbol on the printed map requires a translation from the RGB color model to the CMYK color model. As yet there is no exact translation between the two, and therefore the color map looks different when printed (Slocum 1999).

KEY CONCEPTS AND TERMS

Cathode ray tube (CRT) screen: A display device with a built-in fine mesh of pixels, each of which has three phosphors (red, green, and blue) and can emit light when struck by electrons from electron guns.

Chart map: A map that uses charts such as pie charts or bar charts as map symbols.

Choropleth map: A map that applies shading symbols to data or statistics collected for administrative units such as counties or states.

Chroma: The richness or brilliance of a color. Also called *saturation* or *intensity.*

CMYK: A color model in which colors are specified by the four process colors of cyan (C), magenta (M), yellow (Y), and black (K).

Contrast: A basic element in map design that enhances the look of a map or the figure-ground relationship by varying the design of size, width, color, and texture.

Dasymetric map: A map that uses statistics and additional information to delineate areas of homogeneous values, rather than following administrative boundaries.

Dot map: A map that uses uniform point symbols to show spatial data, with each symbol representing a unit value.

Figure-ground relationship: A tendency in visual perception to separate more important objects (figures) in a visual field from the background (ground).

Flow map: A map that displays different quantities of flow data by varying the width of line symbols.

Font: A complete set of all variants of a given typeface and size.

General reference map: One type of map used for general purpose such as USGS topographic maps.

Graduated color map: A map that uses a progressive color scheme such as light red to dark red to show the variation in spatial data.

Graduated symbol map: A map that uses different sized symbols such as circles, squares, or triangles to represent different magnitudes.

HSV: A color model in which colors are specified by their hue (H), saturation (S), and value (V).

Hue: The quality that distinguishes one color from another, such as red from blue. Hue is the dominant wavelength of light.

Interposition: A tendency for an object to appear behind another because of its incomplete outline.

Layout: The arrangement and composition of map elements on a map.

Map design: The process of developing a visual plan to achieve the map purpose.

Point: Measurement unit of type, with 72 points to an inch.

RGB: A color model in which colors are specified by their red (R), green (G), and blue (B) components.

Sans serif: Without serifs.

Serif: Small, finishing touches added to the end of line strokes.

Spline text: A text string aligned along a curved line.

Sub-divisional organization: A map design principle that groups map symbols at the primary and secondary levels according to the intended visual hierarchy.

Thematic map: One type of map used to emphasize the spatial distribution of a theme, such as a map that shows the distribution of population densities by county in a state.

Typeface: A particular style or design of type.

Type weight: Relative blackness of a type such as bold, regular, or light.

Type width: Relative width of a type such as condensed or extended.

Value: The lightness or darkness of a color.

Visual hierarchy: The process of developing a visual plan to introduce the 3-D effect or depth to maps.

APPLICATIONS: DATA DISPLAY AND CARTOGRAPHY

This applications section consists of three tasks. The first task guides you through the process of making a choropleth map in ArcView, first in black-and-white and then in color. Task 2 involves type design, graduated symbol, choice of color combination, and use of highway shield symbols. Task 3 focuses on the placement of text.

Task 1: Making a Choropleth Map

What you need: *us.shp,* a U.S. theme showing population change by state between 1990 and 1998. The theme is projected onto the Albers equal-area projection and is measured in meters.

Choropleth maps display statistics by administrative units. For this task you will map the rate of population change between 1990 and 1998 by state. The emphasis of Task 1 is on the use of graduated color in ArcView.

1. Start ArcView, open a new view, and add *us.shp* to view.
2. Select Properties from the View menu. Change the name to "Population Change By State, 1990–98," change the Map Units to meters, and click OK. The name of the view will be the title on your map. Select Properties from the theme menu. Change the theme name to "Percentage Change," and click OK. The name of the theme will appear as the description for the theme legend.
3. Double-click the theme name to invoke the Legend Editor. Select Graduated Color from the Legend Type pull-down menu. Select "Zchange," which contains the percent change data between 1990 and 1998, for the Classification Field. Click Apply. The default map uses five classes and the red monochromatic color scheme.
4. The default data classification uses the natural break method and five classes. You can change the number of classes or the classification method by clicking on Classify in the Legend Editor to open the classification dialog. The scroll bars for Type and Number of Classes let you change the classification method and the number of classes, respectively.
5. You can also change the class breaks in the Legend Editor. To change the class breaks, click on a value cell, type in the class breaks you want, and press Enter. Make sure the class breaks you type in do not have gaps between them; otherwise, you may find some

states are not symbolized in the map. Cartographers recommend round numbers such as 5.0 instead of odd numbers such as 5.3 in data classification. Cartographers also recommend logical breaks such as 0 in data classification so that negative values and positive values are clearly separated. There are four negative values, that is, decreases in population from 1990 to 1998, in the states theme (Connecticut, District of Columbia, North Dakota, and Rhode Island).

6. Click on the scroll list button for Color Ramps in the Legend Editor. The list contains color schemes for mapping quantitative data. If you select Gray Monochromatic and click Apply, you should see a black-and-white U.S. map with gray symbols. To continue with the rest of this task, select Orange Monochromatic. If you have grouped the states with negative values into the first class (i.e., -13.8 to 0.0), it might be a good idea to change its color symbol. Double click the symbol for the first class to open the Symbol Palette. Press the Color Palette, and click on a color like green. You should see the green color symbol has become the symbol for the first class. If you click Apply, the U.S. map is redrawn with this new color scheme.
7. If you still have the Color Palette on the screen, you can try the custom color option. Click on Custom on the Color Palette to open the Specify Color dialog, which is based on the HSV (Hue/Saturation/Value) color model. Each color dimension ranges in value from 0 to 255. You can make your own color by selecting a value for each color dimension. Click OK if you like a color in display. This color of your specification becomes a symbol in the color scheme.
8. Close the Legend Editor and the Color Palette. You are now ready to compose your map. Select Layout in the View menu.
9. The Template Manager has five default templates. Click on Portrait and then OK. You should see the layout of the U.S. map

with the title, legend, bar scale, and north arrow.

10. Before manipulating the map elements in the layout, change the layout grid to a finer grid because the grid controls the alignment of the map elements. Select Properties from the Layout menu. In the Layout Properties dialog, set the horizontal and vertical grid spacing to 0.1".

11. You can manipulate an active graphic element in several ways. You can move it by dragging it to where you want it in the layout. You can reduce or enlarge it by dragging on its handles. Or you can select Size and Position from the Graphics menu and specify the height and/or width of the graphic element and its relative position in the layout, all in inches, in the Graphic Size and Position dialog. The only map element that is difficult for reduction or enlargement is the bar scale.

12. If you cannot select a map element, you may need to use the Bring to Front button and Send to Back buttons.

13. To add text such as your name on your map, press the Text tool and click the mouse at the location where you want to start the text string. In the Text Properties dialog, type the text, select the Horizontal Arrangement if it is needed, and then click OK. You can reduce, enlarge, or move the text after the initial placement.

14. To add a neat line such as a map border to your map, press the Draw Point tool and select the rectangle symbol. Then place the cursor at the lower left corner of the intended neat line and drag the cursor to its upper right corner. The pull-down menu of the Draw Point tool contains other graphic devices that can be used for map design purposes, such as point, line, and circle. The default line width of the graphic device is usually too thin for its visual effect. You can change the line width by selecting Show Symbol Window from the Window menu and changing the width of Outline in the Fill Palette.

15. Several styles may be used for the N-arrow. To see the choices, select the N-arrow and double click on it to open a menu with different design options. Select a design and click OK.

16. When you are ready to plot your map, select Print from the File menu. Or select Export and save your layout as a graphics file.

Task 2: Graduated Symbol, Line Symbol, and Type Design

What you need: *idlcity.shp,* a theme showing the 10 largest cities in Idaho; *idhwy.shp,* a theme showing interstates, U.S., and state highways in Idaho; and *idoutl.shp,* an outline map of Idaho.

Task 2 lets you experiment with type design, use the graduated symbol, and try the highway shield symbol in ArcView. The task also involves choice of color symbols and map design in general.

1. Start ArcView, open a new view, and add *idoutl.shp, idhwy.shp,* and *idlcity.shp* to view. Select Properties from the View menu. Change the name of the view to a proper map title, change the Map Units to feet, and click OK. Select Properties from the Theme menu for each of the three themes, and change the name to an appropriate legend description.

2. The only decision to be made about *idoutl.shp* is its color, the background color for the state, which should be contrasted with the point symbols for *idlcity.shp* and the highway symbol. Double-click *idoutl.shp* to invoke the Legend Editor and choose a color symbol for *idoutl.shp.*

3. Double-click *idhwy.shp* to open the Legend Editor. Select Unique Value as the Legend Type, and Route_desc as the Value Field. You should see three default symbols assigned to interstate, state, and U.S. highways. Double-click the default symbol for the interstate to open the Pen Palette. You can change the line symbol (for example, to a double-line symbol) and its color. Repeat the same for state and U.S. highways.

4. For the city theme, you will first show the city population using the graduated symbol. Double click *idlcity.shp* to open the Legend Editor. Select Graduated Symbol from the Legend Type pull-down menu. Select Population for the Classification Field. The default classification divides the cities into 5 classes. For this map you can reduce the number of classes to 3 by clicking Classify and changing the number to 3. Also change the class breaks to 14302–30000, 30000–45000, and 45000–125659.

5. Now work with Symbol and Size Range at the bottom of the Legend Editor. Symbol determines the symbol type, for example, open circle, filled circle, filled circle with outline, and so forth. Size Range determines the size variation of the graduated symbols. The default for Symbol is the filled circle. By double-clicking Symbol, you can change the symbol type using the Marker Palette. You can also use the Color Palette to change the color of the point symbol. Size Range is measured in point size, with 72 points to an inch. The default for Size Range is 4 to 12, a range with fairly small circle sizes on an 8.5-by-11-inch map. You can experiment with different size ranges and see which range works out best on your map. Remember that two cities are close to Boise; a large circle for Boise will cover up its neighboring cities. Click on Apply after you have chosen the symbol type and the size range.

6. The next step is to label the cities. As explained in the chapter, you can label the cities interactively or by using the automatic labeling method. For this task, you will use the interactive method. (You will use the automatic labeling method in Task 3.) Make sure *idlcity.shp* is an active theme. Select Properties from the Theme menu to open the Theme Properties dialog, and click the Text Labels icon on the left. Choose city_name as the Label Field. Uncheck Scale Labels, meaning that you opt not to use scalable labels in your design. (The size of scalable labels will change as you zoom in and out of the view.) Click OK.

7. Label cities in each class with a different type design. Select Show Symbol Window from the Window menu, and click on the Font Palette to open it. Choose the type variations, including typeface, size, and style (called form in the chapter), for the largest class. (It is a good idea to write down your type selection for reference later.) Click on the Label tool and click on a city that belongs to the largest class. You should see the name of the city appears with its handles. The handles indicate the name is now an active graphic element so that you can move it to where you see fit on the map. Do the same for other cities in the largest class. Then repeat the same process for the middle and smallest class. To achieve harmony in text design, stay with the same typeface for all three classes but vary type size and style.

8. After labeling every city, you may still want to change the type design or move a name. You can do the following: click on the Pointer tool, click the name you want to work with (you should see the handles around it), and then move the name or change its type design. While depressing the Pointer tool, click outside the handles to make a name inactive.

9. ArcView has a feature that allows you to add a label with a leader line. This feature is useful for labeling cities close together such as Boise-Nampa-Caldwell. Press the Label tool and select the Labeling with a Leader Line tool from the pull-down menu. Click on the city you want to have a leader line and drag the cursor to where you want to place the label.

10. ArcView also has a feature that allows you to label a road with a highway shield. You need to go through a couple of steps to use this feature. First, make sure *idhwy.shp* is active and choose Properties from the Theme menu. In the Theme Properties dialog, click the Text Labels icon, specify Minor1 (the field with

the highway number) as the Label Field, select On for Alignment Relative to Line and Midway for Alignment Along Length, and uncheck Scale Labels. Click OK to dismiss the Theme Properties dialog. Next, select Text and Label Defaults from the Graphics menu. In the Default Setting dialog, choose the highway icon tool and set up the parameters for the tool. The parameters include two parts: one is the type design for the highway number, and the other is the shield symbol. Typically, you will set the size to be the same for the number and the symbol.

11. To label a highway, press the Label tool and select the icon tool you have set up in the previous step. Position and click the cursor on a highway where you want to place the highway shield. Repeat the same process for other highways in the same class. To label a different class of highways, repeat the previous step before labeling.

12. Map design is an iterative process; therefore, you may have to repeat the above steps to create the map you really want. When you are satisfied with the look of your map in view, go to the layout to compose your map. The default legend in the layout consists of all theme descriptions as a graphic element, which often needs modification. For example, the legend includes a description of Idaho for the outline theme, which is really unnecessary. To delete the outline theme description, first make the legend an active graphic element and select Simplify from the Graphics menu to separate the legend into graphic elements. Then select the legend box and description for the outline theme and delete them. If you want to re-group the graphic elements into two legend parts—one for the cities and the other for the highways—you can use the Group function. First, depress the Shift key while dragging the cursor to make a frame around all the legend boxes and descriptions that belong to *idlcity.shp*. Then select Group from the

Graphics menu. Now you can treat the *idlcity.shp* legend as a separate graphic element in manipulation. You can do the same for the *idhwy.shp* legend.

13. When you are ready to plot your map, select Print from the File menu. Or you can select Export and save your layout as a graphics file.

Task 3: Text Labeling

What you need: *charlie.shp,* a theme showing Santa Creek and its tributaries in north Idaho.

Task 3 lets you try the automatic labeling method in ArcView. Although the method can label all features on a map and remove duplicate name, it requires adjustments on some individual labels and overlapped labels. Therefore, Task 3 also requires you to use the Spline Text tool.

1. Start ArcView, open a new view, and add *charlie.shp* to view. Make the view window as large as possible. Select Properties from the Theme menu to open the Theme Properties dialog. In the dialog, click the Text Labels icon on the left, choose name as the Label Field, select Above for Alignment Relative to Line, and select Midway for Alignment along Length. Uncheck Scale Labels, and click OK to dismiss the dialog.

2. The next step is to select type variations for the stream names. Select Show Symbol Window from the Window menu, and open the Font Palette. Use a serif type such as Times New Roman, a point size of 10, and the italic style.

3. Select Auto-label from the Theme menu to open the Auto-label dialog. In the dialog, opt for Find Best Label Placement, check both Allow Overlapping Labels and Remove Duplicates, and click OK.

4. The view now has labels for the streams, but it has a couple of problems. Bob Cr. and Brown Cr. are in green, meaning that they are either overlapped with other labels or outside the map. The labeling of Pamas Cr. does not follow the course of the creek.

5. The labeling problems with Brown Cr. and Pamas Cr. are actually caused by the placement of Fagan Cr. Therefore the first step in correcting the problems is to re-label Fagan Cr. further upstream. Depress the Pointer tool and then select Fagan Cr. When the handles appear around the label, select Delete Graphics from the Edit menu to delete the label. Now re-label Fagan Cr. Press the Text tool and choose Spline Text from the pull-down menu. Position and click the cursor where you want the label to start. Click a couple of more points to make a curved line, and double click to exit. Type Fagan Cr. in the Spline Text Properties dialog and click OK. Fagan Cr. should now appear on the map. Repeat the same procedure if you are not satisfied with the re-labeling. You can use the same procedure to correct the labeling problems with Brown Cr. and Pamas Cr. For Bob Cr., you need to re-label it below the stream. Remember the names you have re-labeled are now individual graphic elements. The Remove Labels selection from the Theme menu will not remove them.

REFERENCES

Antes, J.R., K. Chang, and C. Mullis. 1985. The Visual Effects of Map Design: An Eye Movement Analysis. *The American Cartographer* 12: 143–55.

Antes, J.R., and K. Chang. 1990. An Empirical Analysis of the Design Principles for Quantitative and Qualitative Symbols. *Cartography and Geographic Information Systems* 17: 271–77.

Brewer, C.A. 1994. Color Use Guidelines for Mapping and Visualization. In A.M. MacEachren and D.R.F. Taylor (eds.). *Visualization in Modern Cartography.* Oxford: Pergamon Press, pp. 123-47.

Chang, K. 1978. Measurement Scales in Cartography. *The American Cartographer* 5: 57–64.

Cuff, D.J. 1972. Value Versus Chroma in Color Schemes on Quantitative Maps. *Canadian Cartographer* 9: 134–40.

Dent, B.D. 1999. *Cartography: Thematic Map Design.* 5th ed. Dubuque, Iowa: WCB/McGraw-Hill.

Kraak, M.J., and F.J. Ormeling. 1996. *Cartography: Visualization of Spatial Data.* Harlow, England: Longman.

Mersey, J.E. 1990. Colour and Thematic Map Design: The Role of Colour Scheme and Map Complexity in Choropleth Map Communication. *Cartographica* 27(3): 1–157.

Monmonier, M. 1996. *How to Lie With Maps.* 2d ed. Chicago: Chicago University Press.

Mower, J.E. 1993. Automated Feature and Name Placement on Parallel Computers. *Cartography and Geographic Information Systems* 20: 69–82.

Robinson, A.H., J.L. Morrison, P.C. Muehrcke, A.J. Kimerling, and S.C. Guptill. 1995. *Elements of Cartography,* 6th ed. New York: John Wiley & Sons.

Slocum, T.A. 1999. *Thematic Cartography and Visualization.* Upper Saddle River, NJ: Prentice Hall.

9

DATA
EXPLORATION

9.1 INTRODUCTION

Starting data analysis in a GIS project can be overwhelming. The GIS database may have dozens of map layers and hundreds of attributes. Where do you begin? What attributes do you look for? What data relationships are there? One way to ease into the analysis phase is data exploration. Broadly defined, **data exploration** is data-centered query and analysis. It allows the user to examine the general trends in the data, to take a close look at data subsets, and to focus on possible relationships between data sets. The purpose of data exploration is to better understand the data and to provide a starting point in formulating research questions and hypotheses.

An important component of data exploration consists of interactive and dynamically linked visual tools. Maps (both vector- and raster-based), graphs, and tables are displayed in multiple windows and dynamically linked so that selecting a record or records from a table, for example, will automatically highlight the corresponding features in a graph or a map. Data exploration, therefore, allows data to be viewed from different perspectives, making it easier for information processing and synthesis. Windows-based GIS packages, which can work with maps, graphs, and tables simultaneously, are well-suited for data exploration.

Chapter 9 is organized into five sections. Section 1 discusses interactive data exploration. Section 2 covers vector data query via attribute data, spatial data, or both. Raster data query using the cell value or graphic method is covered in Section 3. Section 4 briefly explains the use of charts and statistics in data exploration. And Section 5 deals with geographic visualization and different methods for manipulating and displaying GIS data. Throughout its sections, Chapter 9 emphasizes the dynamic linkage between spatial data and attribute data in GIS. ArcView and, to a limited extent, ARC/INFO are used as examples.

9.2 INTERACTIVE DATA EXPLORATION

Statisticians have traditionally used charts and diagrams such as histograms and scatterplots to explore data structure and to discover data patterns. In a histogram, data are grouped into classes and bars are used to show the number of values falling in each class. A scatterplot is a chart in which markings are used to plot the values of two variables along the x- and y-axis. Data exploration in statistics has expanded to include exploratory data analysis (Tukey 1977; Tufte 1983) and dynamic graphics (Cleveland and McGill 1988; Cleveland 1993). **Exploratory data analysis** advocates the use of a variety of techniques for examining data more effectively as the first step in statistical analysis. **Dynamic graphics** enhances exploratory data analysis by using multiple and dynamically linked windows and by letting the user directly manipulate data points in charts and diagrams. Because the views are linked, highlighting in one view causes the immediate highlighting of corresponding data points in other views. This kind of interaction and experimentation with the data set helps the user recognize data patterns and unusual data points called outliers. Exploratory data analysis is therefore a precursor to more formal and structured data analysis.

Common methods of data manipulation in dynamic graphics include selection, deletion, rotation, and transformation of data points. **Brushing**, for example, is a method for selecting and highlighting a data subset in multiple views (Becker and Cleveland 1987). This method allows a user to select a subset of points from a scatterplot and view related data points that are highlighted in other scatterplots. Brushing can also be used in the geographic context by including maps (Monmonier 1989).

Some charts are designed for special uses. The box plot (Tukey 1977), for example, shows five statistics (median, first quartile, third quartile, minimum, and maximum) of a data set in a plot. The variogram cloud (Cressie 1993) is a scatterplot of squared differences in data values for every pair of data points.

Data exploration in GIS is similar to exploratory data analysis: it allows the user to focus on a data subset of interest. But the medium for data exploration has expanded to include both vector- and raster-based maps and map features. As maps are dynamically linked to charts, diagrams, and tables, they provide a new interface to the database and offer a spatial view to data analysis. At the same time, the use of maps has called for specialized methods of exploratory spatial data analysis that consider the special nature of spatial data (Anselin 1999).

A special nature of data exploration in GIS is the linkage between spatial data and attribute data. For example, a GIS user working with soil conditions would want to know how much of the study area is rated poor, but also where those poor soils are distributed. The linkage must be dynamic and simultaneous: highlighting a subset of attribute data should simultaneously highlight the corresponding map features in the view window and vice versa. Windows-based GIS packages can fulfill this requirement easily by displaying the spatial data and attribute data of a theme in separate but dynamically linked windows.

The linkage between spatial and attribute data is an important characteristic of the vector data model. Using the feature attribute table, a vector-based map can be joined or linked with other attribute tables in a relational database for data query. The linkage is not as important for the raster data model primarily for two reasons. First, a raster model uses cells with fixed locations. Second, a raster model with floating-point cell values does not have a value attribute table, thus limiting its use in data query. Therefore, this chapter separates vector data from raster data in the discussion of data exploration.

9.3 VECTOR DATA QUERY

9.3.1 Attribute Data Query

Attribute data query retrieves a data subset from a map by working with its attribute data. The selected data subset may be visually inspected or saved for further processing. Attribute data query requires the use of expressions, which must be interpretable by a GIS or a database management system. These expressions are often different from one system to another. This section covers attribute data query using ArcView, and SQL (Structured Query Language) used with a relational database.

9.3.1.1 Logical Expressions

Data queries in ArcView follow Boolean algebra (after the English logician George Boole, 1815–1864) and consist of logical expressions and Boolean connectors. A **logical expression** contains operands and a logical operator. For example, 'class = 2' is a logical expression, in which 'class' and '2' are operands and '=' is a logical operator. In this example, class is the name of a field, 2 is the field value used in the query, and the logical expression selects those records that have the class value of 2. Operands may be a field, a number, or a string. Logical operators may be equal to ($=$), greater than ($>$), less than ($<$), greater than or equal to ($>=$), less than or equal to ($<=$), or not equal to ($<>$).

Boolean connectors are AND, OR, XOR, and NOT, which are used to connect two or more logical expressions in a query statement. The connector AND connects two expressions in this example: (class = 2) AND (age $>$ 100). Records selected from the statement must satisfy both (class = 2) and (age $>$ 100). If the connector is changed to OR in the example, then records that satisfy either one or both of the expressions are selected. If the connector is changed to XOR, then records that satisfy one and only one of the expressions are selected. The connector NOT negates an expression, meaning that a true expression is changed to false and vice versa. The statement, NOT (class = 2) AND (age $>$ 100), for example, selects those records whose class is not equal to 2 and whose age is greater than 100.

Boolean connectors of NOT, AND, and OR are actually keywords used in the operations of COMPLEMENT, INTERSECT, and UNION on sets in probability. The operations are illustrated in Figure 9.1, with A and B representing two subsets of a universal set.

- The COMPLEMENT of A contains elements of the universal set that do NOT belong to A.
- The INTERSECT of A and B is the set of elements that belong to both A AND B.
- The UNION of A and B is the set of elements that belong to A OR B.

9.3.1.2 Type of Operation

Attribute data query in ArcView begins with a complete data set. Through query, this data set is divided into two subsets: one containing selected records and the other unselected records. A basic query

(a)
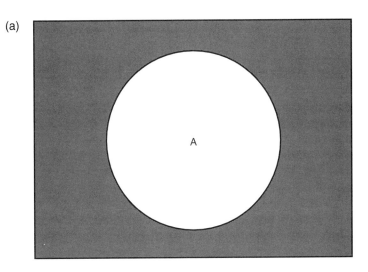

(b)
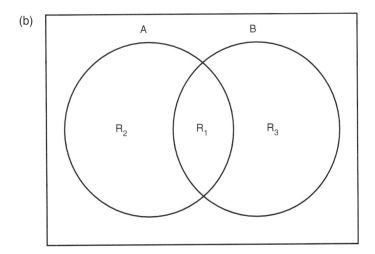

Figure 9.1

The shaded portion in (a) represents the complement of data subset A. R_1 in (b) represents the intersection of data subsets A and B. The union of A and B is the combined area of R_1, R_2, and R_3.

operation, called New Set, is to select a data subset. Given a data subset, three types of operation can act on the subset. First, an operation called Add To Set can add more records to the subset. Second, an operation called Select From Set can reduce it further to a smaller subset. Third, an operation called Switch Selection can switch between the selected subset and the unselected subset. Select All or Select None returns the whole data set by clearing the selected records. ARC/INFO, a command-driven package, has similar query functions as ArcView (Box 9.1).

9.3.1.3 Examples of Query Operation

The following examples show different query operations using data from Table 9.1, which has 10 records and three fields.

Example 1: Select a data subset and then add more records to it

[New Set] cost $>=$ 5 AND soiltype = "Ns1"
- 0 of 10 records selected
[Add To Set] soiltype = "N3"
- 3 of 10 records selected

Example 2: Select a data subset and then switch selection
[New Set] cost $>=$ 5 AND soiltype = "Tn4" AND area $>=$ 300
- 2 of 10 records selected
[Switch Selection]
- 8 of 10 records selected

Example 3: Select a data subset and then select a smaller subset from it
[New Set] cost $>$ 8 OR area $>$ 400
- 4 of 10 records selected
[Select From Set] soiltype = "Ns1"
- 2 of 10 records selected

Box 9.1   **Query Operations in ARC/INFO**

ARC/INFO has the same types of query operations as ArcView. The command RESELECT in Tables performs the same function as 'New Set,' that is, selecting a data subset. ASELECT can add more records to the subset, while NSELECT switches between the selected subset and the unselected subset. There are two ways to return the whole data set: (1) ASELECT without an expression, and (2) SELECT to select the INFO table.

TABLE 9.1   **A Data Set for Query Operation Examples**

Cost	Soiltype	Area	Cost	Soiltype	Area
1	Ns1	500	6	Tn4	300
2	Ns1	500	7	Tn4	200
3	Ns1	400	8	N3	200
4	Tn4	400	9	N3	100
5	Tn4	300	10	N3	100

9.3.1.4 Relational Database Query

Relational database query is performed using a relational database, which consists of many separate but interrelated tables. A query of a table in a relational database not only selects a data subset in the table but also selects records related to the subset in other tables. This feature is desirable in data exploration because it allows the user to examine related data characteristics from multiple tables.

To use a relational database, one must first be familiar with the overall structure of the database, the designation of keys in relating tables, and a data dictionary listing the fields in each table. To perform a query, one can choose either Join or Link in ArcView. Join combines attribute data from two or more tables into a single table. Link keeps tables separate but dynamically linked. To establish a link, the user must designate one table as the source table and the other as the destination table. After the link is established, selecting a record in the destination table will automatically select the corresponding record or records in the source table. The link, however, only exists in the destination table. In other words, selecting a record in the source table will not automatically select the corresponding record or records in the destination table. This kind of directional linking can be limiting and confusing in data exploration. One way to solve this problem is to establish two-way links between tables (Figure 9.2).

As discussed in Chapter 6, the National Map Unit Interpretation Record (MUIR) database is a relational database maintained by the Natural Resources Conservation Service (NRCS). The data-base contains over 80 estimated soil properties, interpretations, and performance data for each polygon on a soil map. A major challenge for using this database is to sort out where each soil attribute resides and how tables are linked. Suppose we ask the question: What types of plants, in their common names, are found where annual flooding frequency is rated as either frequent or occasional? To answer the question, we need the following three MUIR tables: the map unit components table or comp.dbf, which contains data on annual flood frequency; the woodland native plants table or forest.dbf, which has data on forest types; and a look-up table called plantnm.dbf, which has common plant names. The next step is to find the keys that can link the three tables (Figure 9.3). Finally, we need to link the theme table of the soil map to comp.dbf by using musym (map unit symbol) as the key.

After two-way links are established between the tables, we can issue the following query statement to comp.dbf:

([anflood] = "frequent") OR ([anflood] = "occas")

Evaluation of the logical expression selects not only records in comp.dbf that meet the criteria but also the corresponding records in forest.dbf, plantnm.dbf, and the theme table and the corresponding soil polygons in the map. This dynamic selection is possible because the tables are interrelated in MUIR and are dynamically linked to the map. A detailed description of relational database query is included in the applications section.

9.3.1.5 Use SQL to Query a Database

SQL (Structured Query Language) is a data manipulation language designed for relational databases. IBM developed SQL in the 1970s, and many commercial database management systems such as ORACLE have adopted the language. To access data in a database that uses SQL, a GIS user must prepare query statements in SQL.

The basic syntax of SQL is

> **select** <attribute list>
> **from** <relation>
> **where** <condition>

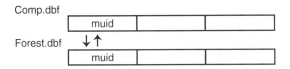

Figure 9.2
To make sure data query can be performed from either comp.dbf or forest.dbf, two-way linkages must be established between the two tables.

Soil theme table

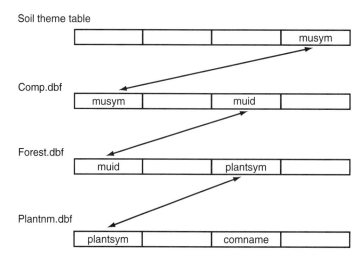

Figure 9.3
The keys relating three dBASE files in MUIR and the theme table. The field comname in plantnm.dbf contains the common plant names.

A simple SQL statement querying the sale date of the parcel, P101, in Figure 9.4 is

> **select** sale date
> **from** parcel
> **where** PIN = "P101"

The following SQL statement, querying the owner name of P104 (Smith), involves two relations (tables) and the AND connector:

> **select** owner name
> **from** parcel, owner
> **where** parcel. PIN = owner. PIN
> and PIN = "P104"

More examples of SQL statements are included in Box 9.2. GIS packages such as ARC/INFO, ArcView, MGE, and MapInfo allow GIS users to use SQL in data query. ArcView has a functionality called SQL Connect, which uses a dialog to connect to an external database. The dialog also includes the keywords of select, from, and where and the spaces so that the user can type in the SQL statement. The selected records from query can then be saved into a table in ArcView. ARC/INFO uses the CONNECT command to link to an external database. After the database is connected in ARC/INFO, query statements can be typed at the command lines.

9.3.1.6 Descriptive Statistics of Attribute Data

Descriptive statistics such as minimum, maximum, range, mean, and standard deviation summarize the data values of an attribute. The **range** is the difference between the minimum and maximum values. The **mean** is the average of data values. And the **standard deviation** is the square root of the average of the squared deviations of each data value about the mean. One can incorporate these descriptive statistics into query statements such as selecting records that are above the mean, or at least 3 standard deviations above the mean. ArcView has a menu selection that can compile the descriptive statistics of a selected field from an attribute table.

9.3.2 Spatial Data Query

Spatial data query refers to the process of retrieving data from a map by working with map features. One may select map features using a cursor, a

Relation 1: Parcel

PIN	Sale date	Acres	Zone code	Zoning
P101	1-10-98	1.0	1	residential
P102	10-6-68	3.0	2	commercial
P103	3-7-97	2.5	2	commercial
P104	7-30-78	1.0	1	residential

Relation 2: Owner

PIN	Owner name
P101	Wang
P101	Chang
P102	Smith
P102	Jones
P103	Costello
P104	Smith

Figure 9.4

The key PIN relates the parcel and owner tables and allows use of SQL with both tables.

Box 9.2   More Examples of SQL Statements

A national forest in Idaho uses Oracle to manage a centralized database, which is accessible to its ranger districts through phone lines. Although SQL statements used for querying the national forest database look complicated because of many attributes involved, the syntax structure remains the same. The following shows two query examples. The first code queries a table called stand_activities for records that have values in the sale_name column:

select sale_name, sale_cntr_no, stand_id, sale_unit_no_1, sale_unit_no_2
from stand_activities

where sale_name is not null

The second code queries two tables, sa (stand_activities) and sn (sale_names), for those records that must satisfy all three expressions about stand_id, activity_code, and accomp_year:

select sale_name, stand_id, activity_code, activity_units, accomp_year
from sa, sn
where stand_id between '23200000' and '29400000'
    and activity_code between 4431 and 4432
    and accomp_year is null

graphic, or the spatial relationships between map features. Spatial data query is the geographic interface to the database and is therefore useful for tasks that cannot be easily accomplished by attribute data query, such as selecting contiguous areas. As in attribute data query, results of spatial data query may be visually inspected or saved as new maps for further processing.

9.3.2.1 Feature Selection by Cursor

The simplest spatial data query is to select a map feature by pointing at the feature itself. An alternative is to select map features within a specified area extent, defined by two opposite points of the extent.

9.3.2.2 Feature Selection by Graphic

This query method uses a graphic such as a circle, a box, a line, or a polygon to select map features, which fall inside or are intersected by the graphic object (Figure 9.5). The graphic to be used for selection, called the selector graphic, can be drawn by using the mouse in real time. Feature selection by graphic is similar to brushing in exploratory data analysis. Like the brush, the selector graphic can be moved around the computer screen and can be made in any size or shape. Examples of query by graphic include selecting restaurants within a one mile radius of a hotel, selecting land parcels that intersect a proposed

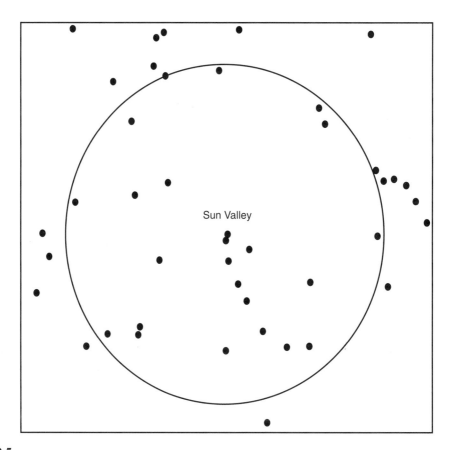

Figure 9.5
A circle with a specified radius is drawn around Sun Valley. The circle is then used as a graphic object to select point features within the circular area.

highway, and finding owners of land parcels within a proposed nature reserve.

9.3.2.3 Feature Selection by Spatial Relationship

This query method selects map features based on their spatial relationships to other features. Map features to be selected may be in the same map as map features used for selection. Or, they may be in different maps. An example of the first type of query is to find roadside rest areas within a radius of 50 miles of a selected rest area; in this case, map features to be selected and used for selection are in the same map. An example of the second type of query is to find rest areas within each county. Two maps are required for this query: one map showing county boundaries and the other roadside rest areas.

Spatial relationships used in query include the following:

- **Containment**—selects features that fall completely within features used for selection. Examples include finding schools within a

selected county, and finding state parks within a selected state.

- **Intersect**—selects features that intersect features used for selection. Examples include selecting land parcels that intersect a proposed road, and finding settlements that intersect an active fault line.

- **Proximity**—selects features that are within a specified distance of features used for selection. Examples include finding state parks within 10 miles of an interstate highway, and finding pet shops within one mile of selected streets. If features to be selected and features used for selection share common boundaries and if the specified distance is 0, then proximity becomes **adjacency**. Examples of spatial adjacency include selecting land parcels that are adjacent to a flood zone, and finding vacant lots that are adjacent to a new theme park.

The use of spatial relationship for spatial data query in ArcView is shown in Box 9.3.

Box 9.3   **Expressions of Spatial Relationship in ArcView**

ArcView handles feature selection by spatial relationship through the Select by Theme dialog. The dialog requires the user to identify features used for selection first and then features to be selected. A theme (selector theme) chosen from a pull-down menu shows where features used for selection come from, and the active theme (target theme) gives the source from which features are to be selected. Although the two sets of map features are usually from two different themes, they can be from the same theme. Six expressions of spatial relationship connect features to be selected and used for selection: (1) are completely within, (2) completely contain, (3) have their center in, (4) contain the center of, (5) intersect, and (6) are within distance of. The first four expres-

sions deal with the containment relationship, and the last two the relationships of intersect, proximity, and adjacency. An input box is added to the dialog whenever the "are within distance of" relationship is specified.

A completed query statement may read: "Select features of active themes that are within distance of the selected features of idroads [name of selector theme]," with a selection distance of "2" miles. If no features are previously selected from the selector theme, then all features in the selector theme are used for selection. After a query statement is set up, the user can then choose the query option of New Set, Add To Set, or Select From Set These query options are the same as for attribute data query.

9.3.2.4 Combining Attribute and Spatial Data Queries

So far we have approached data exploration through attribute data query or spatial data query. In many cases data exploration requires both types of query. For example, both types of query are needed to find gas stations that are within one mile of a freeway exit in southern California and have annual revenue of $2 million. Assuming that the maps of gas stations and freeway exits are available, there are at least two ways to solve the question.

1. Locate all freeway exits in the study area, and draw a circle around each exit with a 1-mile radius. Select gas stations within the circles through spatial data query. Then use attribute data query to find gas stations that have annual revenues exceeding $2 million.
2. Locate all gas stations in the study area, and select those stations with annual revenues exceeding $2 million through attribute data query. Next, use spatial data query to narrow the selection of gas stations to those within 1 mile of a freeway exit.

The first option queries spatial data and then attribute data. The process is reversed with the second option. Assuming that there are many more gas stations than freeway exits, the first option may be a better option, especially if the gas station map must be linked to other attribute tables for getting the revenue data. As shown in the above example, the combination of spatial and attribute data queries opens wide the possibilities of data exploration. Some GIS users might even consider this kind of data exploration as data analysis because that is what they need to do to solve most of their tasks.

9.4 RASTER DATA QUERY

Raster data may be queried using the cell value or the graphic method. An integer grid in ARC/INFO and ArcView has a value attribute table, which functions like the feature attribute table of a vector-based map in data management and query. Without a value attribute table, a floating-point grid is more limited in query functions.

9.4.1 Query by Cell Value

A common method to query a grid is to use a logical expression involving cell values. As examples, the expression, [road] = 1, can query an integer road grid and the expression, [elevation] >= 1243.26, can query a floating-point elevation grid. Because a floating-point elevation grid contains continuous values, querying a specific value is not likely to find any cell in the grid. Therefore, using numeric ranges in query expressions is recommended for floating-point grids. A query statement separates cells of a grid that satisfy the statement from cells that do not.

Raster data query can also use the Boolean connectors of AND, OR, and NOT to string together separate expressions. These separate expressions can apply to different attributes of an integer grid. For example, the statement, ([anflood] = "freq") AND ([wtdepth] < 1), selects cells that have frequent annual flood and water table depth of less than 1 meter from a soils grid. This kind of query is not possible with a floating-point grid.

A compound statement with separate expressions can apply to multiple grids, which may be integer, or floating point, or a mix of both types. For example, the statement, ([grid1] = 13) AND ([grid2] = 3) AND ([grid3] = 4), selects cells that have the value of 13 in grid1, 3 in grid2, and 4 in grid3. Querying multiple grids by cell value is unique with raster data. For vector data, attributes to be used in a compound expression must be all present in the same attribute table.

9.4.2 Query Using Graphic Method

A grid may be queried by pointing the cursor at a cell to identify its cell location and value. An alternative to the cursor is a graphic object such as a circle, a square, or a polygon. The values of those cells that fall within the graphic object can then be displayed in a histogram. ArcView provides a histogram button and displays a histogram with unique cell values if the grid is an integer grid and in numeric ranges if the grid is a floating-point grid.

9.5 CHARTS

Charts are the basic tools for exploratory data analysis. Prepared from an attribute table, charts can be displayed in separate windows from maps and tables. Because charts are dynamically linked to maps and tables, a query of attribute data will simultaneously highlight the selected records in the table, the corresponding features in the map, and the positions of the selected records in the chart.

ArcView has limited capabilities with charts. First, the default maximum number of records is 100 for a chart with one field (e.g., a histogram) and 50 for a chart with two fields (e.g., a scatterplot). In addition, the choices of charts are limited to the display of one or two data sets. ArcView cannot be used for displays of multi-dimensional data. Finally, interactivity with charts is limited to two tools, erase and erase with polygon. The erase tool allows the user to remove data points in a chart, one at a time. Available only on scatterplots, the erase with polygon tool allows the user to remove multiple data points within a specified polygon.

Commercial statistical analysis packages such as SAS, SPSS, SYSTAT, S-PLUS (now Insightful), and Excel offer more tools to work with charts than does a GIS package. Therefore, one way to explore data with charts is to export data files from a GIS to a statistical analysis package (Scott 1994). S-PLUS for ArcView is a software product from MathSoft, Inc. (http://www.insightful.com/), which allows GIS users to move tabular data from ArcView to S-PLUS and then returns graphics and analytical results to ArcView. Recent studies have also shown that one can enhance exploratory data analysis in GIS by using macro programs in a GIS (Batty and Xie 1994) or by linking a GIS to software specifically written for spatial data analysis (Walker and Moore 1988; Haslett et al. 1990; Anselin 1999).

9.6 GEOGRAPHIC VISUALIZATION

MacEachren (1995) used the term **geographic visualization** to describe the use of maps for setting up a context for visual information processing, which can then lead to formulation of research questions or hypotheses. Geographic visualization therefore has the same objective as exploratory data analysis. This section covers four methods for geographic visualization: data classification, spatial aggregation, multiple maps in a view, and map comparison.

9.6.1 Data Classification

Data classification is probably the most common method for map manipulation. The method is useful for data visualization as well as for creating new attribute data and new maps.

9.6.1.1 Data Classification for Visualization

Classification aggregates data and map features using a classification method and a number of classes (Box 9.4). By changing the classification method and the number of classes, the same data can produce different looking maps, each with its own spatial pattern. Of the different map types, choropleth maps are most affected by classification. Cartographers often make several versions of the choropleth map from the same data and choose one—typically one with a good spatial organization of classes—for final map production. For data exploration, classification is most useful in separating spatial data by some descriptive statistics.

Suppose we want to explore the rate of unemployment by state in the U.S in 1997. To get a preliminary look at the data, we may place rates of unemployment into two classes based on the national average: above and below the national average (Figure 9.6a). Though much generalized, the map breaks the country into contiguous regions, which may suggest factors that can explain why some states have done better than others.

To further isolate those states that are way above or below the national average, we can classify rates of unemployment by their mean and

Data classification methods available in GIS packages may include the following six methods: equal interval, equal frequency, equal area, mean and standard deviation, natural breaks, and user-defined. Equal interval uses a constant class interval in classification. Equal frequency, also called quantile, divides the total number of data values by the number of classes and ensures that each class contains the same number of data values. Equal area divides the map area by the number of classes and ensures that each class contains the equal proportion of area. Mean and standard devi-ation sets the class breaks at units of the standard deviation (0.5, 1.0, etc.) above or below the mean. The method of natural breaks considers the grouping of data values in classification. Typically, the natural breaks method uses a computing algorithm to minimize differences between data values in the same class and to maximize differences between classes. The user-defined method lets the user choose the appropriate or meaningful class breaks. For example, in mapping rates of population change by state, the user may choose zero or the national average as a class break.

standard deviation. Figure 9.6b shows the classification result using the mean and standard deviation method. We can now focus our attention on states that are more than one standard deviation above the mean.

As a tool for data exploration, data classification can offer additional information by linking a classified map with other attributes that are not used in classification. Suppose percent change in median household income by state is also available as an attribute. We can see whether states that are grouped with low unemployment rates tend to have higher rates of income growth, and vice versa.

9.6.1.2 Data Classification for Creating New Data

As described in Chapter 6, new attribute data for a vector-based map can be created by selecting a data subset, that is, records that fall within a class according to the classification scheme, and assigning a value to the selected data subset.

Creating a new raster model by classification is often referred to as **reclassification**, recoding, or transforming through look-up tables (Tomlin 1990). Two methods of data reclassification may be used. The first method is a one-to-one change, meaning a cell value in the input grid is assigned a new value in the output grid. For example, irri-gated cropland in a land use grid is assigned a value of 1 in the output grid. The second method assigns a new value to a range of cell values in the input grid. For example, cells with population densities between 0 and 25 persons per square mile in a population density grid are assigned a value of 1 in the output grid.

Both methods of reclassification can be applied to integer grids. Floating-point grids, on the other hand, can only be reclassified by assigning a new value to a range of cell values in the input grid. For example, a value of 2 is assigned to cells with slope values between 10.0 and 20.0%. Reclassification of a floating-point grid results in an integer grid.

Raster data reclassification is an important function in data exploration and data analysis. It has the following applications:

- Data Simplification. Reclassification can group continuous slope values, for example, into a set of classes, for example, 1 for slopes of 0.0–10.0%, 2 for 10.0–20.0%, and so forth.
- Data isolation. Reclassification can create a new grid that contains a unique category or value such as irrigated cropland, or a range of values such as slopes of 10.0–20.0%.
- Data Ranking. Reclassification can create a new grid that shows the result of the ranking of cell values in the input grid. For example,

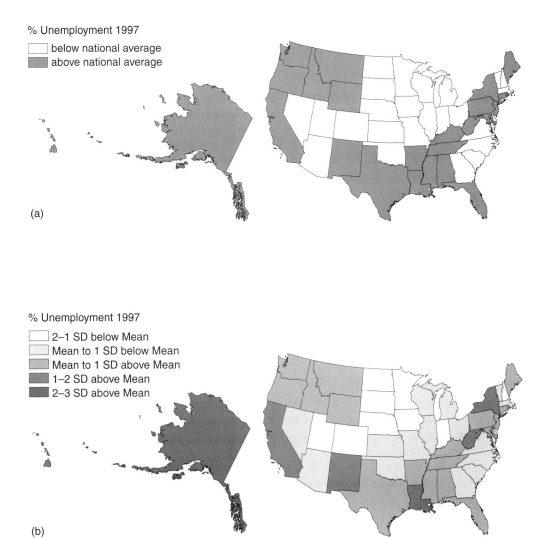

Figure 9.6
The top map shows rate of unemployment in 1997 as either above or below the national average of 4.9%. The bottom map uses the mean and standard deviation (SD) for data classification.

a reclassified grid can show the ranking of 1 to 5, with 1 being least suitable and 5 being most suitable.

The RECLASSIFY function in ArcView performs reclassification. The dialog offers the two methods of reclassification mentioned above and a third method that uses values of a field in the input grid as a look-up table.

9.6.2 Spatial Aggregation

Spatial aggregation is similar to data classification except that it groups data spatially. Figure 9.7

shows percent population change in the United States by state and by region. Regions are aggregates of states and therefore give a more general view of population growth in the country. Figure 9.7 uses two levels of geographic units defined by the U.S. Census Bureau for data collection; other levels are county, census tract, block group, and block. These levels of geographic units form a hierarchical order, and we can aggregate data from one lower level to a higher level. Displaying aggregated data at different levels offers a view of the effect of spatial scaling.

If distance is the primary factor in a study, spatial aggregation can be based on distance measures from points, lines, or areas. An urban area, for example, may be aggregated into distance zones away from the city center or from its streets (Batty and Xie 1994). Different from the geography of census, these distance zones require additional data processing. First, distance zones must be established by proximity or buffering. Then, distance zones must be overlaid with the base maps of census data and data must be re-compiled for different distance zones.

Spatial aggregation for raster data means preparing a coarser-resolution grid and computing values of the larger cells. For example, cells of a grid may be aggregated by a factor of 3 to produce a generalized grid. Each cell in the output grid is made of a 3×3 matrix in the input grid, and the cell value is a computed statistic from the nine input cell values such as mean, median, minimum, maximum, or sum.

9.6.3 Map Comparison

Map comparison helps a GIS user to sort out the relationship between different thematic data presented in maps. For example, the display of wildlife locations on a vegetation map may reveal the association between the wildlife species and a vegetation type (Figure 9.8). When the maps to be compared consist of only point or line features, the maps can be coded in different colors and superimposed on one another in a single view. But this process does not work if the maps to be compared

consist of polygon features or raster data. A map showing the rate of unemployment cannot be placed on top of a map showing the rate of income change unless both maps are prepared on semi-transparent films.

We have three options for comparing polygon or raster maps. The first option is to place all polygon and raster maps, along with point and line maps, onto the screen but to turn on and off polygon and raster maps so that only one of them is viewed at a time.

The second option is to open two or more views so that more than one polygon or raster map can be viewed simultaneously. This option is useful for observing temporal changes. For example, land cover maps of 1970, 1980, and 1990 for a study area can be placed in separate views so that the user can observe the changes over time. The idea of using multiple views is similar to the use of a scatterplot matrix (Cleveland 1993). The difference is that one map can only show one set of data, whereas a scatterplot can plot two or more variables. With maps for many points in time, one can run map animation showing continuous changes (DiBiase et al. 1992; Weber and Buttenfield 1993; Peterson 1995).

The third option is to use map symbols that can show data from two polygon themes. One method for this option is to use cartographic point symbols such as bar charts. A bar chart is placed in each polygon, and each bar chart has two or more bars representing, for example, amounts of timber harvest from two or more years for comparison. Another method is the bivariate choropleth map (Meyer et al. 1975). A bivariate choropleth map combines two color schemes, one for each variable. For example, a yellow-to-red color scheme for rate of unemployment may be combined with a yellow-to-blue color scheme for rate of income change to produce an unemployment/income change bivariate map. But if each variable is grouped into four classes, the bivariate map will have 16 color symbols, and the lack of logical progression between the mixed color symbols can be a problem (Olson 1981). An obvious improvement to the readability of a bivariate map is to re-

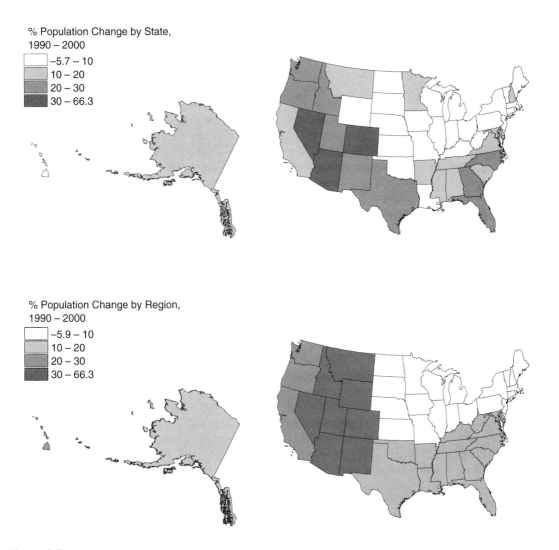

Figure 9.7
The top map shows percent population change by state, 1990–2000. The darker the symbol, the higher the percent increase. The bottom map shows percent population change by region.

duce the number of classes for each variable. Figure 9.9, for example, classifies the rate of unemployment and the rate of income change into either above or below the national average and then shows the combinations of the two variables on a single map.

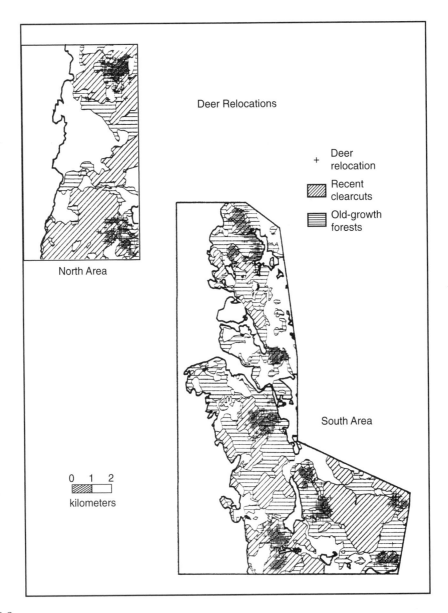

Figure 9.8
An example of using multiple maps in data exploration. In this view of deer relocations in SE Alaska, the focus is on the distribution of deer relocations along the clearcut/old forest edge.

KEY CONCEPTS AND TERMS

Adjacency: A spatial relationship that can be used to select features that share common boundaries.

Attribute data query: The process of retrieving data from a map by working with its attribute data.

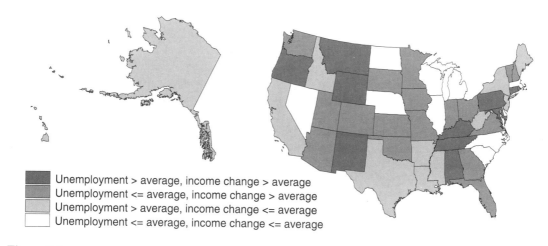

Figure 9.9
A bivariate map showing the combinations of (1) rate of unemployment in 1997, either above or below the national average, and (2) rate of income change 1996–98, either above or below the national average.

Boolean connector: A keyword such as AND, OR, XOR, or NOT that is used to construct complex logical expressions.

Brushing: A data exploration technique for selecting and highlighting a data subset in multiple views.

Containment: A spatial relationship that can be used in data query to select features that fall completely within specified features.

Data exploration: Data-centered query and analysis.

Dynamic graphics: A data exploration method in which the user can directly manipulate data points in charts and diagrams that are displayed in multiple and dynamically linked windows.

Exploratory data analysis: Use of a variety of techniques such as charts, diagrams, and scatterplots to examine data as the first step in statistical analysis.

Geographic visualization: A method that is used to display geographic data in visual representations and to set up a context for visual information processing, leading to formulation of research questions or hypotheses.

Intersect: A spatial relationship that can be used in data query to select features that intersect specified features.

Logical expression: A combination of a field, a value, and a logical operator, such as 'class = 2', from which an evaluation of True or False is derived.

Mean: The average of all values in a data set.

Proximity: A spatial relationship that can be used to select features that are within a distance of specified features.

Range: The difference between the minimum and maximum values in a data set.

Reclassification: The process of reclassifying cell values of an input grid to create a new grid.

Relational database query: Query in a relational database, which not only selects a data subset in a table but also selects records related to the subset in other tables.

Spatial data query: The process of retrieving data from a map by working with map features.

Standard deviation: A measure of data dispersion, which is defined as the square root of the average of the squared deviations of each data value about the mean.

Structured Query Language (SQL): A data query and manipulation language designed for relational databases.

APPLICATIONS: DATA EXPLORATION

This applications section covers five tasks. Task 1 presents an overview of data exploration in Arc-View, including use of summary statistics, selecting features by graphics, and making charts. Both Tasks 2 and 3 deal with attribute data query. Task 4 combines spatial and attribute data queries. Task 5 covers raster data query.

Task 1. An Overview of Data Exploration in ArcView

What you need: *idcities.shp*, with 654 places in Idaho; *idcounty.shp*, showing county boundaries in Idaho; and *snowsite.shp*, with 206 snow stations in Idaho and the surrounding states.

Task 1 is loosely organized around the question of finding a good skiing spot within 40 miles of Sun Valley, Idaho. This task is intended to familiarize you with several aspects of data exploration using ArcView: getting summary statistics, selecting map features by graphic, and creating charts. To keep it organized, each aspect of data exploration is included in a sub-section.

- **Derive summary statistics**

1. Start ArcView, open a new view, and add *idcities.shp*, *idcounty.shp*, and *snowsite.shp* to view. All three themes are based on the Idaho statewide coordinate system and measured in meters. Select Properties from the View menu. In the View Properties dialog, set the Map Units as meters and the Distance Units as miles.
2. Make *snowsite.shp* active, and choose Table from the Theme menu to open the attribute table of *snowsite.shp*. In the *snowsite.shp* theme table, click on the field Swe-max to make it active. Then select Statistics from the Field menu. Swe-max stands for maximum snow water equivalent.

- **Select map features with graphics**

1. Make *idcities.shp* the active theme. Click the Query Builder button. Set the query expression as [City_name] = "Sun Valley" and click New Set. Use the Zoom To Select button and then the Zoom In button to zoom in the area around Sun Valley.
2. Press and hold the Draw Point tool, and choose Draw Circle from the pull-down menu.
3. This step is to draw a circle around Sun Valley and to use the circle as a graphic to select snow stations within the circle. Click on Sun Valley as closely as possible at its location and drag the cursor to make a circle with a radius of 40 miles. The message at the bottom of the computer screen shows you how closely the circle radius is to 40 miles. If you cannot get exactly 40 miles interactively, you can select Size/Position from the Graphics menu and type 40 miles as the radius value.
4. Now select snow stations that fall within the circle. Make sure *snowsite.shp* is active, and click the Select Features Using Graphics button. Those selected snow stations are highlighted.
5. Select Table from the Theme menu. Click Promote to move to the top of the table the highlighted records, that is, records for the selected snow stations.

- **Make charts**

1. While the *snowsite.shp* table with its selected records is still active, click the Create Chart button. This part of Task 1 shows you how you can make charts using attribute data of the selected snow stations from the previous section.
2. In the Chart Properties dialog, do the following: rename the chart as Swe-max, click Swe-max as the Field, click Add to move Swe-max to the Group, change Label Series to Swe-max, and click OK. The chart

is basically a histogram depicting the maximum snow water equivalent of each selected snow station. Make another chart using the field of Elev. Elev lists the elevation of the snow station in feet.

3. To see the relationship between Swe-max and Elev, you can make a scatterplot. Click the Create Chart button to open a new chart. Add both Swe-max and Elev to Groups.

4. Click the *xy* Scatter Chart Gallery button to open options of scatterplots. Select the option in the upper left with the linear scaling of *x* and *y* and click OK. The other options are based on the logarithmic scaling of *x*, or *y*, or both *x* and *y*. The scatterplot shows a positive relationship between Swe-max and Elev: the higher the snow station is, the more the maximum snow water equivalent is expected.

Task 2. Attribute Data Query

What you need: *wp.shp*, a vegetation stand coverage; *wpdata.dbf*, a dBASE file containing stand data for *wp.shp*.

As explained in the text, query or data selection is the most important element of data exploration. Approached from either attribute data or spatial data, results of query are displayed in the linked windows of view, table, and chart. Task 2 focuses on attribute data query.

1. Start ArcView, open a new view, and add *wp .shp* to view. Open the theme table of *wp.shp*.

2. Activate the Project window. Click Tables and Add to open the Add Table dialog. Make sure the file type is dBASE. Double-click on *wpdata.dbf* to select it.

3. At this point, you have opened two tables: the theme table (Attributes of *wp.shp*) and *wpdata.dbf*. To join the data from *wpdata.dbf* to the theme table, do the following: click on Id in *wpdata.dbf* , click on Id in the *wp.shp* theme table, and then click the Join button to join the two tables. Id is the key relating the two tables.

4. Make sure the *wp.shp* theme table includes attribute data from *wpdata.dbf* and is active. Click the Query Builder button to open the Query Builder dialog. The top part of the dialog box shows, from left to right, fields in the attributes of *wp.shp* table, logical operators and Boolean connectors, and values of the selected field. Notice that the name of each field is enclosed in a pair of square brackets. The bottom part of the dialog box has the display area of logical expressions on the left and three buttons for different query methods on the right. The buttons are New Set, Add To Set, and Select From Set. New Set selects a new data subset from the theme table. Add To Set adds a new data subset to the previously selected records. Select From Set selects a new data subset from the previously selected records.

5. Double-click the field of Origin in the Query Builder dialog, click the > operator, and double-click the value of 0. A logical expression, ([Origin] > 0), is now shown in the display area. This is the first logical expression. Click the connector AND, double-click the field of Origin, click the <= operator, and type 1900 to complete the second logical expression. The completed logical expressions should read: ([Origin] > 0) AND ([Origin] <= 1900). Now click on New Set. Records in the theme table that satisfy the logical expressions are highlighted. The upper left corner of the ArcView window shows "175 of 856 selected." Do not dismiss the Query Builder dialog because you will use it again for refining the query operation.

6. The field Origin represents the origin of trees in a stand, expressed in the year trees were planted. The value 0 means that the origin is unknown. Therefore, the logical expressions in step 5 selected stands with trees known to be at least 100 years old. Click the Promote button to bring the selected records to the top of the theme table. Examine the Origin

values of the selected records to see if any of them are after 1900. Now view the map. Highlighted polygons correspond to the selected records.

7. Finally, narrow the selected records by including aspect as an additional criterion. Return to the Query Builder dialog. Drag and highlight the logical expressions between the outer parentheses and delete them. Construct the following logical expressions: ([As] = "N") OR ([As] = "NE") OR ([As] = "NW"). Then click on Select From Set. The number of records selected, as shown in the upper left corner of the ArcView window, is reduced from 175 to 44. The reduced data subset shows only old-growth stands that have the aspects of north, northeast, and northwest. Again, you can verify that the selected records do meet both the origin and aspect criteria. View the map to see where those stands are located.

Task 3. Relational Database Query

What you need: *mosoils.shp*, a soil coverage; *comp.dbf*, *forest.dbf*, and *plantnm.dbf*, three dBASE files from the National Map Unit Interpretation Record (MUIR) database maintained by the Natural Resources Conservation Service (NRCS).

Task 3 lets you work with the MUIR database. By linking the tables in the database properly, you can explore many soil attributes in the database from any table. And, because the tables are linked to the soil map, you can also see where selected records are located.

1. Start ArcView, open a new view, and add *mosoils.shp* to view. Open the theme table of *mosoils.shp*.

2. Next, add the dBASE files to the computer screen. Activate the Project window. Click on Tables and Add. Navigate to the three dBASE files and add them as tables. You should now have four tables and the soil map on the monitor. Arrange them so that you can work with each one of them.

3. The next step is to link the tables. The idea is to keep the four tables separate but dynamically linked rather than joined. To link two tables, you need to know which fields to use as keys. As illustrated in the chapter, musym can link the theme table and *comp.dbf*, muid can link *comp.dbf* and *forest.dbf*, and plantsym can link *forest.dbf* and *plantnm.dbf*. Linking tables is directional, that is, from the source table to the destination table. In data exploration, you want to be able to search soil attributes from any table. Therefore, you need to perform link twice between every two tables. To link the theme table to *comp.dbf*, click on musym in the theme table and musym in *comp.dbf*, and select Link from the Table menu. Then, repeat the same process in the opposite direction: click on musym in *comp.dbf*, musym in the theme table, and select Link from the Table menu. Now, you have completed the two-way linking between the theme table and *comp.dbf*. Do the same between *comp.dbf* and *forest.dbf*, and between *forest.dbf* and *plantnm.dbf*.

4. At this point, the four tables are linked in both directions. The chapter asked a question about what types of plants are found in areas where annual flooding frequency is rated as either frequent or occasional. You can now answer the question by doing the following. Make *comp.dbf* active. Click on the Query Builder button. In the Query Builder dialog, prepare the query expression as ([Anflood] = "FREQ") OR ([Anflood] = "OCCAS"), and click New Set. Records in *comp.dbf* that meet the criteria are highlighted as the corresponding records in the other three tables and the corresponding polygons in the soil map. You can now see where areas with frequent or occasional flooding are located, the forest types in these areas, and the common names of plant species.

5. You can try another query with the tables. You probably want to first clear the selected records by clicking the Select None button. Now, make *plantnm.dbf* active and click the

Query Builder button. Prepare the query statement as ([Comname] = "lupine"). Click New Set. The selected record in *plantnm.dbf* and its corresponding records in the other tables are highlighted. You can also see where lupine can be found on the map.

Task 4. Combining Spatial and Attribute Data Queries

What you need: *thermal.shp*, a coverage with 899 thermal wells and springs; *idroads.shp*, showing major roads in Idaho.

Task 4 assumes that you are asked by a company to locate potential sites for a hot spring resort in Idaho. You are given two criteria for selecting potential sites:

- The hot spring must be within 2 miles of a major road.
- The temperature of the hot spring must be greater than 60˚C.

The field type in *thermal.shp* uses *s* to denote springs and *w* to denote wells. The field temp shows the water temperature in ˚C.

1. Start ArcView, open a new view, and add *thermal.shp* and *idroads.shp* to view. Select Properties from the View menu. In the View Properties dialog, set the Map Units as meters and the Distance Units as miles. Both *thermal.shp* and *idroads.shp* have meters as the map units.
2. Activate *thermal.shp* theme. Click Select By Theme from the Theme menu. In the dialog, set the query statement to read: "Select features of active themes that Are Within Distance Of the selected features of *idroads.shp*," and the Selection Distance to be 2 miles. Click New Set. Those highlighted thermal springs and wells are within 2 miles of major roads in Idaho.
3. Next, narrow the selection of map features by using the second criterion. Select Tables from the Theme menu. Use Promote to move the selected records to the top of the table. Click

on the Query Builder button. Prepare the query expression as: ([Type] = "s") AND ([Temp] > 60). Because you want to select from the previously selected records, click on Select From Set. Dismiss the Query Builder dialog.

4. Again, use Promote to move the selected records to the top. The Type value of the 15 selected records should be all *s* for springs, and the Temp value should be all above 60. In fact, one of the selected records has the name of "Zim's Resort," a hot spring that has already been developed into a resort area. The map shows you where those 15 hot springs are located.
5. As explained in the chapter, this task can also be solved by first selecting hot springs with water temperatures above 60˚C through attribute data query and then selecting those springs that are within 2 miles of major roads through spatial data query. The final answer should be the same.

Task 5. Raster Data Query

What you need: *slope_gd*, a slope grid; and *aspect_gd*, an aspect grid.

Task 5 shows you different methods for querying a single grid or multiple grids.

1. Start ArcView, and load Spatial Analyst. Open a new view, and add *slope_gd* and *aspect_gd* to view. *Slope_gd* has the following slope classes in degree: 1 (0–10), 2 (10–20), 3 (20–30), and 4 (30–40). *Aspect_gd* has the following aspect classes: 1 (flat), 2 (north), 3 (east), 4 (south), and 5 (west).
2. Select Properties from the View menu, and set the Map Units to meters and the Distance Units to kilometers. First query *slope_gd* using the graphic method. Click the Draw Circle tool. Then click a point in *slope_gd* and drag the cursor to make a circle with a radius of 1.5 kilometers. If you cannot get exactly 1.5 kilometers interactively, select Size/Position from the Graphics menu and type 1.5 kilometers as the radius value.

3. Click on the Histogram button. The histogram shows the cell values and their frequencies within the circular area. You can also click on the Identify tool and then the bar graph of a cell value to find its exact frequency (count). To remove the circle, make it active and select Delete Graphics from the Theme menu.

4. Map Query is the tool for querying a grid by its cell values. Select Map Query from the Analysis menu. In the *Map Query 1* dialog, set the query expression as: ([Slope_gd] = 2.AsGrid). The request of AsGrid is automatically added to the expression. Click Evaluate. *Map Query 1* in the Table of Contents shows areas with slopes between 10 and 20 degrees as True.

5. Map Query can also query both *slope_gd* and *aspect_gd* to find areas having slopes between 10 and 20 degrees and the south aspect. Select Map Query from the Analysis menu. In the *Map Query 2* dialog, set the query expression as: ([Slope_gd] = 2. AsGrid) AND ([Aspect_gd] = 4.AsGrid). Click Evaluate. *Map Query 2* shows areas that satisfy the logical expression.

6. To save the result of a map query, first activate the output to be saved and then select Save Data Set from the Theme menu. In the next dialog, specify the name and the path for saving the data set.

REFERENCES

Anselin, L. 1999. Interactive Techniques and Exploratory Spatial Data Analysis. In P.A. Longley, M.F. Goodchild, D.J. MaGuire, and D.W. Rhind (eds.). *Geographical Information Systems*, 2ᵈ ed. New York: John Wiley & Sons, pp. 253–66.

Batty, M., and Y. Xie. 1994. Modelling Inside GIS: Part I. Model Structures, Exploratory Spatial Data Analysis and Aggregation. *International Journal of Geographical Information Systems* 8: 291–307.

Becker, R.A., and W.S. Cleveland. 1987. Brushing Scatterplots. *Technometrics* 29: 127–42.

Cleveland, W.S. 1993. *Visualizing Data*. Summit, NJ: Hobart Press.

Cleveland, W.S., and M.E. McGill, eds. 1988. *Dynamic Graphics for Statistics*. Belmont, CA: Wadsworth.

Cressie, N.A.C. 1993. *Statistics for Spatial Data*, revised edition. New York: John Wiley & Sons.

DiBiase, D., A.M. MacEachren, J.B. Krygier, and C. Reeves. 1992. Animation and The Role of Map Design in Scientific Visualization. *Cartography and Geographic Information Systems* 19: 201–14, 265–66.

Haslett, J., G. Wills, and A. Unwin. 1990. SPIDER— An Interactive Statistical Tool for The Analysis of Spatially Distributed Data. *International Journal of Geographical Information Systems* 3: 285–96.

MacEachren, A.M. 1995. *How Maps Work: Representation, Visualization, and Design*. New York: The Guilford Press.

Meyer, M.A., F.R. Broome, and R.H.J. Schweitzer. 1975. Color Statistical Mapping by the U.S. Bureau of the Census. *American Cartographer* 2: 100–17.

Monmonier, M. 1989. Geographic Brushing: Enhancing Exploratory Analysis of the Scatterplot Matrix. *Geographical Analysis* 21: 81–84.

Olson, J. 1981. Spectrally Encoded Two-Variable Maps. *Annals, Association of American Geographers* 71: 259–76.

Peterson, M.P. 1995. *Interactive and Animated Cartography*. Englewood Cliffs, NJ: Prentice Hall.

Scott, L.M. 1994. Identification of GIS Attribute Error Using Exploratory Data Analysis. *The Professional Geographer* 46: 378–86.

Tomlin, C.D. 1990. *Geographic Information Systems and Cartographic Modeling*. Englewood Cliffs, N.J.: Prentice Hall.

Tufte, E.R. 1983. *The Visual Display of Quantitative Information*. Cheshire, CT: Graphics Press.

Tukey, J.W. 1977. *Exploratory Data Analysis*. Reading, Mass: Addison-Wesley.

Walker, P.A., and D.M. Moore. 1988. SIMPLE—An Inductive Modeling and Mapping Tool for Spatially-Oriented Data. *International Journal of Geographical Information Systems* 2: 347–63.

Weber, C.R., and B.P. Buttenfield. 1993. A Cartographic Animation of Average Yearly Surface Temperatures for the 48 Contiguous United States; 1897–1986. *Cartography and Geographic Information Systems* 20: 141–50.

10

VECTOR DATA ANALYSIS

10.1 INTRODUCTION

The scope of GIS analysis varies among disciplines that use GIS. GIS users in hydrology will likely emphasize the importance of topographic analysis and hydrologic modeling, while GIS users in wildlife management will be more interested in analytical functions dealing with wildlife point locations and their relationship to the environment. This is why GIS companies have taken two general approaches in packaging their products. One is to prepare a set of analytical tools used by most GIS users, and the other is to prepare modules or extensions designed for specific applications such as hydrologic modeling. This chapter, the first of several chapters on data analysis, covers basic analytical tools for vector data analysis.

The vector data model uses points and their x-, y-coordinates to construct spatial features of points, lines, and polygons. Therefore, vector data analysis is based on the geometric objects of point, line, and polygon and the accuracy of analysis results depends on the accuracy of these objects in terms of location and shape. Vector data may be topology-based or non-topological. Therefore, topology can also be a factor for some vector data analysis.

This chapter is grouped into four sections. Section 1 covers buffering and its applications. Section 2 discusses map overlay, types of overlay, and problems with overlay. Tools for measuring distances between points and between points and lines are included in Section 3. Section 4 includes tools for map manipulation. Whenever necessary, ArcView and ARC/INFO are used as examples.

10.2 BUFFERING

Based on the concept of proximity, **buffering** separates a map into two areas: one area that is within a specified distance of selected map features and the other area that is beyond. The area that is within the specified distance is called the buffer zone.

Selected map features for buffering may be points, lines, or areas (Figure 10.1). Buffering around points creates circular buffer zones. Buffer-

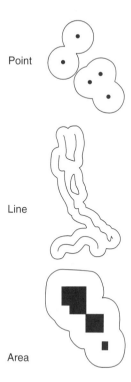

Figure 10.1
Buffering around points, lines, and areas.

ing around lines creates a series of elongated buffer zones. Buffering around polygons creates buffer zones extending outward or from the polygon boundaries.

There are several variations from Figure 10.1 in buffering. The buffer distance does not have to be constant; it can vary according to the values of a given field. For example, streams may be buffered with 200 meters along major streams and 100 meters along tributaries (Figure 10.2). A map feature may have more than one buffer zone. For example, a nuclear power plant may be buffered with distances of 5, 10, 15, and 20 miles, thus forming multiple rings around the plant (Figure 10.3). Buffering around line features does not have to be on both sides of the lines; it can be on either the left side or the right side of the line given that line topology is present. Boundaries of buffer zones may remain intact so that each buffer zone is a separate polygon. Or these boundaries may be dissolved so that there are no overlapped areas between buffer zones (Figure 10.4).

Regardless of its variations, buffering uses distance measurements from map features to create the buffer zones. The GIS user must therefore know the measurement unit of map features, for example, meters or feet, and, if necessary, input that information prior to buffering. ArcView, for example, can use the map unit as the default for the distance unit or allow the user to select a different distance unit, for example, miles instead of feet. Because buffering uses distance measurements from map features, the positional accuracy of map features determines the accuracy of buffer zones.

10.2.1 Applications of Buffering

Buffering creates a buffer zone map, which sets the buffering operation apart from spatial data query. Spatial data query can select map features that are located within a certain distance of other features but cannot create a buffer zone map. A buffer zone is often treated as a protection zone and is used for planning or regulation. Many examples can be cited as follows:

Figure 10.2
Buffering with different buffer distances.

- A city ordinance may stipulate that no liquor stores or pornographic shops shall be within 1000 feet of a school or a church.
- Government regulations may stipulate that logging operations must be at least 2 miles away from any stream to minimize the sedimentation problem and set the 2-mile buffer zones of streams as the exclusion zones.
- A national forest may restrict oil and gas well drilling within 500 feet of roads or highways; within 200 feet of trails; within 500 feet of streams, lakes, ponds, or reservoirs; or within 400 feet of springs.
- Urban planning may set aside land along the edges of streams to reduce the effects of

nutrient, sediment, and pesticide runoff; to maintain shade to prevent the rise of stream temperature; and to provide shelter for wildlife and aquatic life (Thibault 1997).

A buffer zone may be treated as a neutral zone and as a tool for conflict resolution. In controlling the protesting mass, police may require protesters to be at least 300 feet from a building. Perhaps the best-known neutral zone is the demilitarized zone separating North Korea from South Korea along the 38° N parallel.

Sometimes buffer zones may represent the inclusion zones in GIS applications. For example, the siting criteria for an industrial park may stipulate that a potential site must be within 1 mile of a heavy

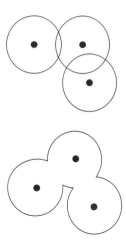

Figure 10.4
Buffer zones not dissolved (top) or dissolved (bottom).

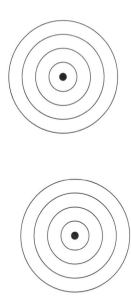

Figure 10.3
Buffering with four rings.

duty road. In this case, the 1-mile buffer zones of all heavy duty roads become the inclusion zones.

Rather than used simply as a screening device, buffer zones themselves may become the object for analysis. A forest management plan may define areas that are within 200 feet of streams as riparian zones. Under the plan, riparian zones are separate from non-riparian areas and are managed differently. Another example comes from urban planning in developing countries, where urban expansion typically occurs near existing urban areas and major roads. Management of future urban growth should therefore concentrate on the buffer zones of existing urban areas and major roads.

Buffering with multiple rings can be useful as a sampling method. Schutt et al. (1999), for example, buffered stream networks in 10-meter increments to a distance of 300 meters so that they could analyze the composition and pattern of woody vegetation as a function of distance from the stream network. One can also apply incremental banding to other studies such as land use change around urban areas.

10.3 MAP OVERLAY

Map overlay combines the geometry and attributes of two feature maps to create the output (Figure 10.5). One of the two maps is called the input map and the other the overlay map. The geometry, or the spatial data, of the output represents the geometric intersection of map features from the input and overlay maps. Therefore, the number of map features on the output map is not the sum of map features on the input and overlay maps but is usually much larger than the sum. Each map feature on the output contains a combination of attributes from the input and overlay maps, and this combination differs from its neighbors.

Feature maps to be overlaid must be spatially registered, that is, based on the same coordinate system. In the case of the UTM coordinate system or the State Plane Coordinate system, they must also be based on the same zone. As GIS users are migrating from NAD27 to NAD83, they must also check to ensure that the datum is the same between overlay maps.

Map overlay can only work with two polygon maps at a time. Therefore, if three polygon maps are to be overlaid, the operation begins with the first two and then uses the output from them with the third. The process, especially the tracking of the intermediate outputs, can be tedious. The problem of using only two maps in an overlay operation disappears with the regions data model, which allows overlaying more than two layers in a single operation (Chapter 15).

10.3.1 Feature Type and Map Overlay

In practice, the first consideration of map overlay is feature type. The input map may contain points, lines, or polygons; the overlay map must be a polygon map; and the output has the same feature type as the input. Map overlay can therefore be grouped by feature type into point-in-polygon, line-in-polygon, and polygon-on-polygon.

In a **point-in-polygon** operation, the same point features in the input are included in the output but each point is assigned with attributes of the polygon within which it falls (Figure 10.6). An example of the point-in-polygon overlay is to find the association between wildlife locations and vegetation types.

In a **line-in-polygon** operation, the output contains the same line features as in the input but each of them is dissected by the polygon boundaries on the overlay map (Figure 10.7). The output, therefore, has more arcs than does the input. Each arc on the output combines attributes from the line map and the polygon within which it falls. An example of the line-in-polygon overlay is to find soil data for a proposed road. The input map includes the proposed road. The overlay map is a soil map.

And the output shows a dissected proposed road, each road segment corresponding to a different set of soil data from its adjacent segments.

The most common overlay operation is **polygon-on-polygon,** involving two polygon maps. The output combines the polygon boundaries from the input and overlay maps to create a new set of polygons (Figure 10.8). Each new polygon carries attributes from both maps, and these attributes differ from those of adjacent polygons. An example of the

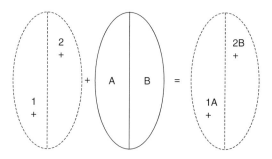

Figure 10.6
Point in polygon overlay. The input is a point theme (the dashed lines are for illustration only and are not part of the point theme). The output is also a point theme, which has attribute data from the overlay polygon theme.

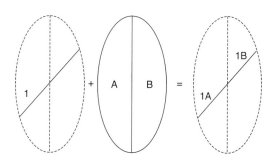

Figure 10.7
Line in polygon overlay. The input is a line theme (the dashed lines are for illustration only and are not part of the line theme). The output is also a line theme. But the output differs from the input in two aspects: the line is broken into two segments, and the line segments have attribute data from the overlay polygon theme.

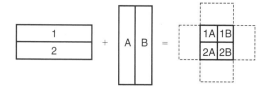

Figure 10.5
Map overlay combines the geometry and attribute data from two maps into a single map. Intersect is the map overlay method in this illustration. The output from intersect include areas that are common to both input maps. The dashed lines are not included in the output.

polygon-on-polygon overlay is to analyze the association between elevation zones and vegetation types.

10.3.2 Map Overlay Methods

If the input map has the same area extent as the overlay map, then that area extent also applies to the output. But, if the input map has a different area extent than the overlay map, then the area extent of the output may vary depending on the overlay method used. The three common map overlay methods are called UNION, INTERSECT, and IDENTITY (Box 10.1).

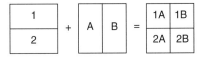

Figure 10.8
Polygon-on-polygon overlay. In the illustration, the two themes to be overlaid have the same area extent. The output combines the geometry and attribute data from the two themes into a single polygon theme.

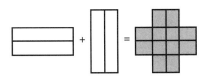

Figure 10.9
The UNION method keeps all areas of the two input themes in the output.

UNION preserves all map features from the input and overlay maps by combining the area extents from both maps (Figure 10.9). It is the Boolean operation that uses the keyword OR, that is, (input map) OR (overlay map). The output therefore corresponds to the area extent of the input map, or the overlay map, or both. UNION requires both the input and overlay maps be polygon maps.

INTERSECT preserves only those features that fall within the area extent common to both the input and overlay maps (Figure 10.10). INTERSECT is the Boolean operation that uses the keyword AND, that is, (input map) AND (overlay map). The output must correspond to the area extent of both the input and overlay maps. INTERSECT is often a preferred method of overlay because any map feature on its output has attribute data from both of its inputs. For example, a forest management plan may call for an inventory of vegetation types within riparian zones. INTERSECT will be a more efficient overlay method than UNION in this case because the output contains

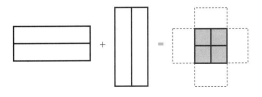

Figure 10.10
The INTERSECT method preserves only the area common to the two input themes in the output. The dashed lines are for illustration only; they are not part of the output.

Box 10.1   Methods of Map Overlay

Different GIS packages offer different map overlay methods, although the methods are all based on the Boolean operations. ArcView offers UNION and INTERSECT. ARC/INFO provides UNION, INTERSECT, and IDENTITY. MGE has the overlay methods of UNION, INTERSECT, MINUS, and DIFFERENCE. MINUS is functionally similar to IDENTITY, and DIFFERENCE uses the Boolean operation of XOR.

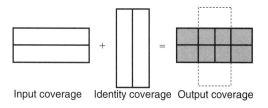

Input coverage   Identity coverage   Output coverage

Figure 10.11
The IDENTITY method produces an output that has the same extent as the input coverage. But the output includes the geometry and attribute data from the identity coverage.

From\To	Point	Line	Polygon
Point	nearest	nearest	inside
Line	nearest	part of	inside
Polygon	n/a	n/a	inside

Figure 10.12
The Assign Data by Location method in ArcView involves two themes: the theme to assign data from, and the theme to assign data to. Each theme can be a point, line, or polygon theme. Data assignment is based on the spatial relationship of nearest, inside, or part of. Two cells (from point to polygon and from line to polygon) are not applicable.

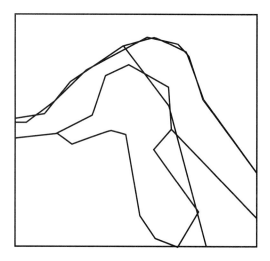

Figure 10.13
The top boundary shows a series of slivers. The slivers are formed between the coastline from two maps used in an overlay operation. If the coastlines register perfectly between the two coverages, then slivers will not be present.

only riparian zones and vegetation types within the zones. The input map for INTERSECT may contain points, lines, or polygons.

IDENTITY preserves only map features that fall within the area extent defined by the input map (Figure 10.11). Expressed in a Boolean expression, IDENTITY represents the operation: [(input map) AND (overlay map)] OR (input map). Features of the overlay map that fall outside the area extent of the input map are left out of the output. The input map for IDENTITY may contain points, lines, or polygons.

ArcView offers the overlay methods of UNION and INTERSECT (Box 10.2). The methods are set up the same way as in ARC/INFO except that the input map for INTERSECT may only be a line or polygon map. ArcView users can overlay a point map with a polygon map using the Assign Data By Location method, also called Spatial Join (Figure 10.12). The point map is the map to assign data to, and the polygon map is the map to assign data from. The Spatial Join operation joins attributes from the polygon map to those of the point map by the spatial correspondence between points and polygons. If, for example, a point corresponds to a polygon with an attribute value of agriculture, then agriculture will be joined with the point's record.

10.3.3 Slivers

A common error from overlaying two polygon maps is **slivers,** very small polygons along correlated or shared boundary lines of the two input maps (Figure 10.13). The existence of slivers often results from digitizing errors. Because of the high precision of manual digitizing or scanning, the shared boundaries on the input maps are not right on top of one another. When the two maps are overlaid, the digitized boundaries intersect to

Union and INTERSECT are functionally the same between ArcView and ARC/INFO. But the map overlay operations differ between the two GIS packages in two ways. ArcView uses shapefiles, so it is a good idea to convert ARC/INFO coverages into shapefiles before running map overlay in ArcView. Second, ArcView does not automatically update the area and perimeter values of polygons created by a map overlay operation. This shortcoming can be frustrating to ArcView users who need to know, for example, the total area of some selected polygons from the output. The ArcView Help document provides a sample script (calcapl.ave) that can calculate feature geometry values, including area and perimeter values. One of the tasks in the Applications section uses calcapl.ave on an overlay output.

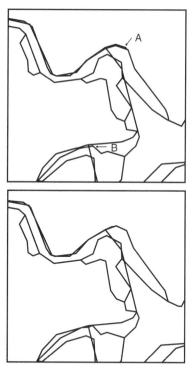

Figure 10.14
Points (and lines) are snapped together if they fall within the specified fuzzy tolerance. Many slivers along the top boundary (A) are removed by use of a fuzzy tolerance. The fuzzy tolerance can also snap arcs that are not slivers (B).

form slivers. Other causes of slivers include errors in the source map or errors in interpretation. Polygon boundaries on soil and vegetation maps are usually interpreted from field survey data, aerial photographs, and satellite images. A wrong interpretation can create erroneous polygon boundaries.

Most GIS packages incorporate the fuzzy tolerance in map overlay operations to remove slivers. The **fuzzy tolerance** forces points that make up the lines to be snapped together if the points fall within the specified distance (Figure 10.14). The fuzzy tolerance is either defined by the user or based on the default value in the GIS package. Slivers that remain on the output of a map overlay operation are those that were not removed by the built-in fuzzy tolerance.

One way to eliminate these slivers is to increase the fuzzy tolerance value. But because the fuzzy tolerance applies to the entire map, a large tolerance value will snap shared boundaries as well as lines that are not shared on the input maps, thus creating distorted map features on the overlay output.

A better option for removing slivers is to apply the concept of minimum mapping unit. The **minimum mapping unit** represents the smallest area unit that will be managed by a government agency or an organization. For example, if a national forest adopts 5 acres as its minimum mapping unit, then any slivers smaller than 5 acres can be removed by combining them with adjacent polygons.

10.3.4 Error Propagation in Map Overlay

Slivers are examples of errors in digital maps that can propagate to the output of a map overlay operation. The generation of errors due to the inaccuracies of the input maps is called **error propagation.** Error propagation usually involves two types of errors: positional and identification errors (MacDougall 1975; Chrisman 1987). Slivers represent positional errors in map overlay if they are resulted from the inaccuracies of boundaries due to digitizing or interpretation errors. Error propagation can also be caused by the inaccuracies of attribute data, or identification errors, such as the incorrect coding of polygon values. Every map overlay product will have some combinations of positional and identification errors.

How serious can error propagation in map overlay be? The answer depends on the number of input maps and the spatial distribution of errors in the input maps. The accuracy of a composite map tends to decrease as the number of input maps increases and the spatial correspondence or coincidence of errors in the input maps decreases.

An error propagation model proposed by Newcomer and Szajgin (1984) considers the spatial coincidence of errors by computing a conditional probability. Assuming that two input maps have square polygons, the model calculates the probability of the event that both maps are correct on the composite map, or $Pr(E_1 \cap E_2)$, using the equation

$$Pr(E_1 \cap E_2) = Pr(E_1)Pr(E_2|E_1) \qquad (10.1)$$

where $Pr(E_1)$ is the probability of the event that the first input map is correct, and $Pr(E_2|E_1)$ is the probability of the event that the second input map is correct given that the first input map is correct. This model suggests that the highest accuracy that can be expected of a composite map is equal to the accuracy of the least accurate individual input map, and the lowest accuracy is equal to

$$1 - \sum_{i=1}^{n} Pr(E_i') \qquad (10.2)$$

where n is the number of input maps in map overlay and $Pr(E_i')$ is the probability that the input map i is incorrect. In other words, the lowest accuracy of the composite map is 1 minus the sum of percent inaccuracies from all input maps.

For example, if a map overlay operation is conducted with three input maps. The accuracy levels of these maps are known to be 0.9, 0.8, and 0.7 respectively. According to Newcomer and Szajgin's model, the accuracy of the composite map is between 0.4 (1 − 0.6) and 0.7.

Newcomer and Szajgin's model is a simple model, quite different from the real world operations. First, the model is based on square polygons, a much simpler geometry than real maps used in map overlay. Second, the model applies only to the Boolean operation of AND in map overlay, for example, input map 1 is correct AND input map 2 is correct. The Boolean operation of OR is different because it requires only that one of the input maps is correct, for example, input map 1 is correct OR input map 2 is correct. Therefore, the probability of the event that the composite map is correct actually increases as more input maps are overlaid (Veregin 1995). Third, Newcomer and Szajgin's model applies only to binary data, meaning that an input map is either correct or incorrect. The model does not work with interval or ratio data and cannot measure the magnitude of errors. Modeling error propagation with numeric data is more difficult than binary data (Arbia et al. 1998; Heuvelink 1998).

10.4 DISTANCE MEASUREMENT

Distance measurement refers to measuring straight-line distances between points, or between points and their corresponding nearest points or lines. Distances can be used directly in data analysis. For example, distance measures were used to test whether deer relocation points were closer to an old-growth/clear-cut edge than random points located within deer's relocation area (Chang et al. 1995). Distances can also be used as input to data analysis. For example, the Gravity Model, a spatial interaction model commonly used in migration studies and business applications, uses distance measures between points as the input.

ArcView covers distance measurement under the Assign Data By Location method. The map to assign data to must be a point map. The map to assign data from may be a point or line map. And the spatial relationship between the two maps is nearest. The Assign Data By Location method joins attributes of the closest point or line to the appropriate record of the point attribute table (Figure 10.12). In addition, a new field called distance, showing the distance measurement, is added to the table.

NEAR and POINTDISTANCE are two ARC/INFO commands that measure distances. NEAR calculates distances between points on a point coverage and their nearest points or lines on another coverage. POINTDISTANCE measures distances for each point on a point coverage to all points on another coverage. POINTDISTANCE does not use the nearest relationship.

10.5 MAP MANIPULATION

Many GIS packages provide tools for manipulating and managing maps in a database. Map manipulation is often part of data analysis or is needed for data preprocessing. Most tools covered in this section involve two maps and are therefore classified as map overlay methods by some GIS users. This chapter, however, limits map overlay to only those methods that can combine spatial and attribute data from two maps into a single map, thus ruling out tools for map manipulation as map overlay methods. Because ArcView and ARC/INFO offer different sets of tools for map manipulation, this section covers the two packages separately.

(a)	(b)

Figure 10.15
Dissolve removes boundaries of polygons that have the same attribute value and create a simplified map (b).

10.5.1 Map Manipulation in ArcView

The GeoProcessing extension to ArcView has Dissolve, Clip, and Merge as map manipulation tools. Dissolve removes boundaries between polygons that have the same value of a selected attribute (Figure 10.15). The main purpose of Dissolve is simplification. Probably the most common use of Dissolve comes after attribute data classification. Classification groups values of a selected attribute into classes and makes obsolete boundaries of adjacent polygons, which have different values initially but are now grouped into the same class. Dissolve removes these unnecessary boundaries and creates a new, simpler map with the classification results as its attribute values.

Clip creates a new map that includes only those features of the input map that falls within the area extent of the clip map (Figure 10.16). Clip is a useful tool, for example, for cutting a map that was acquired elsewhere to fit a study area. The input may be a point, line, or polygon map, but the clip map must be a polygon map.

Merge creates a new map by piecing together two or more maps (Figure 10.17). For example, Merge can put together a map from four input maps, each corresponding to the area extent of a USGS 7.5-minute quad. The output can then be used as a single map for data query or display. But the boundaries separating the input maps still remain on the output and divide a feature into separate features if the feature crosses the boundary.

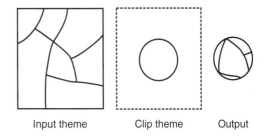

Input theme	Clip theme	Output

Figure 10.16
Clip creates an output that contains only those features of the input map that fall within the area extent of the clip map. The dashed lines are for illustration only; they are not part of the clip map.

Figure 10.17
Merge pieces together two adjacent maps into a single map. Merge does not remove the shared boundary between the maps.

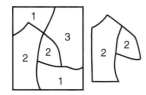

Figure 10.18
RESELECT creates a new coverage with selected map features from the input coverage.

Merge in ArcView is therefore different from edge-matching in ARC/INFO, which can create a seamless coverage.

10.5.2 Map Coverage Manipulation in ARC/INFO

ARC/INFO has more tools for map coverage manipulation than ArcView. Besides DISSOLVE, CLIP, and MERGE, ARC/INFO offers RESELECT, ELIMINATE, UPDATE, ERASE, and SPLIT.

RESELECT creates a new coverage with map features selected from a user-defined logical expression (Figure 10.18). For example, a high canopy closure coverage can be created by selecting stands that have 60 to 80% closure from a vegetation coverage.

ELIMINATE creates a new coverage by removing from the input coverage map features that meet a user-defined logical expression (Figure 10.19). For example, polygons that are smaller than the user-defined minimum mapping unit in a coverage can be eliminated by merging them with other polygons. ELIMINATE is a command that can implement the minimum mapping unit concept.

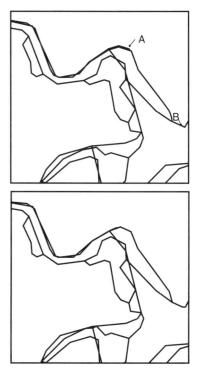

Figure 10.19
ELIMINATE can remove polygons that are smaller than a specified size. Slivers along the top boundary (A) are therefore eliminated. This illustration uses the KEEPEDGE option, which preserves the edge at (B) even though polygons making up the edge are smaller than the specified size.

UPDATE uses a "cut and paste" operation to replace the input coverage by the update coverage and its map features (Figure 10.20). As the name suggests, UPDATE is useful for updating an existing coverage with new map features in limited areas of the coverage. It saves the GIS user from re-digitizing the coverage.

ERASE removes from the input coverage those map features that fall within the area extent of a coverage named as the erase coverage (Figure 10.21). Suppose a suitability analysis stipulates potential sites must be at least 300 meters from any stream. A stream buffer coverage can be used in this case as the erase coverage to remove itself from further consideration.

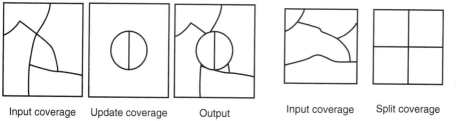

Input coverage   Update coverage   Output

Figure 10.20
UPDATE replaces the input coverage with the update coverage and its map features.

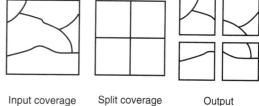

Input coverage   Split coverage   Output

Figure 10.22
SPLIT uses the geometry of the split coverage to divide the input coverage into four separate coverages.

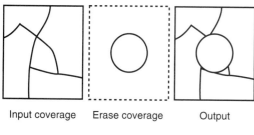

Input coverage   Erase coverage   Output

Figure 10.21
ERASE removes from the input coverage map features that fall within the area extent of the erase coverage. The dashed lines are for illustration only; they are not part of the erase coverage.

SPLIT is opposite to MERGE. Rather than piecing together a single coverage, SPLIT divides the input coverage into two or more coverages (Figure 10.22). A split coverage, that is, a coverage showing area subunits, is used as the template for dividing the input coverage. For example, a national forest can use SPLIT to divide a vegetation coverage by district so that each district office can have its own vegetation coverage.

KEY CONCEPTS AND TERMS

Buffering: A GIS operation in which areas that are within a specified distance of selected map features are separated from areas that are beyond.

Error propagation: The generation of errors in the map overlay output due to inaccuracies of the input maps.

Fuzzy tolerance: A distance tolerance used in a GIS package to force points and lines to be snapped together if they fall within the specified distance.

Identity: An overlay method that preserves only features that fall within the area extent defined by the input map.

Intersect: An overlay method that preserves only those features that fall within the area extent common to the input and overlay maps.

Line-in-polygon overlay: A GIS operation, in which a line map is dissected by the polygon boundaries on the overlay map, and each arc on the output combines attributes from the line map and the polygon within which it falls.

Map overlay: A GIS operation that combines the geometry and attributes of two digital maps to create the output.

Minimum mapping unit: The smallest area unit that is managed by a government agency or an organization.

Point-in-polygon overlay: A GIS operation, in which each point of a point map is assigned attribute data of the polygon within which it falls.

Polygon-on-polygon overlay: A GIS operation, in which the output combines the polygon boundaries from the input and overlay maps to create a new set of polygons and each new polygon carries attributes from both maps.

Slivers: Very small polygons found along the shared boundary of the two input maps in map overlay. Slivers often result from digitizing errors.

Union: A polygon-on-polygon overlay method that preserves all features from the input and overlay maps.

APPLICATIONS: VECTOR DATA ANALYSIS

This applications section covers two tasks. Task 1 covers the basic functions of vector data analysis such as buffering, map overlay, attribute data manipulation, and dissolve. Task 1 also uses an Avenue script to update the area and perimeter values of the composite map from overlay. Task 2 covers distance measurement.

Task 1: Buffering, Overlay, and Use of Avenue Script

What you need: shapefiles of *landuse*, *soils*, and *sewers*.

Task 1 simulates GIS analysis for a real-world project and introduces common vector-based analyses such as buffer, map overlay, dissolve, and tabular data manipulation as well as use of a sample Avenue script. The task is to find a suitable site for a new university aquaculture lab using the following selection criteria:

- Preferred land use is brushland (i.e., lucode = 300 in *landuse.shp*)
- Choose soil types suitable for development (i.e., suit >= 2 in *soils.shp*)
- Site must be within 300 meters of sewer lines

1. Start ArcView and check the GeoProcessing extension.
2. Add *sewers.shp*, *landuse.shp*, and *soils.shp* to view.
3. Select Properties from the View menu. In the View Properties dialog, set the Map Units and Distance Units as meters.

4. The first operation is to buffer *sewers.shp* with a buffer distance of 300 meters. Make *sewers.shp* active, and select Create Buffers from the Theme menu. In the Create Buffers dialog, make sure you want to buffer the features of *sewers.shp*. Click Next. Set a specified distance of 300 meters. Click Next. Click Yes to dissolve barriers between buffers. Save the buffers in a new theme and call the new theme *sewerbuf.shp*. Click Finish. *Sewerbuf.shp* appears in the Table of Contents as Buffer1 of *sewers.shp*. Select Properties from the Theme menu, and rename Buffer1 of *sewers.shp sewerbuf.shp*.

5. The next operation is to overlay *soils.shp* and *landuse.shp*. Select GeoProcessing Wizard from the View menu. In the GeoProcessing dialog, check "Union two themes." Click Next. In the following dialog, select *soils.shp* as the input theme to union, select *landuse.shp* as the polygon overlay theme to union, and specify the output as *landsoil.shp*. Click Finish.

6. Now you need to overlay *landsoil.shp* and *sewerbuf.shp*. (Remember you can overlay only two themes at a time.) Choose "Intersect two themes" as the geoprocessing operation. Select *sewerbuf.shp* as the input theme to intersect, select *landsoil.shp* as the overlay theme, and specify the intersect output as *finalcov.shp*. Click Finish.

7. *Finalcov.shp* has all the attribute data for the suitability analysis. What follows is the query

operation. Open the theme table of *finalcov.shp*. Start editing the table. Add a new field called suitable (Number, 2, 0) to the table. Use the Query Builder to select polygons that meet the criteria: Lucode = 300 and suit >= 2. Click the Calculate button to display the Field Calculator dialog. Click inside the expression box in the lower left of the dialog and type 1 (so that the completed expression reads suitable = 1). Click OK.

8. Click the Promote button to bring polygons that meet the selection criteria to the top of the table. Notice that their suitable values are all 1s.

9. Do one more thing with the *finalcov.shp* theme table: delete one set of area and perimeter from the table. The table contains two sets of area and perimeter from the overlay operation. Highlight the first area field and press delete. Then highlight the first perimeter field and press delete. Now you can save the edits you have made to the table.

10. Because shapefiles do not use topology, the area values in the *finalcov.shp* theme table have not been updated after the map overlay operations. To calculate the correct area values, use an Avenue script available in the ArcView Help document. Navigate to "Sample scripts and extensions\Sample scripts\Views\Data conversion/alteration" and click "Calculates feature geometry values." At the top of the help topic, click on Source Code to open a window with the source code in it. Choose Copy from the Options menu in the source code window.

11. In the Project window, click Scripts and New to open a script window. Then click the Paste button. This action copies the sample script for calculating feature geometry values to the script window. Click the Compile button to compile the script.

12. Activate the View and then activate the script window. (The sample script requires the view document to be the active document before running the script.) Click the Run button to run the sample script. Answer yes to

questions on updating area and perimeter. Return to the *finalcov.shp* theme table. The area values have now been updated. To get the total area of the potential sites, do the following. First, use the Query Builder to select records from the *finalcov.shp* theme table with suitable = 1. Second, press the Area field to highlight it in the theme table. Third, select Statistics from the Field menu. The Statistics dialog shows the sum and other statistics about the Area field for the selected records.

13. *Finalcov.shp* includes polygons with the suitable values of 1 and 0. Use Dissolve on the theme to create a new theme containing only the potential sites. Select GeoProcessing Wizard from the View menu again. Choose the Dissolve operation this time. Click Next. In the following dialog, select *finalcov.shp* as the theme to dissolve, select suitable as the attribute to dissolve, and specify the dissolve output as *finaldis.shp*. No more fields or operations are needed for the output. *Finaldis.shp* has only the polygons selected as the potential sites.

Task 2: Distance Measurement

What you need: *deer.shp* and *edge.shp*.

Task 2 asks you to use the Assign Data by Location method to measure each deer location in *deer.shp* to its closest old-growth/clear-cut edge in *edge.shp*.

1. Start ArcView and load Geoprocessing. Open a new view, and add *deer.shp* and *edge.shp* to view. Select Properties from the View menu, and specify the Map Units as meters.

2. Select Geoprocessing Wizard from the View menu. In the next dialog, click the radio button next to Assign Data by Location. Click Next. Then, specify *deer.shp* as the theme to assign data to and *edge.shp* as the theme to assign data from. Click Finish.

3. Activate *deer.shp*, and open its attribute table. Notice Distance and other fields have been added to the theme table. Click the Query

Builder button. Prepare the query expression as: [Distance] <= 50. Those deer locations that are within 50 meters of their closest edge are highlighted in the theme table as well as in the view.

REFERENCES

Arbia, G., D.A. Griffith, and R.P. Haining. 1998. Error Propagation Modeling in Raster GIS: Overlay Operations. *International Journal of Geographical Information Science* 12: 145–67.

Chang, K., D. L. Verbyla, and J.J. Yeo. 1995. Spatial Analysis of Habitat Selection by Sitka Black-Tailed Deer in Southeast Alaska, USA. *Environmental Management* 19: 579–89.

Chrisman, N.R. 1987. The Accuracy of Map Overlays: A Reassessment. *Landscape and Urban Planning* 14: 427–39.

Heuvelink, G.B.M. 1998. *Error Propagation in Environmental Modeling with GIS*. London: Taylor and Francis.

MacDougall, E.B. 1975. The Accuracy of Map Overlays. *Landscape Planning* 2: 23–30.

Newcomer, J.A., and J. Szajgin. 1984. Accumulation of Thematic Map Errors in Digital Overlay Analysis. *The American Cartographer* 11: 58–62.

Schutt, M.J., T.J. Moser, P.J. Wigington, Jr., D.L. Stevens, Jr., L.S. McAllister, S. S. Chapman, and T.L. Ernst. 1999. Development of Landscape Metrics for Characterizing Riparian-Stream Networks. *Photogrammetric Survey and Remote Sensing* 65: 1157–67.

Thibault, P.A. 1997. Ground Cover Patterns Near Streams for Urban Land Use Categories. *Landscape and Urban Planning* 39: 37–45.

Veregin, H. 1995. Developing and Testing of An Error Propagation Model for GIS Overlay Operations. *International Journal of Geographical Information Systems* 9: 595–619.

RASTER DATA ANALYSIS

11.1 INTRODUCTION

The raster data model uses a regular grid to cover the space and the value in each grid cell to correspond to the characteristic of a spatial phenomenon at the cell location. This simple data structure of a grid with fixed cell locations not only is computationally efficient but also allows a large variety of data analyses.

In contrast to vector data analysis, which is based on the geometric objects of point, line, and polygon, raster data analysis is based on cells and grids. Raster data analysis can be performed at the level of individual cells, or groups of cells, or cells within an entire grid. Some raster data operations use a single grid, while others use two or more grids. An important consideration in raster data analysis is the type of cell value. Statistics such as mean and standard deviation are designed for numeric values, while others such as majority (the most frequent cell value) are designed for both numeric and categorical values.

This chapter covers the basic tools for raster data analysis and is organized into six sections. Section 1 describes raster data analysis environment including the parameters of area extent and cell size. Sections 2 through 5 cover four common types of raster data analysis: local operations, neighborhood operations, zonal operations, and distance measures. Section 6 focuses on spatial autocorrelation, a spatial statistic well-suited for raster data.

Raster data analysis operates on software-specific raster data, although these raster data may be imported from the same data sources such as DEMs, satellite images, and ASCII files. This chapter uses ArcView with the Spatial Analyst extension whenever possible. ArcView offers many similar operations as ARC/INFO GRID.

11.2 DATA ANALYSIS ENVIRONMENT

Raster data analysis begins with the set up of an analysis environment including the area extent for analysis and the cell size for the output. The area extent for analysis may correspond to a specific grid, or an area defined by its minimum and maximum x-, y-coordinates, or a combination of grids. Given a combination of grids with different area extents, the area extent for analysis can be based on the union or intersect of the grids. The union option uses an area extent that encompasses all input grids, whereas the intersect option uses an area extent that is common to all input grids.

A **mask grid** can also determine the area extent for analysis. A mask grid limits analysis to cells that do not carry the cell value of No Data. For example, one option to limit analysis of soil erosion to only private lands is to code public lands with No Data. Data query and reclassification can be used to create a mask grid (Box 11.1). It should be noted that No Data is different from zero because zero is a valid cell value. In some cases, No Data simply means the absence of data in a grid. For example, an elevation grid prepared from a DEM may contain cells with no data along its border.

The GIS user can define the output cell size at any scale deemed suitable. Typically, the output cell size is set to be equal to, or larger than, the largest cell size among the input grids. This follows the rationale that the accuracy of the output should correspond to that of the least accurate

---

Box 11.1 **How to Make a Mask Grid**

Making a mask grid in ArcView requires a couple of steps. First, use Map Query to assign a special value such as 99 to cells to be excluded from analysis. Then, use Reclassify to assign No Data to the special value. The output grid from Reclassify is a mask grid. A separate mask grid is required for some raster data operations but not for local operations (see 11.3). For local operations, cells with No Data will automatically be excluded from analysis. To make a mask grid in ARC/INFO GRID also involves two steps: use the SELECT function to assign No Data to the unselected cells, and then use the SETMASK command to declare the output from SELECT as a mask grid.

input grid. For example, if the input grids have cell sizes of 10 and 30 meters, the cell size for output should be 30 meters or larger.

11.3 LOCAL OPERATIONS

Constituting the core of raster data analysis, **local operations** are cell-by-cell operations. A local operation creates a new grid from either a single input grid or multiple input grids, and the cell values of the new grid are computed by a function relating the input to the output.

11.3.1 Local Operations with a Single Grid

Given a single grid as the input, a local operation computes each cell value in the output grid as a mathematical function of the cell value in the input grid. ArcView, for example, offers arithmetic, logarithmic, trigonometric, and power functions for local operations (Figure 11.1).

Converting a floating-point grid to an integer grid, for example, is a simple local operation, which uses the INTEGER function to truncate the cell value at the decimal point on a cell-by-cell basis. Converting a slope grid measured in percent to one measured in degree is also a local operation but requires a more complex mathematical expression. In Figure 11.2, the expression, slope_d = 57.296 * arctan (slope_p / 100), converts slope_p measured in percent to slope_d measured in degree. Because computer packages typically use radian instead of degree in trigonometric functions, the constant 57.296 ($360 / 2\pi$, where π is 3.1416) is needed to change the angle measure from radian to degree.

11.3.2 Local Operations with Multiple Grids

Local operations with multiple grids are also referred to as compositing, overlaying, or superimposing maps (Tomlin 1990). These local operations are similar to vector-based map overlay in combining spatial and attribute data, but are much more efficient (Chrisman 1997). Because cells are the same in size and location among the grids in a local

Arithmetic	+, -, /, *, absolute, integer, floating-point
Logarithmic	exponentials, logarithms
Trigonometric	sin, cos, tan, arcsin, arccos, arctan
Power	square, square root, power

Figure 11.1
The arithmetic, logarithmic, trigonometric, and power functions offered by ArcView for local operations.

(a)

15.2	16.0	18.5
17.8	18.3	19.6
18.0	19.1	20.2

(b)

8.64	9.09	10.48
10.09	10.37	11.09
10.20	10.81	11.42

Figure 11.2
A slope grid may be measured in percent (a) or in degree (b). A local operation can be used to convert between the two measurement systems.

operation, there is no need to work on the geometry of the output grid. In contrast, the computation of intersections between map features is a necessary part of map overlay with vector data.

A greater variety of local operations have multiple input grids than have a single input grid. Besides mathematical functions that can be used on individual grids, other measures based on the cell values or their frequencies in the input grids can also be derived and stored on the output grid. Summary statistics, including maximum, minimum, range, sum, mean, median, and standard deviation, are measures that apply to grids with numeric data. For example, a local operation using the mean statistic can calculate a mean annual precipitation grid from 20 input grids, each of which has annual precipitation data as its cell values.

Other measures that are suitable for grids with numeric or categorical data are statistics such as majority, minority, and number of unique values. For each cell, a majority output grid tabulates the most frequent cell value among the input grids, a minority grid tabulates the least frequent cell value, and a variety grid tabulates the number of different cell values.

Some local operations do not involve statistics or computation. A local operation, called Combine in ArcView and ARC/INFO, assigns a unique output value to each unique combination of input values. Suppose a slope grid has three cell values (0%–20%, 20%–40%, and greater than 40% slope), and an aspect grid has four cell values (north, east, south, and west aspect). The Combine operation creates an output grid with a unique value for each combination of slope and aspect, such as 1 for greater than 40% slope and the south aspect, 2 for (20%–40%) slope and the south aspect, and so on (Figure 11.3).

11.3.3 Local Operations in ArcView

A GIS user can run local operations in ArcView using Cell Statistics or Map Calculator. Cell Statistics is menu-driven and easy to use. The proper use of Map Calculator, however, requires some understanding of Avenue, the object-oriented scripting language for ArcView. Map Calculator uses requests, which are essentially instructions in Avenue. The syntax for requests is [object].request (parameter). For example, the statement ([grid1].int) uses the request int to convert grid1, a floating-point grid, to an integer grid. The statement, (([grid1] / 100).ATan * 57.296), converts the measurement unit of a slope grid from percent to degree. The additional pair of parentheses around

[grid1] / 100 ensures that the ATan request applies to grid1 after it has been divided by 100. Neither int nor ATan uses the parameter option.

The Map Calculator dialog incorporates a large assortment of requests in its menu choices: arithmetic operators, logical operators, Boolean connectors, and mathematical functions (arithmetic, logarithmic, trigonometric, and power functions). ArcView users can prepare a local operation statement by selecting from Map Calculator's menus or by typing directly in the expression box.

Besides the requests in the Map Calculator dialog, additional requests can also be used. Combine is an example. The statement to combine grid1 and grid2 is as follows: ([grid1].combine({[grid2]})). The parameter in the example is a list of grids to be combined with grid1. The pair of curly brackets in the above statement presents a list although the list contains only grid2. Additional requests are used later in this chapter.

11.3.4 Applications of Local Operations

As the core of raster data analysis, local operations have many applications. A change detection study, for example, can use the unique combinations produced by the Combine operation to trace the change of cell values, such as change of vegetation covers. Local operations are perhaps most useful for GIS projects that require mathematical computation.

The Universal Soil Loss Equation (USLE) (Wischmeier and Smith 1978) uses six environmental factors in the equation

$$A = R\,K\,L\,S\,C\,P \qquad\qquad (11.1)$$

where A is the average soil loss in tons, R is the rainfall intensity, K is the erodibility of the soil, L is the slope length, S is the slope gradient, C is the cultivation factor, and P is the supporting practice factor. With each factor prepared as an input grid, the grids can be multiplied through a local operation to produce the output grid of average soil loss.

A study by Mladenoff et al. (1995) uses the logistic regression model for predicting favorable wolf habitat,

3	2	1
2	1	2
1	2	3

(a)

3	2	4
3	2	4
2	4	1

(b)

1	3	6
2	4	5
4	5	7

(c)

(d)

Combine code	1	2	3	4	5	6	7
(slope, aspect)	(3,3)	(2,3)	(2,2)	(1,2)	(2,4)	(1,4)	(3,1)

Figure 11.3
The Combine operation sorts out the unique combinations of cell values in (a) and (b) and assigns a unique value for each combination in (c). The combination codes and their representations are shown in (d).

Logit $(p) = -6.5988 + 14.6189\ R$, and
$$p = 1/(1 + e^{\text{logit}(p)}) \qquad (11.2)$$

where p is the probability of occurrence of a wolf pack, R is road density, and e is the natural exponent. Logit (p) can be calculated in a local operation using a road density grid as the input. Likewise, p can be calculated in another local operation using logit(p) as the input.

Because grids are superimposed in local operations, error propagation can be an issue in interpreting the output. Unlike vector data, raster data do not directly involve digitizing errors. Instead, the main source of errors is the quality of the cell values, which in turn can be traced to other data sources. For example, if raster data are converted from satellite images, statistics for assessing the classification accuracy of satellite images can be used to assess the quality of raster data (Congalton 1991; Veregin 1995). But these statistics are based on binary data (i.e., correctly or incorrectly classified), similar to those used in Newcomer and Szajgin's model (1984) in Chapter 10. It is much more difficult to model error propagation with interval and ratio data (Heuvelink 1998).

11.4 NEIGHBORHOOD OPERATIONS

A **neighborhood operation** involves a focus cell and a set of its surrounding cells. The surrounding cells are chosen for their distance and/or directional relationship to the focus cell. Common neighborhoods include a focus cell and its four immediate neighbors, and a focus cell at the center and its eight adjacent neighbors in a 3-by-3 window (Figure 11.4). Other types of neighborhood use circles, annuluses, and wedges. A *circle* neighborhood extends from the focus cell with a specified radius. An *annulus* or doughnut-shaped neighborhood consists of the ring area between a smaller circle and a larger circle centered at the focus cell. A *wedge* neighborhood consists of a piece of a circle centered at the focus cell.

A neighborhood operation typically uses the cell values within the neighborhood, with or without the focus cell value, in a computation, and then assigns the computed value to the focus cell. Although a neighborhood operation works on a single grid, its process is similar to that of a local operation with multiple grids. Instead of using cell values from different input grids, a neighborhood operation uses the cell values from a defined neighborhood.

The output from a neighborhood operation can show summary statistics, including maximum, minimum, range, sum, mean, median, and standard deviation, as well as tabulation of measures such as majority, minority, and variety. To complete a neighborhood operation on a grid, the focus cell is moved from one cell to another until all cells are visited.

11.4.1 Neighborhood Operations in ArcView

ArcView has a menu selection called Neighborhood Statistics, which is designed for neighborhood operations using a grid or a point feature map as the input. The option of using a vector-based map directly in a raster data operation is convenient. If the input is a point feature map, the value of a field in its attribute table is used for computation. The Neighborhood Statistics dialog

−1,−1	0,−1	1,−1
−1,0	0,0	1,0
−1,1	0,1	1,1

Figure 11.4
The spatial relationship between the focus cell and its neighbors can be based on column and row.

has two inputs: statistic and neighborhood. The statistic input includes the options of minimum, maximum, mean, median, sum, range, standard deviation, majority, minority, and variety. The neighborhood input has the options of rectangle, circle, doughnut, and wedge. Depending on which neighborhood is selected, the dialog displays the input fields for the neighborhood dimension, such as the width and height for a rectangular neighborhood.

ArcView has an operation called BlockStats, which also calculates a statistic for the specified neighborhood or block. The difference is that BlockStats assigns the calculated value to all block cells in the output grid. Therefore, the operation does not move from cell to cell, but from block to block.

11.4.2 Applications of Neighborhood Operations

An important application of neighborhood operations is data simplification. The moving average method, for example, reduces the level of cell value fluctuation in the input grid (Figure 11.5). The method typically uses a 3×3 or a 5×5 rectangle as the neighborhood. As the neighborhood is moved from one focal cell to another, the average of cell values within the neighborhood is computed and assigned to the focal cell. The output grid of moving averages represents a generalization of the original cell values. Another example is a neighborhood operation that uses variety as a measure, tabulates how many different cell values in the neighborhood, and assigns the number to the

(a)

1	2	2	2	2
1	2	2	2	3
1	2	1	3	3
2	2	2	3	3
1	2	2	2	3

(b)

1.56	2.00	2.22
1.67	2.11	2.44
1.67	2.11	2.44

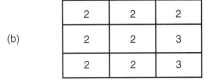

Figure 11.5
The moving average method calculates the mean of cell values within a moving window and assigns the mean to the focal cell. In this illustration, the cell values in (b) are the moving averages of the shaded cells in (a) using a 3 × 3 moving window.

(a)

200	200	110	210	210
200	200	110	210	210
150	150	100	170	170
140	140	130	160	160
140	140	130	160	160

(b)

100	110	110
100	110	110
50	70	70

Figure 11.6
In this illustration, the cell values in (b) are derived for the shaded cells in (a) from a 3 × 3 neighborhood operation using the range statistic.

focal cell. This method can be used to show, for example, the variety of vegetation types or wildlife species in an output grid.

Common in image processing, neighborhood operations are variously called filtering, convolution, or moving window operations for spatial feature manipulation (Lillesand and Kiefer 2000). Edge enhancement, for example, can use a range filter, that is, a neighborhood operation using the range statistic (Figure 11.6). The range measures the difference between the maximum and minimum cell values within the defined neighborhood. A high range value therefore indicates the existence of an edge exists within the neighborhood. The opposite to edge enhancement is a smoothing operation based on the majority measure (Figure 11.7). The majority operation assigns the most frequent cell value to every cell within the neighborhood, thus creating a smoother grid than the original grid.

(a)

1	2	2	2	2
1	2	2	2	3
1	2	1	3	3
2	2	2	3	3
1	2	2	2	3

(b)

2	2	2
2	2	3
2	2	3

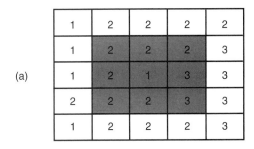

Figure 11.7
In this illustration, the cell values in (b) are derived for the shaded cells in (a) from a 3 × 3 neighborhood operation using the majority statistic.

Another area of study that heavily depends on neighborhood operations is terrain analysis (Chapter 12). Slope, aspect, and surface curvature measures of a cell are all derived from neighborhood operations using elevation values of its adjacent neighbors.

Neighborhood operation can also be important to studies that need to select cells by its neighborhood characteristics. For example, installing a gravity sprinkler irrigation system requires information about elevation drop within a circular neighborhood of a cell. Suppose that a system requires an elevation drop of 130 feet within a distance of 0.5 mile to make it financially feasible. A neighborhood operation on an elevation grid can answer the question by using a circle with a radius of 0.5 mile as the neighborhood and (elevation) range as the statistic. A query of the output grid can show which cells meet the criterion.

Because of its ability in summarizing statistics within a defined area, a neighborhood operation can be used to select sites that meet a study's specific criteria. An example is a study by Crow et al. (1999), which used local operations to select a stratified random sample of 16 plots that represented two ownerships located within two regional ecosystems in Wisconsin.

11.5 ZONAL OPERATIONS

A **zonal operation** works with groups of cells of same values or like features. These groups are called zones. Zones do not have to be connected. A zonal operation may work with a single grid or two grids. Given a single input grid, zonal operations are used to describe the geometry of zones, such as area, perimeter, thickness, and centroid. The area is the sum of the cells that fall within the zone times the cell size. The perimeter of a contiguous zone is the length of its boundary. For a zone consisting of separate regions, the perimeter is the sum of the length of each region. The thickness calculates the radius (in cells) of the largest circle that can be drawn within each zone. The centroid determines the parameters of an ellipse that best approximates each zone, including centroid, major axis, and minor axis. These geometric measures of zones are particularly useful for studies of landscape ecology (Forman and Godron 1986).

Figure 11.8
The zonal operation in this illustration uses the zones of 1, 2, and 3 in (a) and cell values in (b) to calculate the zonal means of 2.17, 2.25, and 4.17. The output grid is shown in (c).

Given two grids in a zonal operation, one input grid and one zonal grid, a zonal operation produces an output grid, which summarizes cell values in the input grid for each zone in the zonal grid. The summary statistics include area, minimum, maximum, sum, range, mean, standard deviation, median, majority, minority, and variety. The last four measures are not available if the input grid is a floating-point grid. Figure 11.8 shows a zonal operation of computing the mean by zone. Figure 11.8a is the zonal grid with three zones, Figure 11.8b is the input grid, and Figure 11.8c is the output grid.

11.5.1 Zonal Operations in ArcView

Map Calculator in ArcView can calculate the geometry of zones in an input grid. This calculation is performed through the ZonalGeometry request with a parameter, which states the type of geometric descriptor to be calculated. The geometric descriptors or measures include area, perimeter, thickness, and centroid. For example, the statement, ([grid1].ZonalGeometry(#GRID_GEOMDESC_THICKNESS)), computes the thickness of each zone in grid1.

Zonal operations with two grids are performed through three menu selections in ArcView: Summarize Zones, Histogram by Zone, and Tabulate Areas. Summarize Zones calculates the summary statistics of area, minimum, maximum, range, mean, standard deviation, sum, variety, majority, minority, and median. Summarize Zones uses the active theme, which can be an integer grid or a polygon feature map, as the zonal grid and the grid selected by the user as the input grid. The output of the operation is a table that lists for each zone the number of cells, area, and statistics. The user can also select a summary statistic to make a histogram with zones and the statistic along the two axes. Histogram by Zone does not perform calculation but plots the cell values of the input grid and the zonal grid in a histogram. Tabulate Areas generates tables of areas sorted by the cell values of the input grid and the zonal grid.

11.5.2 Applications of Zonal Operations

Area, perimeter, thickness, and centroid are part of a large group of geometric measures used in landscape ecology (McGarigal and Marks 1994). Other geometric measures can often be derived from area and perimeter. For example, a shape index called the areal roundness is defined as 354 times the square root of a zone's area divided by its perimeter (Tomlin 1990). Essentially, the areal roundness compares the shape of a zone to a circle. The area roundness value approaches 100 for a circular shape and 0 for a highly distorted shape. The measure should be used on a grid with contiguous zones only. We can use the following expression in ArcView to compute the areal roundness:

$$((([grid1].ZonalGeometry(\#GRID\_GEOMDESC\_AREA)).Sqrt * 354) /$$
$$([grid1].ZonalGeometry(\#GRID\_GEOMDESC\_PERIMETER)))$$

where grid1 is the input grid and Sqrt is the square root function. Perhaps the most important thing about using this long expression is to make sure that parentheses are paired correctly. The same task can be accomplished with a much simpler expression in ARC/INFO GRID (Box 11.2).

Zonal operations with two grids are useful in generating descriptive statistics for comparison purposes. For example, to compare topographic characteristics of different soil textures, we can use a soil grid that contains the categories of sand, loam, and clay as the zonal grid and slope, aspect, and elevation as the input grids. By running a series of zonal operations, we can then summarize the slope, aspect, and elevation characteristics by the three soil textures.

Box **11.2**   **Calculate the Areal Roundness in ARC/INFO GRID**

The statement to calculate the areal roundness of each zone in grid1 in ARC/INFO GRID is as follows:

outgrid = 354 * sqrt(zonalarea(grid1)) / zonalperimeter(grid1)

where outgrid is the output grid, and zonalarea and zonalperimeter are the ARC/INFO GRID functions for computing a zone's area and perimeter. The statement is simpler than the expression required by ArcView.

11.6 DISTANCE MEASURE OPERATIONS

Distance measure operations calculate distances away from cells designated as the source cells. The cells, for which distances are measured, and the source cells are in the same grid. An example of this type of operation is to calculate for each cell the distance to the closest stream in a stream grid. Distance measure operations are also called extended neighborhood operations (Tomlin 1990) or global operations because they cover the entire study area.

Distance measures in a grid follow the node-link relationship (Figure 11.9). A node represents the center of a cell, and a link—either a lateral link or a diagonal link—connects the node to its adjacent cells. Distances are calculated along the links in cells: a lateral link is 1.0 cell, and a diagonal link is 1.4142 cells.

Distances in raster data analysis may be expressed as physical or Euclidean distance or cost distance. The **physical distance** measures the distance by summing the links between two cells and multiplying the sum by the cell size, whereas the **cost distance** measures the cost for traversing the physical distance. A truck driver, for example, is more interested in the time or the fuel cost for covering a route than its physical distance. The cost distance in this case is determined not only by the physical distance but also the speed limit and road condition.

Some distance measure operations to be discussed later also involve direction. Direction measures in a grid follow the eight principal directions from a focal cell to its eight neighboring cells.

Expressed in degree and in a clockwise direction, they are 0° or 360°, 45°, 90°, 135°, 180°, 225°, 270°, and 315°. When direction measures are included in a grid, they are coded in numbers from 0 to 8 rather than in degrees (Figure 11.10).

11.6.1 Physical Distance Measure Operations

A **physical distance measure operation** uses cells as units in measurement. There are two types of physical distance measure operations. The first is to buffer the source cells with continuous distances, thus creating a series of wave-like distance zones over the entire grid (Figure 11.11). The second operation determines for each cell in a grid its closest source cell by the physical distance measure (Figure 11.12).

11.6.2 Cost Distance Measure Operations

A **cost distance measure operation** uses the cost or impedance to move through each cell as distance unit. Cost distance measure operations are much more complex than physical distance measure operations. To begin with, a cost distance operation requires another grid defining the cost or impedance to move through each cell. The cost for each cell in the cost grid is often the sum of different costs. For example, the cost for constructing a pipeline may include construction and operational costs as well as the potential costs of environmental impacts (Box 11.3). Given a cost grid, the cost distance of a lateral link is the average of the costs in the linked cells.

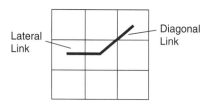

Figure 11.9
Distance measures in a grid follow the links, which connect the cells at their centers. A lateral link connects two direct neighbors, whereas a diagonal link connects two diagonal neighbors.

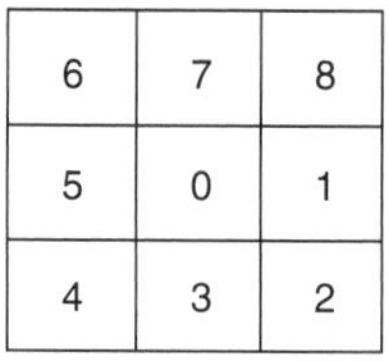

Figure 11.10
Direction measures in a direction grid are numerically coded. The focal cell has a code of 0. The numeric codes 1 to 8 represent the direction measures of 90°, 135°, 180°, 225°, 270°, 315°, 360°, and 45° in a clockwise direction.

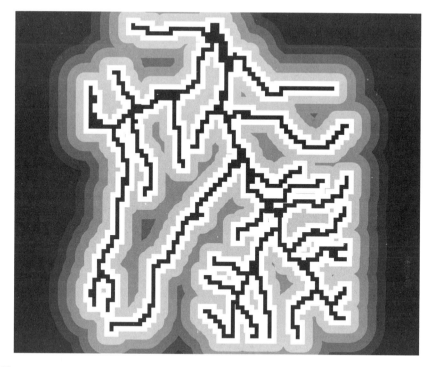

Figure 11.11
An example of continuous distance measures from a stream network.

Box · 11.3   **Cost Grid for a Site Analysis of Pipelines**

A site analysis of a pipeline project must consider the construction and operational costs. Some of the variables that can influence the costs include the following:

- distance from source to destination;
- topography, such as slope and grading;
- geology, such as rock and soils;
- number of stream, road, and railroad crossings;
- right-of-way costs; and
- proximity to population centers.

In addition, the site analysis should consider the potential costs of environmental impacts during construction and liability costs that may result from accidents after the project has been completed. Environmental impacts of a proposed pipeline project may involve the following:

- cultural resources;
- land use, recreation, and aesthetics;
- vegetation and wildlife;
- water use and quality; and
- wetlands.

Each of the above variables must be evaluated and measured in actual or, more likely, relative cost values. Relative costs are ranked values. For example, costs may be ranked from 1 to 5, with 5 being the highest rank value. A grid is made for each cost variable after it has been evaluated. The final step is to make the (total) cost grid by summing the individual cost grids.

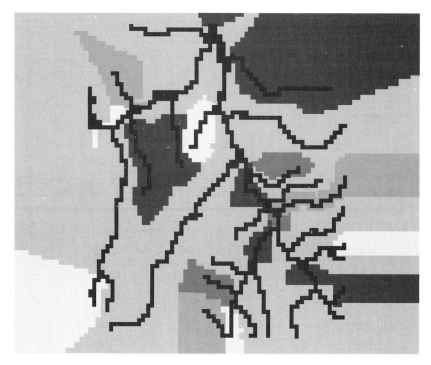

Figure 11.12
An example of assigning each cell to its closest stream.

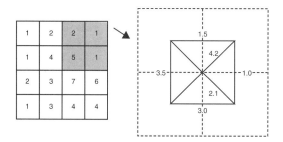

Figure 11.13
The cost distance of a lateral link is the average of the costs in the linked cells, (e.g.), $(1 + 2) / 2 = 1.5$. The cost distance of a diagonal link is the average cost times 1.4142, (e.g.), $1.4142 \times ((1 + 5) / 2) = 4.2$.

The cost distance of a diagonal link is the average cost times 1.4142 (Figure 11.13).

The objective of cost distance measure operations is no longer to calculate the distance from each cell to the closest source cell but to find the path with the least accumulative cost. The inclusion of cost creates many paths connecting a cell and the source cell, each with a different accumulative cost. The algorithm for finding the least accumulative cost follows an iterative process, which begins by activating cells adjacent to the source cell and by computing costs to the cells. The cell with the lowest cost distance is chosen from the active cell list, and its value is assigned to the output grid. Next, cells adjacent to the chosen cell are activated and added to the active cell list. Again, the lowest cost cell is chosen from the list and its neighboring cells are activated. Each time a cell is reactivated, meaning that the cell is accessible to the source cell through a different path, its accumulative cost must be re-computed. The lowest

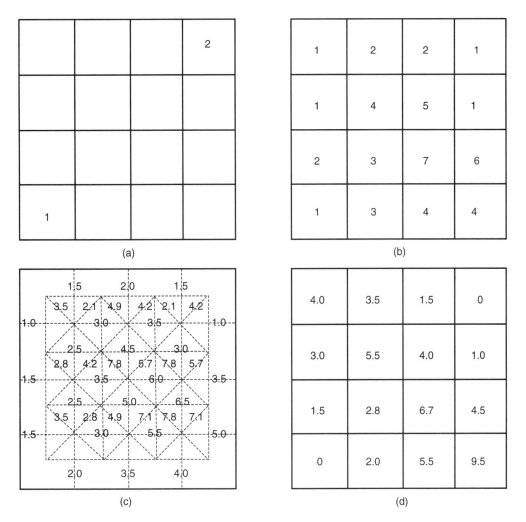

Figure 11.14
Using the grid with two source cells (a) and the cost grid (b), this illustration shows the cost distance for each link (c), and the least accumulative cost distance from each cell to a source cell (d).

accumulative cost is then assigned to the reactivated cell. This process continues until all cells in the output grid are assigned with their least accumulative costs to the source cell.

Figure 11.14 illustrates the cost distance measure operation. Figure 11.14a shows a grid with the source cells at the opposite corners. Figure 11.14b represents a cost grid. To simplify the computation, both grids are set to have a cell size of 1. Figure 11.14c shows the cost of each lateral link

and the cost of each diagonal link. Figure 11.14d shows for each cell the least accumulative cost. The derivation of Figure 11.14d is explained in Box 11.4.

11.6.3 Distance Measure Operations in ArcView

ArcView provides menu selections to run physical distance measure operations and requests to

Box 11.4 **Derivation of the Least Accumulative Cost Grid**

Step 1. Activate cells adjacent to the source cells, place the cells in the active list, and compute cost values for the cells. The cost values for the active cells are as follows: 1.0, 1.5, 1.5, 2.0, 2.8, 4.2.

		1.5	0
		4.2	1.0
1.5	2.8		
0	2.0		

Step 2. The active cell with the lowest value is assigned to the output grid, and its adjacent cells are activated. The cell at row 2, column 3, which is already on the active list, must be reevaluated, because a new path has become available. As it turns out, the new path with a lateral link from the chosen cell yields a lower accumulative cost of 4.0 than the previous cost of 4.2. The cost values for the active cells are as follows: 1.5, 1.5, 2.0, 2.8, 4.0, 4.5, 6.7.

		1.5	0
		4.0	1.0
1.5	2.8	6.7	4.5
0	2.0		

Step 3. The two cells with the cost value of 1.5 are chosen, and their adjacent cells are placed in the active list. The cost values for the active cells are as follows: 2.0, 2.8, 3.0, 3.5, 4.0, 4.5, 5.7, 6.7.

	3.5	1.5	0
3.0	5.7	4.0	1.0
1.5	2.8	6.7	4.5
0	2.0		

Step 4. The cell with the cost value of 2.0 is chosen and its adjacent cells are activated. Of the three adjacent cells activated, two have the accumulative cost values of 2.8 and 6.7. The values remain the same because the alternative paths from the chosen cell yield higher cost

values (5 and 9.1 respectively). The cost values for the active cells are as follows: 2.8, 3.0, 3.5, 4.0, 4.5, 5.5, 5.7, 6.7.

	3.5	1.5	0
3.0	5.7	4.0	1.0
1.5	2.8	6.7	4.5
0	2.0	5.5	

Step 5. The cell with the cost value of 2.8 is chosen. Its adjacent cells all have accumulative cost values assigned from the previous steps. These values remain unchanged because they are all lower than the values computed from the new paths.

	3.5	1.5	0
3.0	5.7	4.0	1.0
1.5	2.8	6.7	4.5
0	2.0	5.5	

Step 6. The cell with the cost value of 3.0 is chosen. The cell to its right has an assigned cost value of 5.7, which is higher than the cost of 5.5 via a lateral link from the chosen cell.

4.0	3.5	1.5	0
3.0	5.5	4.0	1.0
1.5	2.8	6.7	4.5
0	2.0	5.5	

Step 7. All cells have been assigned with the least accumulative cost values, except the cell at row 4, column 4. The least accumulative cost for the cell is 9.5 from either source.

4.0	3.5	1.5	0
3.0	5.5	4.0	1.0
1.5	2.8	6.7	4.5
0	2.0	5.5	9.5

run cost distance measure operations. The menu selection of Find Distance creates an output grid containing continuous distance measures away from the source cells in a grid or the features in a vector-based map. By using either Map Query or Reclassify, one can convert this continuous distance grid to a grid containing discrete distance buffers of the source cells. Task 3 in the Applications section covers this conversion process. The menu selection of Assign Proximity creates a grid, in which each cell is assigned the value of its closest source cell or the ID of the closest feature in a feature map.

The CostDistance request calculates for each cell the least accumulative cost, over a cost grid, to its closest source cell. The request has the options of producing a direction grid, also called a back link grid, showing the direction of the least cost path and an allocation grid showing the assignment of each cell to a source. Using the same data as in Figure 11.14, Figure 11.15a shows the least cost paths of two cells and Figure 11.15b shows the assignment of each cell to the two sources. The darkest cell in Figure 11.15b can be assigned to either one of the two sources. Also, a maximum distance parameter can be used with the CostDistance request to define the threshold for the least accumulative costs. The CostPath request uses the distance and direction grids created by the Cost Distance request to generate the least cost path from any cell or zone (set of cells with the same value). Task 4 in the Applications section uses both CostDistance and CostPath.

11.6.4 Applications of Distance Measure Operations

Like buffering around vector-based features, physical distance measure operations have many applications. An example is a raster model of potential nesting habitat of greater sandhill cranes in northwestern Minnesota (Herr and Queen 1993). The study used a cell resolution of 30 meters and compiled a vegetation map from Landsat Thematic Mapper data. Using continuous distance zones measured from undisturbed vegetation, roads, buildings, and agricultural land, the study categorized potentially suitable nesting vegetation as optimal, sub-optimal, marginal, or unsuitable. Distance measure operations provided the tools to implement the model.

Cost distance measure operations are commonly used for site analysis of roads and pipelines to find the least accumulative cost path

(a)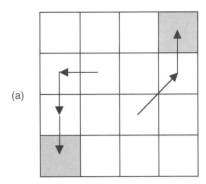

(b)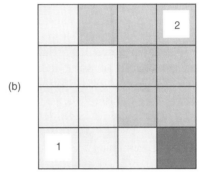

Figure 11.15
Using the same data as in Figure 11.14, this illustration shows the last cost path (a) and the allocation grid (b).

from the cost grid. Cost is not the only measure of impedance. For example, Hepner and Finco (1995) built a model of gas dispersion using an impedance grid that represented the product of surface distance, slope impedance, and wind impedance. The cumulative impedance in their study represented the difficulty of gas travel between the release point (source cell) and any given cell in the study area. And equal steps of the cumulative impedance values supposedly defined the progression over time of a dense gas cloud over the terrain.

Because distance measure operations are based on cells, they are not as precise as buffering around vector-based features. The same is true with directions as direction measures in these operations follow the eight principal directions.

Therefore, a least accumulative cost path often displays a zigzag pattern.

11.7 SPATIAL AUTOCORRELATION

Spatial autocorrelation measures the relationship among values of a variable according to the spatial arrangement of the values (Cliff and Ord 1973). The relationship may be described as highly correlated if like values are spatially close to each other and independent or random if no pattern can be discerned from the arrangement of values. The absence of significant spatial autocorrelation can validate the use of standard statistical tests of hypotheses. On the other hand, the presence of significant spatial autocorrelation should encourage the researcher to incorporate spatial dependency in the analysis (Legendre 1993). Spatial autocorrelation is well-suited to raster data analysis because cells in a grid follow a well-defined spatial arrangement.

Spatial autocorrelation does not belong to any of the above raster data operations. It is probably best classified as a global operation because the computation of spatial autocorrelation uses every cell in a grid and produces a statistic that applies to the entire grid.

One popular spatial autocorrelation measure is **Moran's I**, which can be computed by

$$\frac{\sum\limits_{i=1}^{n} \sum\limits_{j=1}^{m} w_{ij} (x_i - x_m)(x_j - x_m) / \sum\limits_{i=1}^{n} \sum\limits_{j=1}^{m} w_{ij}}{\sum\limits_{i=1}^{n} (x_i - x_m)^2 / n} \quad (11.3)$$

where x_i is the value of cell i, x_j is the value of cell i's neighbor j, x_m is the mean cell value of the grid, w_{ij} is a coefficient, and n is the total number of cells in the grid. The coefficient w_{ij} has a value of 1 if j is one of the four cells directly adjacent to i and a value of 0 for other cells or cells with No Data. Moran's I is positive when nearby areas have similar attribute values, negative when they have dissimilar values, and close to zero when attribute values are arranged randomly. The values Moran's I takes on tend to range between -1 and 1, but are not restricted to the range.

Another popular spatial autocorrelation measure is **Geary's c**, which can be computed by

$$\frac{\sum\limits_{i=1}^{n} \sum\limits_{j=1}^{m} w_{ij} (x_i - x_j)^2 / \sum\limits_{i=1}^{n} \sum\limits_{j=1}^{m} w_{ij}}{\sum\limits_{i=1}^{n} (x_i - x_m)^2 / (n - 1)} \quad (11.4)$$

The notations in the equation are the same as those for Moran's I. Whereas Moran's I uses the covariance, $(x_i - x_m)(x_j - x_m)$, in the computation, Geary's c uses the variance, $(x_i - x_j)^2$. Geary's c has a value of 1 for a random pattern, less than 1 for a positively correlated pattern, and greater than 1 for a negatively correlated pattern. Figure 11.16 shows three patterns and the associated values of Moran's I and Geary's c for comparison.

(a)

Moran's I = −0.6, Geary's c = 0.48

(b)

Moran's I = −0.15, Geary's c = 1.07

(c)

Moran's I = −0.65, Geary's c = 1.29

Figure 11.16
Three sets of spatial autocorrelation statistics: (a) a positively correlated pattern, (b) a random pattern, and (c) a negatively correlated pattern.

Recent developments in spatial statistics have included measures of spatial autocorrelation in multiple distances and local autocorrelation statistics (Getis and Ord 1996; Lam et al. 1996; Lee and Wong 2001). ARC/INFO, IDRISI, GRASS, and ILWIS are all capable of computing spatial autocorrelation.

KEY CONCEPTS AND TERMS

Cost distance: Distance measured by the cost or impedance of moving between cells.

Cost distance measure operation: A distance measure operation that uses the cost or impedance to move through each cell as distance unit.

Distance measure operation: A raster data operation that calculates distances away from cells designated as the source cells.

Geary's c: A spatial autocorrelation statistic that uses the variance in the computation.

Local operation: A cell-by-cell operation in raster data analysis.

Mask grid: A grid that limits raster data analysis to cells that do not carry the cell value of No Data.

Moran's I: A spatial autocorrelation statistic that uses the covariance in the computation.

Neighborhood operation: A raster data analysis operation that involves a focus cell and a set of its surrounding cells.

Physical distance: Distance measured by summing the links between two cells and multiplying the sum by the cell size.

Physical distance measure operation: A distance measure operation that uses cells as units.

Spatial autocorrelation: A spatial statistic that measures the relationship among values of a variable according to the spatial arrangement of the values.

Zonal operation: A raster data analysis operation that involves groups of cells of same values or like features.

APPLICATIONS: RASTER DATA ANALYSIS

This applications section covers the basic operations of raster data analysis. Task 1 covers the local and neighborhood operations. Task 2 uses the zonal operation. Task 3 includes the physical distance measure operation in data query. Task 4 uses the CostDistance and CostPath requests to solve the least accumulative cost distance problem.

Task 1: Local and Neighborhood Operations

What you need: *emidalat*, an elevation grid with a cell resolution of 30 meters.

Task 1 lets you work with two raster data manipulations. The first uses a local operation to convert the elevation values of *emidalat* from meters to feet. The second uses a neighborhood operation to generalize the elevation values of *emidalat*.

1. Start ArcView and load the Spatial Analyst extension. Open a new view, and add *emidalat* to view. Select Properties from the View menu and specify meters as Map Units.

2. Select Map Calculator from the Analysis menu. In the next dialog, double-click *emidalat* in the Layers list, click the multiplication (*) button, and press 3.28 on the keypad. The constant 3.28 is for converting the elevation values from meters to feet. You should see the expression,

([emidalat] * 3.28) in the expression box. Click Evaluate.

3. The output is called *Map Calculation 1*, which shows *emidalat* in feet. *Map Calculation 1* is a temporary grid. You can save the grid by selecting Save Data Set from the Theme menu and specifying the name and path for the grid.

4. To start the second part of Task 1, activate *emidalat*.

5. Select Neighborhood Statistics from the Analysis menu. In the Neighborhood Statistics dialog, make sure the Statistic is the mean, the Neighborhood is rectangle, and both the Width and Height values are 3 (cells). Click OK. The output with moving means as the cell values is named *NbrMean of Emidalat*. Check the box next to the new grid theme to view. Unless you look in detail, you probably cannot tell the difference between the new grid and *emidalat*. Make *NbrMean of Emidalat* active, and select Properties from the Theme menu. Notice in the Theme Properties dialog that the new grid has a cell size of 30 (meters), the same as *emidalat*. The dialog also shows that the new grid has the Temporary status. You can save *NbrMean of Emidalat* by selecting Save Data Set from the Theme menu and specifying its name and path.

Task 2: Zonal Operation

What you need: *precipgd*, a grid showing the average annual precipitation in Idaho; *hucgd*, a watershed grid.

Task 2 asks you to derive annual precipitation statistics by watershed in Idaho. Both *precipgd* and *hucgd* are projected onto the Idaho Transverse Mercator coordinate system and are measured in meters. The precipitation measurement unit for *precipgd* is 1/100 of an inch; for example, the cell value of 675 means 6.75 inches.

1. Start ArcView, and load Spatial Analyst. Open a new view, and add *precipgd* and *hucgd* to view. In this operation, *hucgd* is the zonal grid and *precipgd* is the input grid.

2. Activate *hucgd*. Select Summarize Zones from the Analysis menu. In the next dialog, make sure that *precipgd* is the theme to be summarized.

3. You can select a statistic to chart. Select the Mean to make the chart.

4. The table titled Stats of *Precipgd* within Zones of *Hucgd* lists the following statistics of annual precipitation by watershed: minimum, maximum, range, mean, standard deviation, sum, variety, minority, and majority.

Task 3: Physical Distance Measure

What you need: *strmgd*, a grid showing streams; *elevgd*, a grid showing elevation zones.

Task 3 asks you to locate the potential habitats of a plant species. The cell values in *strmgd* are the ID values of streams. The cell values in *elevgd* are elevation zones 1, 2, and 3. Both grids have the cell resolution of 100 meters. The potential habitats of the plant species must meet the following two criteria:

- Elevation zone 2, and
- Within 200 meters of streams

1. Start ArcView and load the Spatial Analyst extension.

2. Open a new view, and add *strmgd* and *elevgd* to view. Select Properties from the View menu and specify meters as Map Units.

3. First, create distance zones of 200 meters from streams. By selecting Find Distance from the Analysis menu, the Spatial Analyst extension creates a new grid called *Distance to Strmgd*. The new grid comes with a default classification system. To create two distance zones of within 200 meters of streams and more than 200 meters from streams, you can reclassify the grid. Activate *Distance to Strmgd*. Select Reclassify from the Analysis menu. Click Classify in the Reclassify Values dialog. Change the Number of Classes to 2, and click OK. Now click the first cell under Old Values, change the numeric range to 0–200, and enter the range. Change the

second cell under Old Values to 201–2421, and keep No Data for the third cell. Click OK.

4. The Reclassify function adds a new grid called *Reclass of Distance to Strmgd* to view. The new grid only has three values: 1, 2, and No Data. Cells with the value of 1 are within 200 meters of streams.

5. The final step for Task 3 is to query *Reclass of Distance to Strmgd* and *Elevgd*. Select Map Query from the Analysis menu. In the window area of the Map Query 1 dialog, prepare the following query expression: ([Reclass of Distance to Strmgd] = 1.AsGrid) AND ([Elevgd] = 2.AsGrid). You do not have to type the request of AsGrid because it is automatically added. Click Evaluate.

6. The result of the query is shown in the *Map Query 1* grid. Cells with the value of True (1) satisfy both criteria.

7. Task 3 can also be completed without reclassifying the *Distance to Strmgd* grid. In that case, the query expression should be changed to ([Distance to Strmgd] <= 200.AsGrid) AND ([Elevgd] = 2.AsGrid).

Task 4: A Least Accumulative Cost Distance Example

What you need: *sourcegrid* and *costgrid*, the same grids as in Figure 11.14; *pathgrid*, a grid to be used with the CostPath request.

1. Start ArcView and load the Spatial Analyst extension.

2. Open a new view, and add *sourcegrid* and *costgrid* to view. The upper right and lower left corner cells are designated as the source cells in *sourcegrid*. Select Map Calculator from the Analysis menu. Then use the CostDistance request to calculate for each cell in *sourcegrid* the least accumulative cost distance to a source cell over *costgrid*. The syntax for CostDistance is as follows:

aGrid.CostDistance(costGrid, directionFN, allocationFN, maxDistance)

Of the parameters, directionFN is the file name for a direction or back link grid, allocationFN is the file name for an allocation grid, and maxDistance defines the threshold for the least accumulative costs. For this task, use a direction grid (named *dir_gd*) and an allocation grid (named *alloc_gd*) but not a distance threshold. Type the following statement in the Map Calculator dialog's expression box and click Evaluate:

([sourcegrid].CostDistance([costgrid], "dir_gd".AsFileName, "alloc_gd".AsFileName, Nil))

3. The output grid from the CostDistance request is called *Map Calculation 1*. Make *Map Calculation 1* active. Select Properties from the Theme menu, and rename *Map Calculation 1 least_gd*. Use the identify tool to check the cell values in *least_gd*; they should be the same as in Figure 11.14d.

4. The CostDistance request has also created *dir_gd* and *alloc_gd*. Add *dir_gd* and *alloc_gd* to view. *Alloc_gd* is easy to understand: the cells in the grid are divided by their allocation to either source 1 (lower left corner) or source 2 (upper right corner). *Dir_gd* requires some explanation. The cell values in *dir_gd* range from 0 to 8. Each cell value indicates which neighboring cell to move into to get back to the source. The notations of the cell values are shown as follows:

6	7	8
5	0	1
4	3	2

The following diagram shows the cell values in *dir_gd* and the back links the cell values represent:

3	1	1	0
3	5	1	7
3	4	8	7
0	5	5	7

5. Next, use the CostPath request to calculate the shortest cost paths for two cells with values in *pathgrid*. Select Map Calculator from the Analysis menu. The syntax for the CostPath request is as follows:

AGrid.CostPath (distanceGrid, directionGrid, byZone)

Of the parameters, distanceGrid and directionGrid are the output grids created by the CostDistance request. The byZone parameter can be True if the cost paths are to be calculated for each zone or each set of cells with the same value, or False if the cost paths are to be calculated for each cell with a value. To run the CostPath request, type the following statement in the Map Calculator dialog's expression box and click Evaluate:

([pathgrid].CostPath ([least_gd], [dir_gd], False))

The output grid from the CostPath request has three cell values: 1 represents the source cell, 3 is the least cost path for the cell with a value of 1 in *pathgrid*, and 4 is the least cost path for the cell with a value of 2 in *pathgrid*. Make the output grid active and open its theme table. The theme table shows the least accumulative cost for each path and the starting row and column of each path. The numbering of rows and columns starts with the upper left corner cell, which has a row number of 0 and a column number of 0.

REFERENCES

Chrisman, N. 1997. *Exploring Geographic Information Systems*. New York: John Wiley & Sons.

Cliff, A.D., and J.K. Ord. 1973. *Spatial Autocorrelation*. New York: Methuen.

Congalton, R.G. 1991. A Review of Assessing the Accuracy of Classification of Remotely Sensed Data. *Photogrammetric Engineering & Remote Sensing* 37: 35–46.

Crow, T.R., G.E. Host, and D.J. Mladenoff. 1999. Ownership and Ecosystem as Sources of Spatial Heterogeneity in a Forested Landscape, Wisconsin, USA. *Landscape Ecology* 14: 449–63.

Forman, R.T.T., and M. Godron. 1986. *Landscape Ecology*. New York: John Wiley & Sons.

Getis, A., and J. K. Ord. 1996. Local Spatial Statistics: An Overview. In P. Longley and M. Batty (eds.), *Spatial Analysis: Modelling in a GIS Environment*. Cambridge, UK: GeoInformation International, pp. 261–77.

Hepner, G.H., and M.V. Finco. 1995. Modeling Dense Gas Contaminant Pathways over Complex Terrain Using a Geographic Information System. *Journal of Hazardous Materials* 42: 187–99.

Herr, A.M., and L.P. Queen. 1993. Crane Habitat Evaluation Using GIS and Remote Sensing. *Photogrammetric Engineering & Remote Sensing* 59: 1531–38.

Heuvelink, G.B.M. 1998. *Error Propagation in Environmental Modelling with GIS*. London: Taylor and Francis.

Lam, N.S., M. Fan, and K. Liu. 1996. Spatial-Temporal Spread of the AIDS Epidemic, 1982–1990: A Correlogram Analysis of Four Regions of the United States. *Geographical Analysis* 28: 93–107.

Lee, J., and D.W.S. Wong. 2001. *Statistical Analysis with ArcView GIS*. New York: John Wiley & Sons.

Legendre, P. 1993. Spatial Autocorrelation: Trouble or New Paradigm? *Ecology* 74: 1659–73.

Lillesand, T.M., and R.W. Kiefer. 2000. *Remote Sensing and Image Interpretation*, 4th ed. New York: John Wiley & Sons.

McGarigal, K., and B.J. Marks. 1994. *Fragstats: Spatial Pattern Analysis Program for Quantifying Landscape Structure*. Forest Science Department, Oregon State University.

Mladenoff, D.J., T.A. Sickley, R.G. Haight, and A.P. Wydeven. 1995. A Regional Landscape Analysis and Prediction of Favorable Gray Wolf Habitat in the Northern Great Lakes Regions. *Conservation Biology* 9: 279–94.

Newcomer, J.A., and J. Szajgin. 1984. Accumulation of Thematic Map Errors in Digital Overlay Analysis. *The American Cartographer* 11: 58–62.

Tomlin, C.D. 1990. *Geographic Information Systems and Cartographic Modeling*. Englewood Cliffs, N.J.: Prentice Hall.

Veregin, H. 1995. Developing and Testing of An Error Propagation Model for GIS Overlay Operations. *International Journal of Geographical Information Systems* 9: 595–619.

Wischmeier, W.H., and D.D. Smith. 1978. Predicting Rainfall Erosion Losses: A Guide to Conservation Planning. *Agricultural Handbook 537*. Washington, DC: U.S. Department of Agriculture.

12

TERRAIN MAPPING AND ANALYSIS

12.1 INTRODUCTION

The terrain with its undulating, continuous land surface is a familiar phenomenon to GIS users. The land surface has been the object of mapping for hundreds of years. Mapmakers have introduced various techniques for terrain mapping and have developed land surface measures as derivatives of elevation data such as slope and aspect. GIS has made it easier to incorporate terrain mapping and analysis into applications ranging from wildlife habitat analysis to hydrologic modeling.

The land surface is a three-dimensional surface. Most GIS packages treat elevation data, often called the z values, as attribute data at point or cell locations rather than as an additional coordinate to the x-, y-coordinates. In raster format, the z values correspond to cell values. In vector format, the z values are stored in a field of a feature attribute table. Terrain mapping and analysis can use either raster data or vector data as the input.

This chapter is organized into four sections. Section 1 covers two common data sources for terrain mapping and analysis: DEM (Digital Elevation Model) and TIN (Triangulated Irregular Network). Section 2 describes different methods for terrain mapping. Section 3 discusses terrain analysis including slope, aspect, surface curvature, viewshed analysis, and watershed analysis. Section 4 compares DEM and TIN for terrain mapping and analysis.

Although this chapter uses ArcView and ARC/INFO as examples, many GIS packages are capable of performing terrain mapping and analysis (Box 12.1). GIS vendors typically group terrain mapping and analysis functions into a module or an extension. ArcView has the Spatial Analyst and 3-D Analyst extensions, and ARC/INFO has the spatial modeling modules of GRID and TIN. MGE has the Terrain Analyst, SPANS has the Topographer, and PAMAP has the TOPOGRAPHER.

12.2 DATA FOR TERRAIN MAPPING AND ANALYSIS

12.2.1 DEM

A DEM represents an array of elevation points. The quality of a DEM can influence the accuracy of terrain measures, such as slope and aspect. The U.S. Geological Survey (USGS) classifies the quality of 7.5-minute DEMS into four levels, with Level 1 having the poorest quality. Using known sources such as bench marks (vertical control points) and spot elevations as test points, the USGS calculates the root mean squared error (RMSE) of the DEM data. Level 1 accuracy has a RMSE target of 7 meters and a maximum RMSE of 15 meters. Level 2 accuracy has a maximum RMSE of one half the contour interval. Level 3 accuracy has a maximum RMSE of one-third of a contour interval—not to exceed 7

Box 12.1   **A Survey of Terrain Analysis Functions among GIS Packages**

Terrain mapping and analysis is an important component of GIS packages, especially those that are raster-based. The following lists some of the terrain analysis functions covered in this chapter and the GIS packages that carry them:

Slope, aspect: ARC/INFO, ArcView, IDRISI, SPANS, GRASS, ILWIS, PAMAP

Surface curvature: ARC/INFO, IDRISI, GRASS

Viewshed analysis: ARC/INFO, ArcView, GRASS, IDRISI, SPANS, PAMAP, MFworks

Watershed analysis: ARC/INFO, ArcView, GRASS, IDRISI, PAMAP, MFworks

meters. Most USGS DEMs in use are either Level 1 or Level 2.

Errors in USGS DEMs may be classified as either global or relative (Carter 1989). Global errors are systematic errors caused by displacements of the DEM, as evidenced by mismatching elevations along the boundaries of adjacent DEMs. One can usually correct global errors by applying geometric transformations, including translation, rotation, and scaling. Relative errors are local but significant errors relative to the neighboring elevations. Examples of relative errors are artificial peaks and blocks of elevations extending above the surrounding lands, especially along ridgelines. Relative errors can only be corrected by editing the DEM data.

For terrain analysis, point-based DEMs must first be converted to software-specific raster data such as elevation grids for ARC/INFO and ArcView. Elevation points in a DEM thus become cells in an elevation grid. In the case of 3 arc-second DEMs from the USGS, they must also be projected into a real-world coordinate system.

12.2.2 TIN

A TIN approximates the land surface with a series of non-overlapping triangles. Elevation values (z values) along with x-, y-coordinates are stored at nodes that make up the triangles. In contrast to DEMs, TINs are based on an irregular distribution of elevation points. GIS users typically compile the input data to a TIN from a variety of sources: DEMs, surveyed elevation points, contour lines, and **breaklines**. Breaklines are line features that represent changes of the land surface such as streams, shorelines, ridges, and roads.

Because triangles in a TIN vary in size by the complexity of topography, not every point in a DEM needs to be used to create a TIN. Instead, the process is to select points that are more important in representing the terrain. Several algorithms for selecting significant points from a DEM have been proposed in GIS (Lee 1991; Kumler 1994). Here we examine two algorithms: **VIP** (Very Important Points) used by ARC/INFO, and **maximum z-tolerance** used by ARC/INFO and ArcView.

To select points from a DEM, VIP first converts a DEM to a grid and then evaluates the importance of each point (each cell in the elevation grid) by measuring how well its value can be estimated from the neighboring point values (Chen and Guevara 1987). The conversion of a point-based DEM to an elevation grid is necessary for the selection process, which is essentially a neighborhood operation.

Figure 12.1a shows a 3 x 3 moving window in an elevation grid. To assess the importance of P, VIP first estimates the elevation at P by using four pairs of its neighbors: up and down (B–F), left and right (H–D), upper right and lower left (C–G), and upper left and lower right (A–E). Figure 12.1b shows a vertical profile for the case with C–G. P_e is the estimated elevation at P from the elevations at G and C (Z_G and Z_C), and P_h is the actual elevation at P. The difference between P_h and P_e, d, therefore represents the elevation offset at P. But VIP uses s, which measures the distance of the perpendicular line from P_h to line Z_GZ_C, for the offset measure at P. The offset s is a better measure than d in flat or steep slope areas according to Chen and Guevara (1987). After calculating four offset values, one for each pair of neighbors around P, VIP uses their average as an indicator of the significance of P.

Using the above procedure, VIP computes the significance value for each cell in an elevation grid. A frequency distribution of the significance values from a grid usually looks like a normal distribution, with more cells having lower significance values (Chen and Guevara 1987). The selection of points by VIP can be based on either a desired number or a specified significance level.

The maximum z-tolerance algorithm selects points from an elevation grid to construct a TIN such that, for every point in the elevation grid, the difference between the original elevation and the estimated elevation from the TIN is within the specified maximum z-tolerance. This algorithm is used by the LATTICETIN command, which converts an elevation grid to a TIN, in ARC/INFO and by ArcView's 3-D Analyst.

(a)

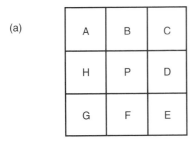

(b)

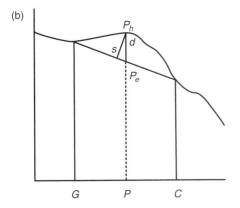

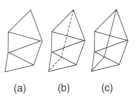

(a)   (b)   (c)

Figure 12.2
The dashed line in (b) represents a breakline, which subdivides the triangles in (a) into a series of smaller triangles in (c).

Figure 12.1
VIP evaluates the significance of an elevation point by measuring how well its value can be estimated from the neighboring point values. Figure 12.1b shows the case of using elevations at G and C to estimate the elevation at P. P_e is the estimated elevation, P_h is the actual elevation, and d represents the offset between P_e and P_h. Rather than using d as the measure of significance, VIP uses s.

The maximum z-tolerance algorithm uses an iterative process. The process begins by constructing a candidate TIN. Then, for each triangle in the TIN, the algorithm computes the elevation difference from each point in the grid to the enclosing triangular facet. The algorithm determines the point with the largest difference. If the difference is greater than a specified z-tolerance, the algorithm flags the point for addition to the TIN model. After every triangle in the current TIN is checked, a new triangulation is recomputed with the selected additional points. This process continues until all points in the grid are within the specified maximum z-tolerance.

The construction of a TIN can use other data sources besides a DEM. Additional elevation points may come from photogrammetric survey data or GPS readings. Contour lines can provide points with contour (elevation) readings. Breaklines provide not only points with elevation values but also the physical structure in the form of triangle edges (Figure 12.2).

A TIN connects elevation points from various data sources into a series of non-overlapping triangles. A common algorithm for connecting points is called the **Delaunay triangulation** (Tsai 1993). Triangles formed by the Delaunay triangulation have the following characteristics: all nodes (points) are connected to their nearest neighbors to form triangles; and triangles are as equi-angular, or compact, as possible.

Unlike compact triangles derived from the Delaunay triangulation, triangles along the border of a TIN are often stretched and elongated, thus distorting the landform features derived from those triangles. One way to solve the problem is to include elevation points beyond the border of a study area for processing and then to clip the study area from the larger coverage.

12.3 TERRAIN MAPPING

12.3.1 Contouring

Contouring is the most common method for terrain mapping. **Contour lines** connect points of equal elevation (Box 12.2), and the **contour interval** represents the vertical distance between contour lines. The arrangement and pattern of contour lines reflect the topography. For example, contour lines are closely spaced in steep terrain and are curved in the upstream direction along a stream. With some training and experience, map readers can visualize the terrain by simply studying contour lines. Contour lines can also be used for manually measuring slope and aspect, although the practice is becoming rare with the use of GIS.

Automated contouring follows two basic steps: (1) detecting a contour line intersecting a grid cell or a triangle, and (2) drawing the contour line through the grid cell or triangle (Jones et al. 1986). A TIN is a good example for illustrating automated contouring because it is already triangulated and has elevation readings for all nodes. Given a contour line, every triangle edge is examined to determine if the contour should pass through the edge. If it does, a linear interpolation, which assumes a constant gradient between the end nodes of the edge, can determine the con-

tour's position along the edge. After all the positions for the contour are calculated, they are connected to form the contour line (Figure 12.3). The initial contour line consists of straight-line segments, which can be smoothed by splining, that is, by fitting a mathematical function to points that make up the line. Another way of producing smooth contour lines is to divide a triangle into a series of smaller triangles and to use these smaller triangles for contouring.

Contour lines cannot intersect one another or stop in the middle of a map, although they can close up by themselves in cases of depressions or isolated hills. Contour maps created from a GIS sometimes contain irregularities or even errors (Figure 12.4).

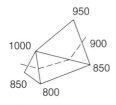

Figure 12.3
The contour line of 900 connects points that are interpolated to have the value of 900 along the triangle edges.

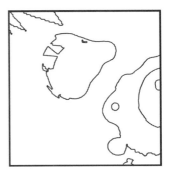

Figure 12.4
Automated contouring may produce contour lines that are highly irregular.

Box 12.2   **Contour Lines**

Cartographers use the term *contour lines* in terrain mapping. A contour line connects points of equal elevations. Contour lines are one type of isolines, which can be used for all kinds of data in mapmaking. For example, isolines can show precipitation or temperature distributions. But GIS software developers have not always observed the distinction between contour lines and isolines. ARC/INFO and ArcView, for example, use the term contour lines in place of isolines.

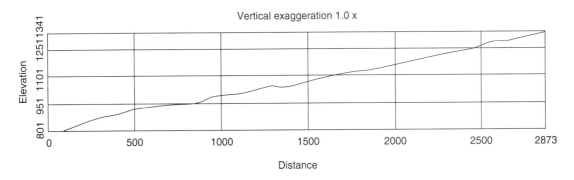

Figure 12.5
A vertical profile showing changes in elevation along a stream tributary. The profile has a vertical exaggeration factor of 1.0, or no vertical exaggeration.

12.3.2 Vertical Profiling

A **vertical profile** shows changes in elevation along a line, such as a hiking trail, a road, or a stream (Figure 12.5). The manual method usually involves the following steps:

- draw a line on a contour map;
- mark each intersection between a contour and the profile line and record its elevation;
- raise each intersection point to a height proportional to its elevation; and
- plot the vertical profile by connecting the points.

Automated profiling follows the same procedure but substitutes the contour map with an elevation grid or a TIN.

12.3.3 Hill Shading

Also known as shaded relief or simply shading, **hill shading** attempts to simulate how the terrain looks with the interaction between sunlight and surface features (Figure 12.6). A mountain slope directly facing incoming light will be very bright, but a slope opposite to the light will be dark. Hill shading helps viewers recognize the shape of landform features on a display. An excellent example is Thelin and Pike's (1991) digital shaded-relief map of the United States (http://www.usgs.gov/reports/

Figure 12.6
An example of hill shading, with the sun's azimuth at 315^0 (NW) and the sun's altitude at 45^0.

misc/Misc._Investigations_Series_Maps_(I_Series)/ I_2206/usa_dem.gif).

Hill shading used to be produced by talented artists. Nowadays the computer can generate high-quality shaded maps. Four factors control the visual effect of hill shading. The sun's azimuth is the direction of the incoming light, ranging from 0^0 to 360^0 in a clockwise direction. The sun's altitude is the angle of the incoming light measured above the horizon between 0^0 and 90^0. The other two factors are the surface's slope and aspect: slope ranges from 0^0 to 90^0 and aspect from 0^0 to 360^0. Using the above four factors, the following equation can compute the relative radiance value for every cell in an elevation grid or for every triangle in a TIN (Eyton 1991):

$$R_f = \cos (A_f - A_s) \sin H_f \cos H_s + \cos H_f \sin H_s \quad (12.1)$$

where R_f is the relative radiance value of a facet (a grid cell or a triangle), A_f is the facet's aspect, A_s is the sun's azimuth, H_f is the facet's slope, and H_s is the sun's altitude. R_f ranges in value from 0 to 1. If multiplied by the constant 255, R_f can be converted to the illumination value (I_f) for display. An I_f value of 255 would result in white and an I_f value of 0 would result in black on a shaded map. Both ARC/INFO and ArcView use I_f for hill shading (Box 12.3).

Relative radiance is similar to another measure called the incidence value (Franklin 1987):

$$\cos (H_f) + \cos (A_f - A_s) \sin (H_f) \cot (H_s) \quad (12.2)$$

The notations in Equation 12.2 are the same as Equation 12.1. The incidence value can also be derived by multiplying the relative radiance value by $\sin (H_s)$. Besides producing hill shading, both relative radiance and incidence can be used in image processing as variables representing the interaction between the incoming radiation and local topography.

12.3.4 Hypsometric Tinting

Also known as layer tinting, **hypsometric tinting** applies color symbols to different elevation zones. The use of well-chosen color symbols can help viewers see the progression in elevation, especially on a small-scale map. Hypsometric tinting can also be used to highlight a particular elevation zone, which may be important, for example, in a wildlife habitat study.

12.3.5 Perspective View

Perspective views are 3-D views of the terrain: the terrain has the same appearance as viewed with an angle from an airplane (Figure 12.7). Four parameters can control the appearance of a 3-D view (Figure 12.8):

Box 12.3 **A Worked Example of Computing Relative Radiance**

Suppose a cell in an elevation grid has a slope value of 10^0 and an aspect value of 297^0 (facing west to northwest), the sun's altitude is 65^0, and the sun's azimuth is 315^0 (from northwest). The relative radiance value of the cell can be computed by

$$R_f = \cos (297–315) \sin (10) \cos (65) + \cos (10) \sin (65) = 0.9623$$

The cell will appear bright with a R_f value of 0.9623. Suppose that the sun's altitude is lowered to 25^0 and

the sun's azimuth remains at 315^0. Then the cell's relative radiance value becomes

$$R_f = \cos (297–315) \sin (10) \cos (25) + \cos (10) \sin (25) = 0.5658$$

The cell will appear in medium gray with a R_f value of 0.5658.

Figure 12.7
An example of a 3-D perspective view.

- **Viewing azimuth** is the direction from the observer to the surface, ranging from 0^0 to 360^0 in a clockwise direction.
- **Viewing angle** is the angle measured from the horizon to the altitude of the observer. A viewing angle is always between 0^0 and 90^0. An angle of 90^0 means viewing the surface from directly above, while an angle of 0^0 means viewing the surface directly ahead. Therefore, the 3-D effect reaches its maximum as the angle approaches 0^0 and its minimum as the angle approaches 90^0.
- **Viewing distance** is the distance between the viewer and the surface. Adjustment of the viewing distance allows the surface to be viewed close up or from a distance.
- **Z-scale** is the ratio between the vertical scale and the horizontal scale. Also called the vertical exaggeration factor, z-scale is useful for highlighting minor landform features.

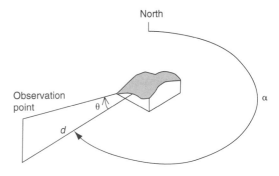

Figure 12.8
The diagram shows three parameters controlling the appearance of a 3-D view. α is the viewing azimuth, measured clockwise from the north. θ is the viewing angle, measured from the horizon. d is the viewing distance, measured between the observation point and the 3-D surface.

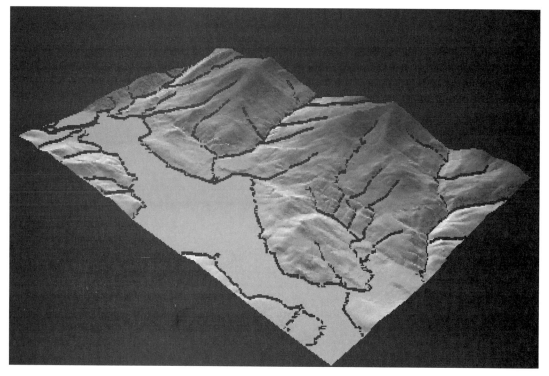

Figure 12.9
An example of 3-D draping. In this case, streams and shorelines are draped on a 3-D surface.

The 3-D Analyst extension to ArcView provides graphical interfaces for manipulating the viewing parameters. Therefore, GIS users can easily rotate the surface, navigate the surface, or take a close-up view of the surface. To make perspective views even more realistic, they can be superimposed with thematic layers such as land cover, vegetation, and roads in a process called **3-D draping** (Figure 12.9). Perspective views represent an exciting method for portraying the land surface.

12.4 TERRAIN ANALYSIS

12.4.1 Slope and Aspect

Slope measures the rate of change of elevation at a surface location, and **aspect** is the directional measure of slope. If we define the elevation (z) of a point on the land surface as a function of the point's position (x and y), then we can define slope (S) at the point as a function of the first-order derivatives of the surface in the x and y directions:

$$S = ((\partial z/\partial x)^2 + (\partial z/\partial y)^2)^{0.5} \qquad (12.3)$$

And we can define the slope's directional angle as

$$A = \arctan ((\partial z/\partial y) / (\partial z/\partial x)) \qquad (12.4)$$

Slope may be expressed as percent slope or degree slope. Percent slope is 100 times the ratio of rise (vertical distance) over run (horizontal distance), whereas degree slope is the arc tangent of the ratio of rise over run (Figure 12.10). Aspect (A) is a directional measure in degree. Aspect starts with 0^0 at the north, moves clockwise, and ends with 360^0 also at the north (Figure 12.11).

Aspect is a circular measure. Therefore, an aspect of 10^0 is closer to 360^0 than to 30^0. GIS users often have to transform aspect measures before they can use aspects in numerical analysis. One common method is to classify aspects into four principal directions (north, east, south, and west) or eight principal directions (north, northeast, east, southeast, south, southwest, west, and northwest) and to treat aspects as categorical data. Another method is to transform aspect values to capture the principal directions (Figure 12.12) (Chang and Li 2000). To capture the N-S principal direction, for example, one can set 0^0 at north, 180^0 at south, and 90^0 at both west and east.

As basic elements for analyzing and visualizing landform characteristics, slope and aspect are important in studies of watershed units, landscape units, and morphometric measures (Moore et al. 1991). When used with other variables, slope and aspect can assist in solving problems in forest inventory estimates, soil erosion, wildlife habitat suitability, site analysis, and many other fields.

12.4.1.1 Computing Algorithms for Slope and Aspect Using Grid

When an elevation grid is used as the data source, slope and aspect are computed for each cell in the grid. One can measure the slope and aspect for a cell by the quantity and direction of tilt of the cell's normal vector—a directed line perpendicular to the cell (Figure 12.13). Given a normal vector (n_x, n_y, n_z), the formula for computing the cell's slope is

$$(n_x^2 + n_y^2)^{0.5} / n_z \qquad (12.5)$$

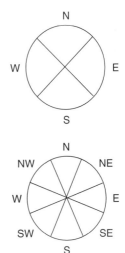

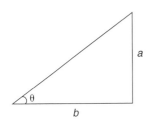

Figure 12.10
Percent slope in the diagram is 100 x (a/b). a is the vertical distance or rise, and b is the horizontal distance or run. Degree slope can be calculated by arctan (a/b).

Figure 12.11
Aspect is a directional measure in degrees. Aspect measures are often grouped into 4 principal directions (top) or 8 principal directions (bottom).

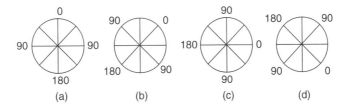

Figure 12.12
Transformation methods to capture the N-S direction (a), the NE-SW direction (b), the E-W direction (c), and the NW-SE direction (d).

And the formula for computing the cell's aspect is

$$\arctan(n_y / n_x) \qquad (12.6)$$

Different approximation methods have been proposed for computing slope and aspect. Here we will examine three common methods. All three methods use a 3 x 3 moving window to estimate the slope and aspect of the center cell, but they differ in the number of neighboring cells used in the estimation and the weight applying to each cell.

The first method, which is attributed to Fleming and Hoffer (1979) and Ritter (1987), uses the four immediate neighbors of the center cell. The slope (S) at C_0 in Figure 12.14 can be computed by

$$S = ((e_1\text{-}e_3)^2 + (e_4\text{-}e_2)^2)^{0.5} / 2d \qquad (12.7)$$

where e_i are the neighboring cell values, and d is the cell size. The n_x component of the normal vector to C_0 is $(e_1\text{-}e_3)$, or the elevation difference in the x dimension. The n_y component is $(e_4\text{-}e_2)$, or the elevation difference in the y dimension. To compute the percent slope at C_0, one can multiply S by 100.

S's directional angle D can be computed by

$$D = \arctan((e_4\text{-}e_2) / (e_1\text{-}e_3)) \qquad (12.8)$$

D is measured in radian and is with respect to the x-axis. To convert D to aspect, that is, measured in degree and from a north base of 0°, the following algorithm (Ritter 1987; Hodgson 1998) can be used:

```
If S <> 0 Then
     T = D x 57.296
     If nx = 0
          If ny < 0 Then
               Aspect = 180
          Else
               Aspect = 360
     ElseIf nx > 0 Then
               Aspect = 90 − T
     Else     'nx < 0
               Aspect = 270 − T
Else     'S = 0
     Aspect = −1 'undefined aspect for flat surface
END If
```

The second method for computing slope and aspect is called Horn's algorithm (1981), an algorithm used in ARC/INFO and ArcView. Horn's algorithm uses eight neighboring cells and applies a weight of 2 to the four immediate neighbors and a weight of 1 to the four corner cells. Horn's algorithm computes slope at C_0 in Figure 12.15 by

$$S = (((e_1 + 2e_4 + e_6) − (e_3 + 2e_5 + e_8))^2 + ((e_6 + 2e_7 + e_8) − (e_1 + 2e_2 + e_3))^2)^{0.5} / 8d \qquad (12.9)$$

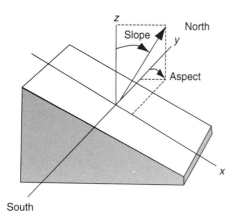

Figure 12.13

The normal vector to the cell is the directed line perpendicular to the cell. The quantity and direction of tilt of the normal vector determine the slope and aspect of the cell. (Redrawn from Hodgson, 1998, CaGIS vol. 25, no. 3, pp. 173–185; reprinted with the permission of the American Congress on Surveying and Mapping.)

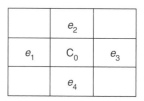

Figure 12.14

Ritter's algorithm for computing slope and aspect at C_0 uses the four immediate neighbors of C_0.

And the D value at C_0 is computed by

$$D = \arctan(((e_6 + 2e_7 + e_8) - (e_1 + 2e_2 + e_3)) / ((e_1 + 2e_4 + e_6) - (e_3 + 2e_5 + e_8))) \quad (12.10)$$

D can be converted to aspect by using the same algorithm for the first method except that $n_x = (e_1 + 2e_4 + e_6)$ and $n_y = (e_3 + 2e_5 + e_8)$ (Box 12.4).

The third method called Sharpnack and Akin's algorithm (1969) also uses eight neighboring cells but applies the same weight to every cell. The formula for computing S is

$$S = (((e_1 + e_4 + e_6) - (e_3 + e_5 + e_8))^2 + ((e_6 + e_7 + e_8) - (e_1 + e_2 + e_3))^2)^{0.5} / 6d \quad (12.11)$$

And the formula for computing D is

$$D = \arctan(((e_6 + e_7 + e_8) - (e_1 + e_2 + e_3)) / ((e_1 + e_4 + e_6) - (e_3 + e_5 + e_8))) \quad (12.12)$$

12.4.1.2 Computing Algorithms for Slope and Aspect Using TIN

The algorithms for computing slope and aspect for a triangle in a TIN also use the bi-directional normal vector, that is, the vector perpendicular to the triangular surface. Suppose a triangle is made of the following three nodes: A (x_1, y_1, z_1), B (x_2, y_2, z_2), and C (x_3, y_3, z_3) (Figure 12.16). The normal vector is a cross product of vector AB, $((x_2 - x_1), (y_2 - y_1), (z_2 - z_1))$, and vector AC, $((x_3 - x_1), (y_3 - y_1), (z_3 - z_1))$. And the three components of the normal vector are:

$$n_x: (y_2 - y_1)(z_3 - z_1) - (y_3 - y_1)(z_2 - z_1)$$
$$n_y: (z_2 - z_1)(x_3 - x_1) - (z_3 - z_1)(x_2 - x_1)$$
$$n_z: (x_2 - x_1)(y_3 - y_1) - (x_3 - x_1)(y_2 - y_1) \quad (12.13)$$

The S and D values of the triangle can be derived from Equations 12.5 and 12.6, and the D value can then be converted to the aspect measured in degree and from a north base of 0^0 (Box 12.5).

12.4.1.3 Factors Influencing Slope and Aspect Measures

Computing algorithms can influence slope and aspect measures. Using a digitized contour map of moderate topography in Australia, Skidmore (1989) compared six algorithms, including the three described in the previous section. He reported that Horn's algorithm and Sharpnack and Akin's algorithm were among the best for estimating both slope and aspect. Comparing five algorithms using a synthetic surface and a study area in Tennessee, Hodgson (1998) found Ritter's algorithm to be consistently more accurate in estimating slope than the methods using eight neighboring cells. Jones (1998) compared eight algorithms using a synthetic surface and a DEM in Scotland and reported that Ritter's algorithm was the best in estimating slope and aspect, followed by Horn's algorithm and Sharpnack and Akin's algorithm.

The accuracy of slope and aspect measures is probably influenced by the resolution and quality of DEM to a greater degree than the computing algorithm. Isaacson and Ripple (1990) reported greater amounts of detail in the slope and aspect maps created from the 7.5-minute DEM than from

e_1	e_2	e_3
e_4	C_0	e_5
e_6	e_7	e_8

Figure 12.15
Horn's algorithm for computing slope and aspect at C_0 uses the eight neighboring cells of C_0. The algorithm also applies a weight of 2 to e_2, e_4, e_5, and e_7, and a weight of 1 to e_1, e_6, e_3, and e_8.

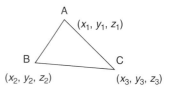

Figure 12.16
The algorithm for computing slope and aspect of a triangle in a TIN uses the x, y, and z values at the three nodes of the triangle.

Box 12.4 **A Worked Example of Computing Slope and Aspect Using Grid**

The following diagram shows a 3 x 3 window of an elevation grid. Elevation readings are measured in meters, and the cell size is 30 meters.

1006	1012	1017
1010	1015	1019
1012	1017	1020

This example first computes the slope and aspect of the center cell using Horn's algorithm:

n_x = (1006 + 2x1010 + 1012) − (1017 + 2x1019 + 1020) = −37

n_y = (1012 + 2x1017 + 1020) − (1006 + 2x1012 + 1017) = 19

$S = ((−37)^2 + (19)^2)^{0.5} / 8x30 = 0.1733$

S_p = 100 x 0.1733 = 17.33

D = arctan (n_y/n_x) = arctan (19/−37) = −0.4744

T = −0.4744 x 57.296 = −27.181

Because $S <> 0$ and $n_x < 0$

Aspect = 270 − (−27.181) = 297.181

For comparison, S_p has a value of 17.16 using the Fleming and Hoffer's algorithm and a value of 17.39 using the Sharpnack and Akin's algorithm. Aspect has a value of 299.06 using the Fleming and Hoffer's algorithm and a value of 296.56 using the Sharpnack and Akin's algorithm.

Box 12.5 **A Worked Example of Computing Slope and Aspect Using TIN**

A triangle in a TIN is made of the following nodes with their *x*, *y*, and *z* values measured in meters.

Node 1: $x_1 = 532260$, $y_1 = 5216909$, $z_1 = 952$

Node 2: $x_2 = 531754$, $y_2 = 5216390$, $z_2 = 869$

Node 3: $x_3 = 532260$, $y_3 = 5216309$, $z_3 = 938$

n_x = (5216390 − 5216909) (938 − 952) − (5216309 − 5216909) (869 − 952) = − 42534

n_y = (869 − 952) (532260 − 532260) − (938 − 952) (531754 − 532260) = −7084

n_z = (531754 − 532260) (5216309 − 5216909) − (532260 − 532260) (5216390 − 5216909) = 303600

S_p = 100 x $(((− 42534)^2 + (− 7084)^2)^{0.5} / 303600$ = 14.20

D = arctan (−7084 / −42534) = 0.165

T = 0.165 x 57.296 = 9.454

Because $S <> 0$ and $n_x < 0$, aspect = 270 − T = 260.546

---

the 1-degree DEM. The 7.5-minute DEM has a sampling interval of 30 meters, whereas the 1-degree DEM has a sampling interval of about 90 meters. Chang and Tsai (1991) found that the accuracy of the estimated slope and aspect decreased with a decreasing DEM resolution from 20 to 80 meters. Gao (1998) reported that the reliability of

both slope and aspect measures decreased with an increasing sampling interval of the DEM from 20 to 60 meters, and the inverse relationship between the reliability of slope measure and the DEM sampling interval was statistically significant.

In a study of assessing the influence of DEM quality, Bolstad and Stowe (1994) compared a

---

Box 12.6   **A Worked Example of Computing Surface Curvature**

1017	1010	1017
1012	1006	1019
1015	1012	1020

e_1	e_2	e_3
e_4	e_0	e_5
e_6	e_7	e_8

The diagram above represents a 3 x 3 window of an elevation grid, with a cell size of 30 meters. This example shows how to compute the profile curvature, plan curvature, and surface curvature at the center cell. The first step is to estimate the coefficients D–H of the quadratic polynomial equation that fits the 3 x 3 window:

$$D = ((e_4 + e_5) / 2 - e_0) / L^2$$

$$E = ((e_2 + e_7) / 2 - e_0) / L^2$$

$$F = (-e_1 + e_3 + e_6 - e_8) / 4L^2$$

$$G = (-e_4 + e_5) / 2L$$

$$H = (e_2 - e_7) / 2L$$

where e_0 to e_8 are elevation values within the 3 x 3 window according to the diagram below, and L is the cell size.

Profile curvature $= -2 ((DG^2 + EH^2 + FGH) / (G^2 + H^2)) = -0.0211$

Plan curvature $= 2 ((DH^2 + EG^2 - FGH) / (G^2 + H^2)) = 0.0111$

Curvature $= -2 (D + E) = -0.0322$

All three measures are based on 1/100 (z-units). The negative curvature value means the surface at the center cell is upwardly concave. The elevation grid above shows the center cell is like a shallow basin surrounded by higher elevations in the neighboring cells.

USGS 7.5-minute DEM and a DEM produced from a SPOT panchromatic stereopair for the same study area. They reported that slope and aspect errors for the SPOT DEM were statistically significant, while slope and aspect errors for the USGS DEM were not significantly different from zero.

Local topography can also influence the accuracy of estimated slopes and aspects. Chang and Tsai (1991) reported that errors in slope estimates were greater in areas of higher slopes, but errors in aspect estimates were greater in areas of lower relief. Carter (1992) attributed errors in the calculation of aspect, especially in areas of low slope, to the problem of data precision, that is, the rounding of elevations to the nearest whole number. Bolstad and Stowe (1994) found larger slope errors on steeper slopes and speculated that the correlation was likely due in part to difficulties in stereo-correlation in forested terrain. Florinsky (1998)

reported higher aspect and slope errors within flat areas due to the data precision problem.

12.4.2 Surface Curvature

GIS applications in hydrological studies often require computation of surface curvature to determine if the surface at a cell location is upwardly convex or concave. A common method, which is used by the CURVATURE command in ARC/INFO, is to fit a 3 x 3 window with a quadratic polynomial equation (Zevenbergen and Thorne 1987; Moore et al. 1991):

$$z = Ax^2y^2 + Bx^2y + Cxy^2 + Dx^2 + Ey^2 + Fxy + Gx + Hy + I \qquad (12.14)$$

The coefficients A–I can be estimated by using the elevation values in the 3 x 3 window and the grid cell size (Box 12.6). Three curvature measures can then be computed from the coefficients:

Profile curvature $= -2\,((DG^2 + EH^2 + FGH)\,/\,(G^2 + H^2))$ (12.15)

Plan curvature $= 2\,((DH^2 + EG^2 - FGH)\,/\,(G^2 + H^2))$ (12.16)

Curvature $= -2\,(D + E)$ (12.17)

Profile curvature estimates along the direction of maximum slope. Plan curvature estimates orthogonal to, or across, the direction of maximum slope. And curvature measures the difference between the two, that is, (profile curvature − plan curvature). A positive curvature value at a cell means that the surface is upwardly convex at the cell. A negative curvature value means that the surface is upwardly concave. And a 0 value means that the surface is flat.

Figure 12.17a
Viewshed analysis divides the study area into not visible from observation points, visible from observation point 1, and visible from observation point 2.

12.4.3 Viewshed Analysis

The **viewshed** refers to areas of the land surface that are visible from an observation point, or observation points (Figure 12.17). The basis for viewshed analysis is the line of sight operation, the operation of determining whether a given target is visible from an observation. Viewshed analysis expands the operation to cover every possible point (or cell using the raster format) in the study area as the target. For more than one observation point, the operation is repeated for each point. The output of viewshed analysis is a binary map: visible and invisible areas.

The selection of observation points is crucial for viewshed analysis. To gain as much visibility as possible, observation points should be located in areas of high elevation. Data query using elevation values, graphics, or the cursor can help GIS users locate areas ideal for observation points. For some viewshed analysis, the physical structure of the observation station can add height to the point location and make it higher than its immediate surroundings. For example, a forest lookout station is usually 15 to 20 meters high. In ArcView, the height of the observation station is added to the elevation of the point location in a field called Spot.

Applications of viewshed analysis include siting forest lookout stations, selecting locations for housing and resort area developments, and evaluating scenic quality along a highway or a river. Viewshed analysis can also help locate antennas for wireless communication. Depending on the size of the grid or TIN and the number of observation points, viewshed analysis can be computationally intensive, especially if observation points are distributed continuously along a linear feature such as a highway.

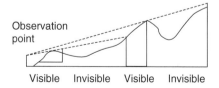

Figure 12.17b
Viewshed analysis is based on the line of sight operation.

12.4.4 Watershed Analysis

A **watershed** is an area that drains water and other substances to a common outlet. A watershed is also called a basin or catchment. Elevation grids and raster data operations are used in **watershed analysis** for derivation of topographic features such as watersheds and stream networks, which are important in characterizing the hydrologic process (Moore 1996).

Watershed analysis usually requires three data sets in raster format: a filled elevation grid, a flow direction grid, and a flow accumulation grid. A **filled elevation grid** is void of depressions. A depression is a cell or cells that are surrounded by higher elevation values, and thus represents an area of internal drainage. Although some depressions are natural, such as quarries or glaciated potholes, many are imperfections in the DEM. Therefore depressions must be removed from an elevation grid. One method for removing a depression is to increase its cell value to the lowest cell value surrounding the depression (Jenson and Domingue 1988).

A **flow direction grid** shows the direction water will flow out of each cell of a filled elevation grid. The most common method for determining a cell's flow direction, which is used by ARC/INFO and ArcView, is to find the steepest distance-weighted gradient to one of its eight surrounding cells (Figure 12.18). One limitation of the method is that it does not allow flow to be distributed to multiple cells (Moore 1996).

A **flow accumulation grid** tabulates for each cell the number of cells that will flow to it (O'Callaghan and Mark 1984). Cells having high accumulation values generally correspond to stream channels, whereas cells having an accumulation value of zero generally correspond to ridge lines (Figure 12.19a). Therefore a fully connected drainage network can be derived from a flow accumulation grid by using some threshold accumulation value (Jenson and Domingue 1988).

Watersheds can be delineated for selected points or for an entire grid (Figure 12.19b). A specific watershed can be derived for a point such as a hydrologic station by following the ascending flow paths from the point. Watersheds can be delineated for an entire grid by using a user-defined minimum size in each watershed and drainage line intersections as the starting points (Jenson and Domingue 1988).

The Hydrologic Modeling extension to ArcView provides the menu interface for GIS users to create a filled elevation grid, a flow direction grid, and a flow accumulation grid from an elevation grid and to delineate watersheds from the data set.

12.5 GRID VERSUS TIN

ARC/INFO and ArcView allow GIS users to use either grids or TINs for land surface mapping and analysis, although the two GIS packages differ in the choice of algorithms and the type of output (Box 12.7). ARC/INFO and ArcView also provide algorithms for conversion from a grid to a TIN, or from a TIN to a grid. Given the options, GIS users might have the question as to which data model to use. There is no easy answer to the question. Grids and TINs differ in two areas: data flexibility and computational efficiency.

A main advantage of using a TIN lies in the flexibility with input data sources. One can construct a TIN using inputs from DEM, breaklines, contour

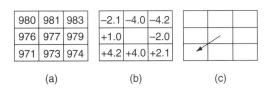

(a)   (b)   (c)

Figure 12.18

The flow direction of the center cell in (a) is determined by first calculating the distance-weighted gradient to each of its eight neighbors. For the four immediate neighbors, the gradient is calculated by dividing the elevation difference between the center cell and the neighbor by 1. For the four corner neighbors, the gradient is calculated by dividing the elevation difference by 1.414. The results in (b) show that the steepest gradient, and therefore the flow direction, is from the center cell to the lower left cell ($+4.2$).

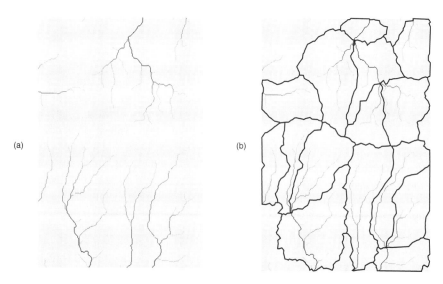

Figure 12.19

A connected drainage network appears in (a), which is a flow accumulation grid. The darkness of the symbol corresponds to the flow accumulation value. The delineation of watersheds is shown in (b). A ridgeline in the middle of (b) separates watersheds to the north from those to the south. Watershed boundaries cannot be determined for some areas along the border of the grid.

Box 12.7   **TINs in ARC/INFO and ArcView**

ARC/INFO and ArcView differ in the construction and processing of TINs. ARC/INFO offers both VIP and maximum z-tolerance for selecting points from an elevation grid to be included in a TIN. VIP is incorporated in the CREATETIN command, whereas the maximum z-tolerance algorithm is incorporated in the LATTICETIN command. ArcView's 3-D Analyst offers only the maximum z-tolerance algorithm.

Perhaps more important to the GIS user is the processing of TINs. ARC/INFO users can convert a TIN to a polygon coverage and then derive slope and as-

pect coverages from the polygon coverage. ArcView users cannot convert a TIN to a polygon coverage because ArcView uses non-topological shapefiles. Although ArcView allows users to derive slope or aspect from a TIN, the output is a grid rather than a polygon coverage. The conversion from a TIN to a grid means loss of data accuracy. Therefore, the quality of slope and aspect grids derived from a TIN in ArcView is questionable. TINs in ArcView are used mainly for 3-D display.

---

lines, GPS data, and survey data. The user can add elevation points to a TIN at their precise locations or add line features, such as streams, roads, ridgelines, and shorelines, to define surface discontinuities. Because the GIS user put together a TIN, the quality of the TIN reflects the user's time and effort.

An elevation grid is fixed with a given cell size. One cannot add new sample points to an elevation grid to increase its surface accuracy. Assuming the production method is the same, the only way to improve the accuracy of a grid is to increase its resolution, for example, from 30 meters to 10 meters.

Researchers, especially those working with small watersheds, have in fact advocated DEMs with a 10-meter resolution (Zhang and Montgomery 1994). But increasing DEM resolution is a costly operation because it requires the re-compiling of elevation data and more computer memory.

Besides data flexibility, TIN is also an excellent data model for terrain mapping and 3-D display. The triangular facets of a TIN better define the land surface than an elevation grid and create a sharper image. Most GIS users seem to prefer the look of a map based on a TIN rather than an elevation grid (Kumler 1994).

Computational efficiency is the main advantage of using grids for terrain analysis. The simple data structure makes it relatively easy to perform local, neighborhood, zonal, and global operations on an elevation grid. Therefore, computations of slope, aspect, surface curvature, relative radiance, and other topographic variables using an elevation grid are fast and efficient. In contrast, the computational load using a TIN can increase significantly as the number of triangles increases. For some operations in ARC/INFO and ArcView, TINs are in fact converted to elevation grids prior to data analysis.

Finally, which data model is more accurate in measuring elevation, slope, aspect, and other land surface parameters? Kumler (1994) made a series of comparisons between two types of DEM, a TIN derived from the VIP algorithm, a TIN derived from the maximum z-tolerance algorithm, and three other TINs derived from contour lines. He concluded that TINs made of points sampled from DEMs were inferior to the full DEM, and the contour-based TINs were not as efficient as DEMs in terrain modeling. But Kumler's comparisons were limited to elevation estimates and did not cover slope, aspect, or other topographic parameters.

A TIN made of sampled points from a DEM obviously will not perform as well as the full DEM in estimating elevations. This is also true with contour-based TINs because they contain much smaller numbers of elevation points than contour-based DEMs. The main advantage of using a TIN for land surface analysis is data flexibility, which allows the GIS user to add point and linear features in areas where the topography changes rapidly. For example, a DEM cannot recognize a stream in a hilly area and its accompanying topographic characteristics if the stream width is smaller than the DEM resolution. A TIN will have no problem of depicting the stream as a breakline.

KEY CONCEPTS AND TERMS

3-D drape: The method of superimposing thematic layers such as vegetation and roads on perspective views.

Aspect: The directional measure of slope.

Breaklines: Line features that represent changes of the land surface such as streams, shorelines, ridges, and roads.

Contour interval: The vertical distance between contour lines.

Contour lines: Lines connect points of equal elevation.

Delaunay triangulation: An algorithm for connecting points to form triangles such that all points are connected to their nearest neighbors and triangles are as compact as possible.

Filled elevation grid: An elevation grid that is void of depressions.

Flow accumulation grid: A grid that shows for each cell the number of cells that will flow to it.

Flow direction grid: A grid that shows the direction water will flow out of each cell of a filled elevation grid.

Hill shade: A graphic method, which simulates how the land surface looks with the interaction between sunlight and landform features. The method is also known as *shaded relief* or *shading*.

Hypsometric tint: A mapping method, which applies color symbols to different elevation zones. The method is also known as *layer tinting*.

Maximum *z*-tolerance: A TIN construction algorithm, which ensures that, for each elevation point selected, the difference between the original elevation and the estimated elevation from the TIN is within the specified maximum *z*-tolerance.

Perspective view: A graphic method that produces 3-D views of the land surface.

Slope: The rate of change of elevation at a surface location, measured as an angle in degrees or as a percentage.

Vertical profile: A chart showing changes in elevation along a line such as a hiking trail, a road, or a stream.

Viewing angle: A parameter for creating a perspective view, which represents the angle measured from the horizon to the altitude of the observer.

Viewing azimuth: A parameter for creating a perspective view, which represents the direction from the observer to the surface.

Viewing distance: A parameter for creating a perspective view, which represents the distance between the viewer and the surface.

Viewshed: Areas of the land surface that are visible from an observation point, or observation points.

VIP: An elevation point selection algorithm, which evaluates the importance of an elevation point by measuring how well its value can be estimated from the neighboring point values.

Watershed: An area that drains water and other substances to a common outlet.

Watershed analysis: Analysis that involves derivation of flow direction, watershed boundaries, and stream networks.

Z-scale: A parameter for creating a perspective view, which is the ratio between the vertical scale and the horizontal scale. It is also called the *vertical exaggeration factor*.

APPLICATIONS: TERRAIN MAPPING AND ANALYSIS

The applications section includes four tasks. Task 1 uses DEM data for land surface mapping and analysis in ArcView. Task 2 performs a viewshed analysis. Tasks 3 and 4 deal with TINs: Task 3 builds and displays a TIN in ArcView, and Task 4 builds a TIN in ARC/INFO and derives a slope and aspect maps from the TIN.

Task 1: Terrain Mapping and Analysis Using DEM

What you need: *plne*, an elevation grid; *streams.shp*, a stream shapefile.

The elevation grid *plne* is imported from a USGS 7.5-minute DEM. The shapefile *streams.shp* shows major streams in the area. Task 1 covers terrain mapping and analysis using Spatial Analyst and 3-D Analyst in ArcView.

1.1 Create a contour theme

1. Start ArcView and load both 3-D Analyst and Spatial Analyst.
2. Open a new view, and add the grid data source, *plne*, and the shapefile, *streams.shp*, to view. Click the check box for *plne* to display the elevation grid. Make *plne* active.
3. Select Create Contours from the Surface menu. Enter 100 as the contour interval and 800 as the base contour. Elevations in *plne* range from 743 to 1986 meters.
4. Draw the newly created contour theme.

1.2 Create a vertical profile

1. Activate *streams.shp*, and open its theme table. Use the Query Builder to select 'usgh_id = 167,' a small tributary to be used for plotting a vertical profile. Because the

selected tributary is a planimetric feature, elevation changes along the tributary must be derived from an elevation grid.

2. Select Convert to 3-D Shapefile from the Theme menu. Select Surface to get Z values from, and *plne* as the surface. In the following dialog, set 30 meters as the sample distance on grid and name the 3-D shapefile as *profile.shp*. The sample distance of 30 meters is the same as the cell size of *plne*.

3. Activate *profile.shp*.

4. Open a new layout, and then click on the Profile Graph tool. Define the area within the layout to draw the vertical profile. Press the cursor at the upper left corner of the intended area and drag the cursor to the lower right, while holding down the mouse button. Release the mouse button.

5. In the Profile Graph Properties dialog, set the vertical exaggeration factor to 10. Click OK. The layout shows the vertical profile of the selected stream.

1.3 Create a hillshade theme

1. Activate *plne*, and select Compute Hillshade from the Surface menu. Take the default values of 315 as the azimuth and 45 as the altitude. In other words, the light source for creating the hillshade theme comes from the northwest, at an angle of 45^0 above the horizon. Click OK.

2. Draw the hill shade theme.

1.4 Create a perspective view

1. ArcView manages 3-D perspective views via the 3-D-Scenes document, which is added by 3-D Analyst. Select the 3-D-Scenes document and double click on the New button. A separate Table of Contents and a 3-D viewer are now opened.

2. Select Add Theme from the 3-D Scene menu. Add the elevation grid *plne* to view. The initial view of *plne* is 2-D. To change it to a 3-D view, select 3-D Properties from the

Theme menu. In the 3-D Theme Properties dialog, check Surface and click OK. The 3-D view has a (1) scale of 1, that is, with no vertical exaggeration.

3. The 3-D viewer has a set of tools that allow you to manipulate the 3-D view. The Navigate tool lets you view the surface from different viewing directions and distances. The Rotate Viewer tool automatically rotates the 3-D view in a counterclockwise direction. You can stop the rotation by clicking on the Stop button. You also have tools to zoom in or zoom out, or to identify or select features.

4. Next, you want to drape streams on the 3-D view. Select Add Theme from the 3-D Scene menu, and select the feature theme *streams.shp*. Because *streams.shp* is a 2-D theme, you must convert it to a 3-D shapefile. Activate *streams.shp* and select Convert to 3-D Shapefile from the Theme menu. In the following dialog, select Surface to get z values from, *plne* as the surface, and 30 meters as the sample distance. Then, name the new shapefile as *strm3d.shp*. Display *strm3d.shp* and streams are draped on the 3-D surface.

5. You can work with the View document alongside the 3-D-Scenes document. For example, you can create a hill shade of *plne* and display it in 3-D view.

1.5 Create a slope theme

1. Activate *plne*, and select Derive Slope from the Surface menu to create a slope theme from *plne*. Display the slope theme by clicking on its check box.

2. The slope theme is a floating-point grid. The default classification divides slope values (in degrees) into nine equal-interval classes. You want to use a legend with meaningful class breaks. Select Reclassify from the Analysis menu. Click on the Classify button. Change the number of classes to 5, and click OK. In the Reclassify Values dialog, change Old

Values to 0–10, 10–20, 20–30, 30–40, and 40–55 respectively and click OK. The new theme, *Reclass of Slope*, classifies slope values with a 10-degree interval. Draw the new theme.

3. Reclassify not only changes the classification but also converts the slope theme from a floating-point grid to an integer grid. One advantage of having an integer grid is that you can display its attribute data. Select Table from the Theme menu to display the attributes of value and count of *Reclass of Slope*. Value represents the re-classed slope value: 1 for 0–10^0, 2 for 10–20^0, and so on. Count shows the number of cells in each slope category. You can derive the area percentage of a slope category by dividing its count by the total count.

1.6 Create an aspect theme

1. Activate *plne*, and select Derive Aspect from the Surface menu to create an aspect theme. The aspect theme is displayed in nine categories: flat, north, northeast, east, southeast, south, southwest, west, and northwest. But the aspect grid is actually a floating-point grid and does not have a theme table. Use Reclassify in the Analysis menu to convert the aspect grid to an integer grid or to change the aspect classification, for example, from the eight to four principal directions.

2. To convert the aspect grid to an integer grid but still keep the eight principal directions, use the following table in the Reclassify Values dialog:

Old Values	−1	0–22.5	22.5–67.5	67.5–112.5	112.5–157.5
New Values	−1	1	2	3	4
Old Values	157.5–202.5	202.5–247.5	247.5–292.5	292.5–337.5	337.5–360
New Values	5	6	7	8	1

The north aspect is made of two components in the table: 0–22.5° and 337.5–360°. Flat areas have a value of −1.

As with slope classes, you can derive the area percentage of different aspects by opening the attribute table of the *Reclass of Aspect* theme.

3. To convert the aspect grid to an integer grid and from the eight to four principal directions, use the following table in the Reclassify Values dialog:

Old Values	−1	0–45	45–135	135–225	225–315	315–360
New Values	−1	1	2	3	4	1

The reclassification groups the study area into flat areas and four principal aspects (north, east, south, and west). The north aspect consists of two sectors: one from 0° to 45°, and the other from 315° to 360°.

Task 2. Viewshed Analysis

What you need: *plne*, an elevation grid; and *lookouts.shp*, a lookout location shapefile.

The elevation grid *plne* is the same as for Task 1. The lookout location theme contains two points, labeled 1 and 2. A viewshed analysis determines areas in *plne* that are visible from the two lookout locations and areas that are not visible.

1. Start ArcView and load Spatial Analyst. Open a new view, and add *plne* and *lookouts.shp* to view.

2. Activate *lookouts.shp*. To label the two lookout locations, select Auto-label from the Theme menu. In the Auto-label dialog, select Id for the Label Field and click OK.

3. Activate both *lookouts.shp* and *plne*, because the viewshed analysis requires both themes. Select Calculate Viewshed from the Surface menu.

4. Draw the *Visibility of Lookouts.shp* theme. The study area is categorized as either visible or not visible. Activate the visibility theme and open its theme table. The first record in the table shows the count of cells not visible from the lookout points. The second record shows the count of cells visible from one lookout point, and the third record the count of cells visible from two lookout points.

You can also create your own lookout shapefile for viewshed analysis by creating a new theme and by adding point(s) to the theme. The following shows you how to create a lookout shapefile:

1. Select New Theme from the View menu, and Point for the Feature Type. Name the new theme *newpoints.shp* and save it.
2. Click on the Draw Point tool. Click two or more points as your choices for the lookout locations. To exit from drawing points, click the Select Graphics tool and click anywhere outside the study area in view. Select Stop Editing from the Theme menu and save the changes.
3. This step is to label the ID of the lookout location. Make *newpoints.shp* active. Select Table from the Theme menu to open the table for *newpoints.shp*. Select Starting Editing from the Table menu. Click ID to make it active. Click on the Edit tool. Click the first cell under ID, and enter an ID value of 1. Repeat this procedure for entering ID's of other cells.
4. Select Stop Editing from the Table menu, and save the edits. Your own lookout shapefile is now ready to be used in a viewshed analysis.

Task 3. Build and Display TIN in ArcView

What you need: *emidalat*, an elevation grid; and *emidastrm.shp*, a stream shapefile.

Task 3 shows you how to construct a TIN from an elevation grid and to modify the TIN with *emidastrm.shp* as breaklines. You will also display the TIN with elevation, slope, and aspect data.

1. Start ArcView and load 3-D Analyst and Spatial Analyst. Add *emidalat* and *emidastrm.shp* to view.
2. The first part of the task is to construct a TIN from *emidalat*. Make *emidalat* active. Select Convert Grid to TIN from the Theme menu. Name the output TIN *emidatin*. Enter 10 (meters) as the z-value tolerance. As explained in the chapter, 3-D Analyst uses the maximum z-tolerance to make a TIN. The input value of 10 meters means that the TIN

cannot differ in height by more than 10 meters from *emidalat* at any cell location.

3. Next, use *emidastrm.shp* to modify *emidatin*. Activate *emidatin*. Select Starting Editing from the Theme menu. The check box next to *emidatin* in the Table of Contents changes to dashed lines, meaning that *emidatin* is ready for modification. Activate *emidastrm.shp*. Select Add Features to TIN from the Surface menu. In the Create new TIN dialog, choose Spot for the Height Source, and Hard Breaklines for the Input. Hard breaklines provide edges of new triangles for reconstructing the TIN. Click OK. You have modified *emidatin*. Activate *emidatin*. Select Stop Editing from the Theme menu, and save edits to *emidatin*.
4. Now you can view *emidatin* in several ways. Open *emidatin*'s legend editor. In the Legend Editor dialog, check the box next to Points and turn off Lines and Faces. The Points legend type includes Single Symbol and Elevation Range. Try Single Symbol first and click Apply. The point symbols show the nodes of *emidatin*. Notice more points are located along streams and in hilly areas. Next try Elevation Range. The point symbols are color coded by elevation range.
5. To see triangles in *emidatin*, turn on Lines with the Single Symbol legend and turn off Points and Faces. Click Apply.
6. The legend type for Faces includes Single Symbol, Elevation Range, Slope, and Aspect. Faces refer to triangular facets in *emidatin*. The legend type allows you to display *emidatin* with symbols representing elevation ranges, slopes, or aspects.
7. You can also query triangles in *emidatin*. Change the Lines legend type to Single Symbol and the Faces legend type to Elevation Range. Click Apply. Then zoom into a small area so that you can clearly see the triangles. Press the Identify tool. Click several points within a triangle. You should see in the Identify Results dialog that the slope and aspect values are constant, but that the elevation values change each time you click a new point.

8. The last part of this task is to view *emidatin* as a 3-D scene. Click 3-D Scenes in ArcView's Table of Contents, and open a new 3-D Scene Viewer. Add *emidatin* as a new theme to view. You should see a 3-D view of *emidatin*. Notice the 3-D view of *emidatin* is clear and sharp because of the triangular facets. Use the tools of the 3-D Viewer to rotate or pan *emidatin*. You can also zoom in and zoom out. The legend editor for the 3-D scene of *emidatin* is the same as in regular view. Therefore, you can change the display of *emidatin* by changing the legend types for points, lines, and faces.

Task 4. Build and Analyze TIN in ARC/INFO

What you need: *emidalat*, an elevation grid; and *breakstrm*, a stream coverage.

Like Task 3, Task 4 uses *emidalat* and *break-strm* to create a TIN and to derive slope and aspect from the TIN. But for Task 4, you will build the TIN in ARC/INFO using the VIP algorithm and derive slope and aspect maps from the TIN as polygon coverages.

1. The VIP algorithm selects "very important" points from an elevation grid to be included in a TIN. ARC/INFO allows users to specify a percentage of elevation points to be selected. The default is 10%.

   Arc: vip emidalat emidavip /*emidavip* is the output point coverage

2. The CREATETIN command in ARC/INFO can create a TIN using data from different sources. For this task, you will use two data sources: (1) elevation points selected by the VIP algorithm and (2) the hard breakline of *emidastrm*. Elevation along streams in *emidastrm* will be derived from the elevation grid *emidalat*. CREATETIN is a dialog command. You will use subcommands in the dialog to enter the input data.

   Arc: createtin emidatin2 /* *emidatin2* is the output tin
   : cover emidavip point /* input elevation points from *emidavip*
   : lattice emidalat breakstrm line hardline /* input *emidastrm* as hardline
   :end /* exit createtin

3. The next step converts *emidatin2* to a polygon coverage. Each triangle in *emidatin2* becomes a polygon with its slope and aspect measures. Slopes can be measured in percent or degree.

   Arc: tinarc emidatin2 emidapoly poly percent /*emidapoly* is the polygon coverage

4. *Emidapoly* contains the items of percent slope and aspect. Typically, percent slope values are grouped into classes so that slopecode 1 = 0–20%, slopecode 2 = 20–40%, and so on. After the slope classification is completed, *emidapoly* can be dissolved into a slope (polygon) map by using slopecode as the dissolve item. Likewise, *emidapoly* can be dissolved into an aspect (polygon) map by using aspectcode as the dissolve item. The following shows the command to dissolve *emidapoly* into a slope map called *emidaslope*:

   Arc: dissolve emidapoly emidaslope slopecode poly

REFERENCES

Bolstad, P.V., and T. Stowe. 1994. An Evaluation of DEM Accuracy: Elevation, Slope, and Aspect. *Photogrammetric Engineering and Remote Sensing* 60: 1327–32.

Carter, J.R. 1989. Relative Errors Identified in USGS Gridded DEMs. *Proceedings, AUTO–CARTO* 9, pp. 255–65.

Carter, J.R. 1992. The Effect of Data Precision on the Calculation of Slope and Aspect Using Gridded DEMs. *Cartographica* 29: 22–34.

Chang, K., and B. Tsai. 1991. The Effect of DEM Resolution on Slope and Aspect Mapping. *Cartography and Geographic Information Systems* 18: 69–77.

Chang, K., and Z. Li. 2000. Modeling Snow Accumulation with a Geographic Information System. *International Journal of Geographical Information Science* 14: 693–707.

Chen, Z.T., and J.A. Guevara. 1987. Systematic Selection of Very Important Points (VIP) from Digital Terrain Model for Constructing Triangular Irregular Networks. *Proceedings, AUTO-CARTO* 8, pp. 50–56.

Eyton, J.R. 1991. Rate-of-Change Maps. *Cartography and Geographic Information Systems* 18: 87–103.

Fleming, M.D., and R. M. Hoffer. 1979. *Machine Processing of Landsat MSS Data and DMA Topographic Data for Forest Cover Type Mapping.* LARS Technical Report 062879. Laboratory for Applications of Remote Sensing, Purdue University, West Lafayette, Indiana, USA.

Florinsky, I.V. 1998. Accuracy of Local Topographic Variables Derived From Digital Elevation Models. *International Journal of Geographical Information Systems* 12: 47–61.

Franklin, S.E. 1987. Geomorphometric Processing of Digital Elevation Models. *Computers & Geosciences* 13: 603–9.

Gao, J. 1998. Impact of Sampling Intervals on the Reliability of Topographic Variables Mapped from Grid DEMs at a Micro-Scale. *International Journal of Geographical Information Systems* 12: 875–90.

Hodgson, M.E. 1998. Comparison of Angles from Surface Slope/Aspect Algorithms. *Cartography and Geographic Information Systems* 25: 173–85.

Horn, B.K.P. 1981. Hill Shading and the Reflectance Map. *Proceedings of the IEEE* 69(1): 14–47.

Isaacson, D.L., and W.J. Ripple. 1990. Comparison of 7.5-Minute and 1-Degree Digital Elevation Models. *Photogrammetric Engineering and Remote Sensing* 56: 1523–27.

Jenson, S.K., and J.O. Domingue. 1988. Extracting Topographic Structure from Digital Elevation Data for Geographic Information System Analysis. *Photo-grammetric Engineering and Remote Sensing* 54: 1593–1600.

Jones, K.H. 1998. A Comparison of Algorithms Used to Compute Hill Slope As a Property of the DEM. *Computers & Geosciences* 24: 315–23.

Jones, T.A., D.E. Hamilton, and C.R. Johnson. 1986. *Contouring Geologic Surfaces with the Computer.* New York: Van Nostrand Reinhold Company.

Kumler, M.P. 1994. An Intensive Comparison of Triangulated Irregular Networks (TINs) and Digital Elevation Models (DEMs). *Cartographica* 31 (2): 1–99.

Lee, J. 1991. Comparison of Existing Methods for Building Triangular Irregular Network Models of Terrain from Grid Digital Elevation Models. *International Journal of Geographical Information Systems* 5: 267–85.

Moore, I.D. 1996. Hydrological Modeling and GIS. In M.F. Goodchild, L.T. Steyaert, B.O. Parks, C. Johnston, D. Maidment, M. Crane, and S. Glendinning (eds.). *GIS and Environmental Modeling: Progress and Research Issues.* Fort Collin, CO: GIS World Books, pp. 143–48.

Moore, I.D., R.B. Grayson, and A.R. Ladson. 1991. Digital Terrain Modelling: A Review of Hydrological, Geomorphological, and Biological Applications. *Hydrological Process* 5: 3–30.

O'Callaghan, J.F., and D.M. Mark, 1984. The Extraction of Drainage Networks from Digital Elevation Data. *Computer Vision, Graphics and Image Processing* 28: 323–44.

Ritter, P. 1987. A Vector-Based Slope and Aspect Generation Algorithm. *Photogrammetric Engineering and Remote Sensing* 53: 1109–11.

Sharpnack, D.A., and G. Akin. 1969. An Algorithm for Computing Slope and Aspect from Elevations. *Photogrammetric Engineering* 35: 247–48.

Skidmore, A.K. 1989. A Comparison of Techniques for Calculating Gradient and Aspect from a Gridded Digital Elevation Model. *International Journal of Geographical Information Systems* 3: 323–34.

Thelin, G.P., and R.J. Pike. 1991. *Landforms of the Conterminous United States: A Digital Shaded-Relief Portrayal.* Washington, D.C.: U.S. Geological Survey. Map I-2206. scale 1:3,500,000.

Tsai, V.J.D. 1993. Delaunay Triangulations in TIN Creation: An Overview and Linear Time Algorithm. *International Journal of Geographical Information Systems* 7: 501–24.

Zevenbergen, L.W., and C.R. Thorne. 1987. Quantitative Analysis of Land Surface Topography. *Earth Surface Processes and Landforms* 12: 47–56.

Zhang, W., and D. R. Montgomery. 1994. Digital Elevation Model Grid Size, Landscape Representation, and Hydrologic Simulations. *Water Resources Research* 30: 1019–28.

SPATIAL INTERPOLATION

13.1 INTRODUCTION

Besides the land surface that was covered in Chapter 12, GIS users also work with another type of surface, which may not be physically present but can be visualized in the same way as the land surface. Cartographers have called the second type of surface the statistical surface (Robinson et al. 1995). Examples of the statistical surface include precipitation, snow accumulation, water table, and population density.

How can a statistical surface be constructed? The answer is similar to that for the land surface except that input data are typically limited to a sample of point data. To make a precipitation map, for example, one will not find a regular array of weather stations like a digital elevation model (DEM). Therefore, a process of filling in data between sample points is required.

Spatial interpolation is a process of using points with known values to estimate values at other points. For example, the precipitation value at a location with no recorded data can be estimated through interpolation from known precipitation readings at nearby weather stations. In GIS applications, spatial interpolation is typically applied to a grid and estimates are made for all cells in the grid. Spatial interpolation is therefore a means of converting point data to surface data.

Spatial interpolation methods are generally grouped into global and local methods (Burrough and McDonnell 1998). The difference between the two groups lies in the use of control points, that is, points with known values, in estimating unknown values. A global method uses every control point available in estimating an unknown value, whereas a local method uses a sample of control points for estimation.

This chapter is divided into three main sections. Section 1 considers the role of known points in spatial interpolation. Section 2 covers global methods. Section 3 provides an overview of local methods and a comparison of local methods. Perhaps more than any other topic in GIS, spatial interpolation depends on the computing algorithm. Worked examples are included in this chapter to show how spatial interpolation is carried out mathematically. ArcView and ARC/INFO are used as examples in this chapter. Many other GIS packages also have spatial interpolation functionalities (Box 13.1).

13.2 CONTROL POINTS

Control points are points with known values. The number and distribution of control points can greatly influence the accuracy of spatial interpolation (Robinson et al. 1995). A basic assumption in spatial interpolation is that the value to be estimated at a point is more influenced by nearby control points than those that are farther away. To be

---

Box 13.1 | **A Survey of Spatial Interpolation among GIS Packages**

Spatial interpolation is usually included in a raster-based GIS package. The following lists some of the interpolation methods covered in this chapter and the GIS packages that carry them.

Trend surface: ARC/INFO, IDRISI, GRASS, ILWIS

Regression: ARC/INFO, IDRISI, GRASS (linear regression)

Inverse distance weighted interpolation: ARC/INFO, ArcView, IDRISI, GRASS, ILWIS, MFworks, SPANS, Vertical Mapper

Spline: ARC/INFO, ArcView, GRASS (regularized splines with tension)

Kriging: ARC/INFO, IDRISI, ILWIS, GRASS, MFworks, SPANS, Vertical Mapper

effective for estimation, control points should be well distributed within the study area. But this ideal situation is rare in real-world applications. A study area often contains data-poor areas, which represent a major problem in estimation.

Figure 13.1 is a map of 105 weather stations and their 30-year average annual precipitation data in Idaho. The map clearly shows two major data-poor areas in the state: the central part (Clearwater Mountains, Salmon River Mountains, and Lemhi Range), and the southwestern corner (Owyhee Mountains). As shown later in this chapter, these data-poor areas can cause problems in spatial interpolation.

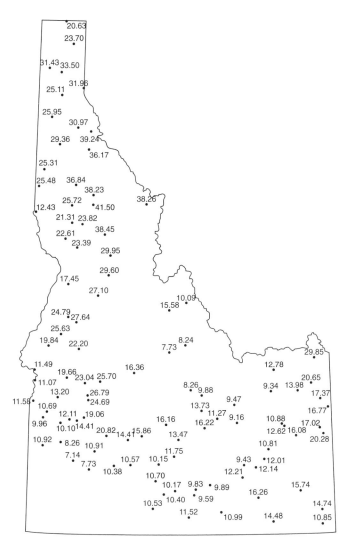

Figure 13.1

A map of 105 weather stations in Idaho and their 30-year average annual precipitation values.

13.3 GLOBAL METHODS

A **global interpolation method** uses every control point available to derive an equation or a model. The model can then be used to estimate unknown values.

13.3.1 Trend Surface Analysis

Trend surface analysis approximates points with known values with a polynomial equation (Davis 1986; Bailey and Gatrell 1995). The equation, which is called the trend surface model, can then be used to estimate values at other points. A linear or first-order trend surface uses the equation

$$z_{x,y} = b_0 + b_1 x + b_2 y \qquad (13.1)$$

where the attribute value z is a function of x and y. The b coefficients are estimated from the control points (Box 13.2). Because the trend surface model is computed by the least squares method, the "goodness of fit" of the surface can be measured by the coefficient of determination (R^2). Also, the deviation or the residual between the observed and the estimated values can be computed for each known point.

The distribution of most natural phenomena is usually more complex than an inclined plane surface from a first-order trend surface. Therefore, higher-order trend surface models are required to approximate more complex surfaces. For example, a cubic or third-order trend surface is based on the equation

$$z_{x,y} = b_0 + b_1 x + b_2 y + b_3 x^2 + b_4 xy + b_5 y^2 + b_6 x^3 + b_7 x^2 y + b_8 xy^2 + b_9 y^3 \qquad (13.2)$$

Unlike a linear surface, a cubic surface includes hills and valleys. Statistical tests (F tests) may be used to compare the fits of alternative models. GIS packages such as ARC/INFO offer up to 12th-order trend surface models. Figure 13.3 shows an isoline map of a third-order trend surface of annual precipitation in Idaho created from 105 data points. The trend surface output is a grid with a cell size of 2000 meters. An obvious problem with the model is the negative values in the data-poor southwest corner of the state.

13.3.2 Regression Models

A **regression model** relates a dependent variable to a number of independent variables in an equation, which can then be used for prediction or estimation. Many regression models use non-spatial attributes such as income and education and are not considered as methods for spatial interpolation. Exceptions, however, can be made for regression models that use spatial variables such as distance to a river (Burrough and McDonnell 1998). Chang and Li (2000), for example, developed a regression model using snow water equivalent (SWE) as the dependent variable and location and topographic variables as the independent variables. One of their watershed models takes the form of

$$SWE = b_0 + b_1 EASTING + b_2 SOUTHING + b_3 ELEV + b_4 PLAN1000 \qquad (13.3)$$

where EASTING and SOUTHING correspond to the column number and the row number in an elevation grid, ELEV is the elevation value, and PLAN1000 is a surface curvature measure. The b coefficients in Equation 13.3 were estimated from a sample of snow courses with known values in the watershed. The regression model was then used to estimate SWE for all cells in the watershed and to produce a continuous SWE surface.

13.3.3 Global Methods in ARC/INFO and ArcView

ARC/INFO's TREND command can run trend surface analysis from first to 12th order. REGRESSION in ARC/INFO runs regression analysis, but REGRESSION does not offer model-selection methods such as stepwise, maximum R^2, and so forth. ArcView does not have menu access to trend surface or regression analysis. ArcView users can run trend surface analysis by using Avenue scripts (Box 13.3).

13.4 LOCAL METHODS

A **local interpolation method** uses a sample of control points in estimating an unknown value.

Box 13.2 A Worked Example of Trend Surface Analysis

Figure 13.2 shows five weather stations with known values around point 0 with an unknown value. The table below shows the x-, y-coordinates of the points, measured in row and column of a grid with the cell size of 2000 meters, and their known values.

Point	x	y	z-Value
1	69	76	20.820
2	59	64	10.910
3	75	52	10.380
4	86	73	14.600
5	88	53	10.560
0	69	67	?

This example shows how we can use Equation 13.1, or a linear trend surface, to interpolate the unknown value at point 0. The least-squares method is commonly used to solve for the coefficients of b_0, b_1, and b_2 in Equation 13.1. Therefore, the first step is to set up three normal equations, similar to those for a regression analysis:

$$\Sigma z = b_0 n + b_1 \Sigma x + b_2 \Sigma y$$
$$\Sigma xz = b_0 \Sigma x + b_1 \Sigma x^2 + b_2 \Sigma xy$$
$$\Sigma yz = b_0 \Sigma y + b_1 \Sigma xy + b_2 \Sigma y^2$$

The equations can be rewritten in matrix form as

$$\begin{bmatrix} n & \Sigma x & \Sigma y \\ \Sigma x & \Sigma x^2 & \Sigma xy \\ \Sigma y & \Sigma xy & \Sigma y^2 \end{bmatrix} \cdot \begin{bmatrix} b_0 \\ b_1 \\ b_2 \end{bmatrix} = \begin{bmatrix} \Sigma z \\ \Sigma xz \\ \Sigma yz \end{bmatrix}$$

Using the values of the five known points, we can calculate the statistics and substitute the statistics into the equation

$$\begin{bmatrix} 5 & 377 & 318 \\ 377 & 29007 & 23862 \\ 318 & 23862 & 20714 \end{bmatrix} \cdot \begin{bmatrix} b_0 \\ b_1 \\ b_2 \end{bmatrix} = \begin{bmatrix} 67.270 \\ 5043.650 \\ 4445.800 \end{bmatrix}$$

We can then solve the b coefficients by multiplying the inverse of the first matrix on the left by the matrix on the right:

$$\begin{bmatrix} 23.210 & -0.163 & -0.168 \\ -0.163 & 0.002 & 0.000 \\ -0.168 & 0.000 & 0.002 \end{bmatrix} \cdot \begin{bmatrix} 67.270 \\ 5043.650 \\ 4445.800 \end{bmatrix} = \begin{bmatrix} -10.094 \\ 0.020 \\ 0.347 \end{bmatrix}$$

Using the coefficients, the unknown value at point 0 can be estimated by

$$P_0 = -10.094 + (0.020)(69) + (0.349)(67) = 14.669$$

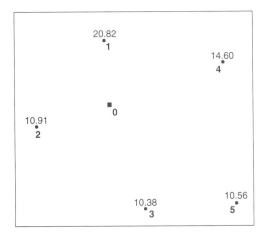

Figure 13.2
The unknown value at Point 0 is interpolated by five surrounding stations with known values.

The sampling of control points is therefore important to local methods. The first issue in sampling is the number of control points to be used in estimation. GIS packages typically allow users to specify the number of control points, or use a default number (e.g., 7 to 12 control points). One might assume that more control points would result in more accurate estimates. The validity of the assumption, however, depends on the distribution of control points relative to the cell to be estimated and the extent of spatial autocorrelation. More control points usually imply more generalized estimations.

After the number of control points is determined, the next task is to search for those control points. A simple option is to use the closest points to the point to be estimated. Another option is to select control points within a radius, the size of

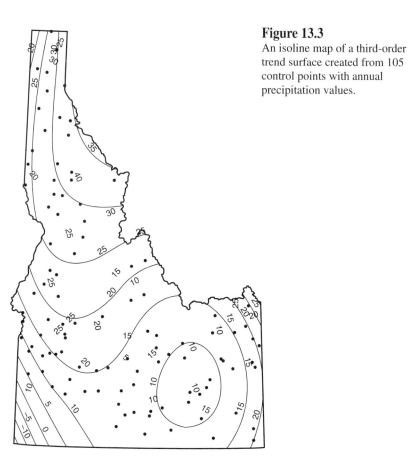

Figure 13.3
An isoline map of a third-order trend surface created from 105 control points with annual precipitation values.

Box 13.3 **Avenue Script for Trend Surface Analysis**

The following code can create an interpolated precipitation grid, which can then be used to make an isoline map similar to Figure 13.3.

‘Add the point theme to view and activate the theme.
theView = av.GetActiveDoc
theTheme = theView.GetActiveThemes.Get(0)

‘Prepare the analysis properties.
theAE = AnalysisPropertiesDialog.Show(theView,FALSE, "Analysys Properties")
theAE.Activate

‘MakeTrend prepares a third-order interpolator.
theInterp = interp.MakeTrend(3, FALSE)

‘Get the map projection of the point theme.
aPrj = theView.GetProjection

‘Set ann_prec as the Z field for interpolation.
theFTab = theTheme.GetFTab
theField = theFTab.FindField("ann_prec")

‘Interpolate a grid using theInterp.
theResult = Grid.MakeByInterpolation(theTheme.Get FTab,aPrj,theField,theInterp,NIL)

‘Make a grid theme from theResult and add it to view.
theGTheme = GTheme.Make(theResult)
theView.AddTheme(theGTheme)

which may have to be adjusted depending on the distribution of control points. ARC/INFO and Arc-View, for example, offer both options. Other search options incorporate the component of direction such as a quadrant or octant requirement (Davis 1986). A quadrant requirement means selecting control points from each of the four quadrants around a cell to be estimated. An octant requirement means using eight sectors.

13.4.1 Thiessen Polygons

Thiessen polygons are constructed around a sample of known points so that any point within a Thiessen polygon is closer to the polygon's known point than any other known points. Thiessen polygons were originally proposed to estimate areal averages of precipitation (Tabios and Salas 1985).

Thiessen polygons require initial triangulation among known points, that is, connecting known points to form triangles. Because different ways of connecting points can form different sets of triangles, the Delaunay triangulation—the same method for constructing Triangulated Irregular Network (TIN)—is often used in preparing Thiessen poly-

gons (Davis 1986). The Delaunay triangulation ensures that each known point is connected to its nearest neighbors, and that triangles are as equilateral as possible. After triangulation, Thiessen polygons can be easily constructed by connecting lines drawn perpendicular to the sides of each triangle at their midpoints (Figure 13.4). Thiessen polygons are also called *Voronoi polygons.*

13.4.2 Density Estimation

Density estimation measures densities in a grid based on a distribution of points and their known values. A simple density estimation method is to place a grid on a point distribution, tabulate points that fall within each cell, sum the point values, and estimate the cell's density by dividing the total point value by the cell size.

Figure 13.5 shows the input and output of an example of simple density estimation. The input is a distribution of sighted deer locations plotted with a 50-meter interval to accommodate the resolution of telemetry. Each deer location has a count value, which measures how many times a deer was sighted at the location. The output is a density grid:

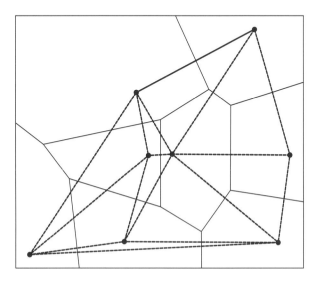

Figure 13.4
The diagram shows control points, Delaunay triangulation in dashed lines, and Thiessen polygons in thin solid lines.

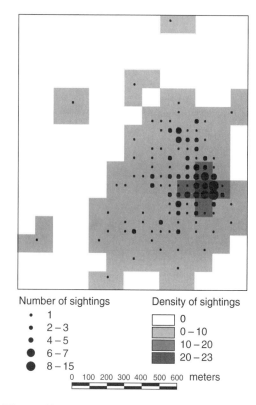

Number of sightings
- 1
- 2 – 3
- 4 – 5
- 6 – 7
- 8 – 15

Density of sightings
- 0
- 0 – 10
- 10 – 20
- 20 – 23

0 100 200 300 400 500 600 meters

Figure 13.5
The simple density estimation method is used to compute the number of deer sightings per hectare from the point data.

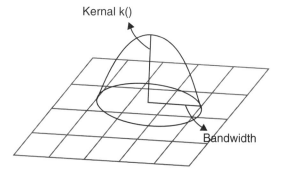

Kernal k()

Bandwidth

Figure 13.6
A kernel function, which represents a probability density function, looks like a "bump" above a grid.

density estimator at point x is then the sum of bumps placed at the observations, x_i, within the bandwidth:

$$\hat{f}(x) = \frac{1}{nh^d} \sum_{i=1}^{n} K\left(\frac{1}{h}(x - x_i)\right) \qquad (13.4)$$

where $K(\ )$ is the kernel function, h is the bandwidth, n is the number of observations within the bandwidth, and d is the data dimensionality. For two-dimensional data ($d = 2$), the kernel function is usually given by

$$K(x) = 3\pi^{-1}(1 - X^T X)^2, \text{ if } X^T X < 1 \qquad (13.5)$$
$$K(x) = 0, \text{ otherwise}$$

By substituting Equation 13.5 for $K(\ )$, Equation 13.4 can be rewritten as

$$\hat{f}(x) = \frac{3}{nh^2\pi} \sum_{i=1}^{n} \left(1 - \frac{1}{h^2}((x - x_i)^2 + (y - y_i)^2)\right)^2 \qquad (13.6)$$

where π is a constant, and $(x - x_i)$ and $(y - y_i)$ are the deviations in x-, y-coordinates between the point x and the observations x_i within the bandwidth.

Using the same input as for the simple estimation method, Figure 13.7 shows the output density grid from kernel estimation. Density values on the grid are expected values rather than probabilities (Box 13.4). Although the difference between Figures 13.7 and 13.5 is not apparent, kernel estimation usually produces a smoother density surface than the simple estimation method.

the cell size is 10,000 square meters or one hectare, and the density measure is the number of sightings per hectare. A circle, rectangle, wedge, or ring based at the center of a cell may replace the cell in the calculation.

Kernel estimation is a different density estimation method, which associates each point or observation with a kernel function (Silverman 1986; Scott 1992; Bailey and Gatrell 1995). Expressed as a bivariate probability density function, a kernel function looks like a "bump," centering at a point and tapering off to 0 over a defined bandwidth or window area (Silverman 1986) (Figure 13.6). The kernel function and the bandwidth determine the shape of the bump, which in turn determines the amount of smoothing in estimation. The kernel

Box **13.4**   **A Worked Example of Kernel Estimation**

Τ his example shows how the value of the cell marked X in Figure 13.7 is derived. The window area is defined as a circle with a radius of 100 meters (h). Therefore, only points within the 100-meter radius of the center of the cell can influence the estimation of the cell density. Using the 10 points within the cell's neighborhood, we can compute the cell density by

$$3/\pi \sum_{i=1}^{10} n_i (1-((x-x_i)^2 + (y-y_i)^2)/h^2)^2$$

where n_i is the number of sightings at point i, x_i and y_i are the x-, y-coordinates of point i, and x and y are the x-, y-coordinates of the center of the cell to be estimated. Because the density is measured per 10,000 square meters or hectare, h^2 in Equation 13.6 is cancelled out. Also, because the output shows an expected value rather than a probability, n in Equation 13.6 is not needed. The computation shows the cell density to be 11.421.

13.4.3 Inverse Distance Weighted Interpolation

The **inverse distance weighted interpolation** method is a local method that assumes that the unknown value of a point is influenced more by nearby control points than those farther away. The method is commonly used in computer-assisted mapping (Monmonier 1982). The degree of influence, or the weight, is expressed by the inverse of the distance between points raised to a power. A power of 1.0 means a constant rate of change in value between points, and the method is called linear interpolation. A power of 2.0 or higher suggests that the rate of change in values is higher near a known point and levels off away from it.

The general equation for the inverse distance weighted method is

$$z_0 = \frac{\sum_{i=1}^{s} z_i \frac{1}{d_i^K}}{\sum_{i=1}^{s} \frac{1}{d_i^K}} \quad (13.7)$$

where z_0 is the estimated value at point 0, z_i is the z value at control point i, d_i is the distance between control point i and point 0, s is the number of control points used in estimation, and k is the specified power.

Figure 13.8 shows an annual precipitation surface in Idaho created by the inverse distance squared method (that is, a power of 2) from 105

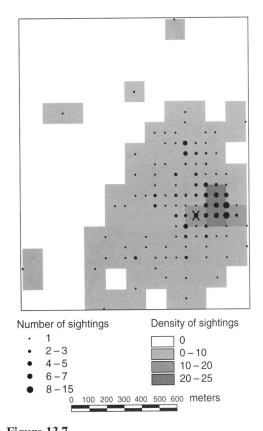

Number of sightings
· 1
· 2 – 3
• 4 – 5
• 6 – 7
• 8 – 15

Density of sightings
☐ 0
0 – 10
10 – 20
20 – 25

0 100 200 300 400 500 600 meters

Figure 13.7
The kernel estimation method is used to compute the number of sightings per hectare from the point data. The letter X marks the cell, which is used as an example in Box 13.4.

point values (Box 13.5). Figure 13.9 shows an iso-hyet (isoline of precipitation) map of the surface. An isoline map is easier to read and compare than a precipitation surface map. Small, enclosed isolines are typical of the inverse distance weighted method. Irregular isolines in the southwest corner of the state are due to the absence of control points.

13.4.4 Thin-plate Splines

Splines for spatial interpolation are conceptually similar to splines for line generalization, except that in spatial interpolation they apply to surfaces rather

than lines. **Thin-plate splines** create a surface that passes through control points and has the least pos-sible change in slope at all points (Franke 1982). In other words, thin-plate splines fit the control points with a minimum-curvature surface. The approxi-mation of thin-plate splines is of the form

$$Q(x,y) = \sum A_i d_i^2 \log d_i + a + bx + cy \qquad (13.8)$$

where x and y are the x-, y-coordinates of the point to be interpolated, $d_i^2 = (x - x_i)^2 + (y - y_i)^2$, and x_i and y_i are the x-, y-coordinates of control point i. Thin-plate splines consist of two components: ($a + bx + cy$) represents the local trend function, which has the same form as a linear or first-order trend surface, and $d_i^2 \log d_i$ represents a basis function,

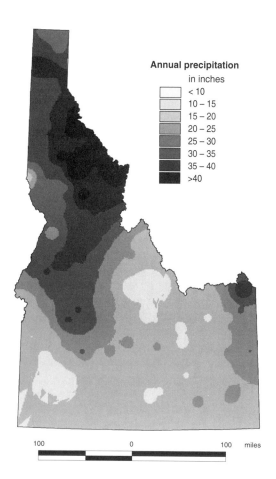

Figure 13.8
An annual precipitation surface map created by the inverse distance squared method.

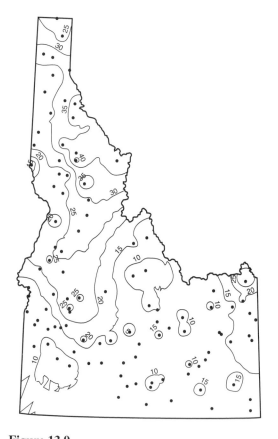

Figure 13.9
An isohyet map created by the inverse distance squared method.

which is designed to obtain minimum curvature surfaces (Watson 1992). The coefficients A_i, and a, b, and c are determined by a linear system of equations (Franke 1982)

$$\sum_{i=1}^{n} A_i d_i^2 \log d_i + a + bx + cy = f_i \quad (13.9)$$

$$\sum_{i=1}^{n} A_i = 0 \quad (13.10)$$

$$\sum_{i=1}^{n} A_i x_i = 0 \quad (13.11)$$

$$\sum_{i=1}^{n} A_i y_i = 0 \quad (13.12)$$

where n is the number of control points, and f_i is the known value at control point i. The estimation of the coefficients requires $n + 3$ simultaneous equations.

Other algorithms may be used for creating minimum-curvature surfaces. For example, instead of the basis function $d^2 \log d$ in Equation 13.9, the biharmonic Green function, $d^2(\log d - 1)$, may be used (Watson 1992; Middleton 2000).

One major problem with thin-plate splines is the steep gradients in data-poor areas, often referred to as overshoots. Different methods for correcting overshoots have been proposed, including thin-plate splines with tension (Franke 1985; Mitas and Mitasova 1988), regularized splines (Mitas and Mitasova 1988), and regularized splines with tension (Mitasova and Mitas 1993).

The approximation of **regularized splines** has the same local trend function as thin-plate splines, but the basis function has a different form

$$\frac{1}{2\pi} \left(\frac{d^2}{4} \left(\ln\left(\frac{d}{2\tau}\right) + c - 1 \right) + \tau^2 \left(K_0\left(\frac{d}{\tau}\right) + c + \right. \right.$$

$$\left. \left. \ln\left(\frac{d}{2\pi}\right) \right) \right) \quad (13.13)$$

where τ is the weight to be used with the splines method, d is the distance between the point to be interpolated and the control point i, c is a constant of 0.577215, and $K_0(d/\tau)$ is the modified zeroth-order Bessel function, which can be approximated by a polynomial equation (Abramowitz and Stegun 1964). The τ value is usually set between 0 and 0.5 because higher τ values tend to result in overshoots in data-poor areas. Both ARC/INFO and ArcView use the default τ value of 0.1.

The **thin-plate splines with tension** method has the following form

$$a + \sum_{i=1}^{n} A_i R(d_i) \quad (13.14)$$

where a represents the trend function, and the basis function $R(d)$ is

$$-\frac{1}{2\pi\phi^2} \left(\ln\left(\frac{d\phi}{2}\right) + c + K_0(d\phi) \right) \quad (13.15)$$

Box 13.5 **A Worked Example of Using the Inverse Distance Weighted Method for Estimation**

This example uses the same data set as in Box 13.2, but interpolates the unknown value at point 0 by the inverse distance squared method. The table below shows the distances in thousands of meters between point 0 and the five known points:

Between Points	Distance
0,1	18.000
0,2	20.880
0,3	32.310
0,4	36.056
0,5	47.202

We can substitute the known values and the distances into Equation 13.7 and estimate z_0:

$$\Sigma\, z_i\, 1/d_i^2 = (20.820)(1/18.000)^2 +$$
$$(10.910)(1/20.880)^2 + (10.380)(1/32.310)^2 +$$
$$(14.600)(1/36.056)^2 + (10.560)(1/47.202)^2 =$$
$$0.1152$$

$$\Sigma 1/d_i^2 = (1/18.000)^2 + (1/20.880)^2 +$$
$$(1/32.310)^2 + (1/36.056)^2 + (1/47.202)^2 =$$
$$0.0076$$

$$z_0 = 0.1152 / 0.0076 = 15.158$$

where ϕ is the weight to be used with the tension method. If the weight ϕ is set close to 0, then the approximation with tension is similar to the basic thin-plate splines method. A larger ϕ value reduces the stiffness of the plate and thus the range of interpolated values so that the interpolated surface resembles the shape of a membrane passing through the control points (Franke 1985). Both ARC/INFO and ArcView use the default ϕ value of 0.1.

Thin-plate splines and their variations are recommended for smooth, continuous surfaces such as elevation and water table. Splines have also been used for interpolating climate data such as mean rainfall (Hutchinson 1995). Figures 13.10 and 13.11 respectively show annual precipitation surfaces created by the regularized splines method and the splines with tension method (Box 13.6). The isolines in both figures are very smooth, compared to isolines in Figure 13.9, which were created by the inverse distance squared interpolation method.

Interpolation by splines in data-poor areas can be excessive, as exemplified by the 45-inch isohyet in north Idaho in Figure 13.10.

13.4.5 Kriging

Kriging (after the South African mining engineer, D.G. Krige) is a geo-statistical method for spatial interpolation. The technique of kriging assumes that the spatial variation of an attribute such as changes in grade within an ore body is neither totally random nor deterministic (Davis 1986; Isaaks and Srivastava 1989; Webster and Oliver 1990; Cressie 1991; Bailey and Gatrell 1995). Instead, the spatial variation may consist of three components: a spatially correlated component, representing the variation of the regionalized variable; a 'drift' or structure, representing a trend; and a random error term. The presence or absence of a drift and the interpretation of the regionalized variable

---

Box 13.6   A Worked Example of Thin-plate Splines

This example uses the same data set as in Box 13.2 but interpolates the unknown value at point 0 by the splines with tension method. The method first involves calculation of $R(d)$ in Equation 13.15 using the distances between the point to be estimated and the control points, distances between the control points, and the ϕ value of 0.1. The following table shows the $R(d)$ values along with the distance values.

Points	0,1	0,2	0,3	0,4	0,5
Distance	18.000	20.880	32.310	36.056	47.202
R(d)	-7.510	-9.879	-16.831	-18.574	-22.834
Points	1,2	1,3	1,4	1,5	2,3
Distance	31.240	49.476	34.526	59.666	40.000
R(d)	-16.289	-23.612	-17.879	-26.591	-20.225
Points	2,4	2,5	3,4	3,5	4,5
Distance	56.920	62.032	47.412	26.076	40.200
R(d)	-25.843	-27.214	-22.868	-13.415	-20.305

The next step is to solve for A_i in Equation 13.14. We can substitute the calculated $R(d)$ values into Equation 13.14 and rewrite the equation and the constraint about A_i in matrix form

$$\begin{bmatrix} 1 & 0 & -16.289 & -23.612 & -17.879 & -26.591 \\ 1 & -16.289 & 0 & -20.225 & -25.843 & -27.214 \\ 1 & -23.612 & -20.225 & 0 & -22.868 & -13.415 \\ 1 & -17.879 & -25.843 & -22.868 & 0 & -20.305 \\ 1 & -26.591 & -27.214 & -13.415 & -20.305 & 0 \\ 0 & 1 & 1 & 1 & 1 & 1 \end{bmatrix} \cdot \begin{bmatrix} a \\ A_1 \\ A_2 \\ A_3 \\ A_4 \\ A_5 \end{bmatrix} = \begin{bmatrix} 20.820 \\ 10.910 \\ 10.380 \\ 14.600 \\ 10.560 \\ 0 \end{bmatrix}$$

The matrix solutions are

$a = 13.203$ $A_1 = 0.396$ $A_2 = -0.226$
$A_3 = -0.058$ $A_4 = -0.047$ $A_5 = -0.065$

Now we can calculate the value at point 0 by

$P_0 = 13.203 + (0.396)(-7.510) + (-0.226)(-9.879) +$
$(-0.058)(-16.831) + (-0.047)(-18.574) + (-0.065)$
$(-22.834) = 15.795$

The splines with tension method is one of the thin-plate splines methods discussed in the text. Using the same data set, the estimation of P_0 by other methods is as follows: 16.350 by thin plate splines and 15.015 by regularized splines (using the τ value of 0.1).

have led to development of different kriging methods for spatial interpolation.

13.4.5.1 Ordinary Kriging

Assuming the absence of a drift, **ordinary kriging** focuses on the spatially correlated component. The measure of the degree of spatial dependence among the sampled known points is the **semivariance,** which is calculated by

$$\gamma(h) = \frac{1}{2n}\sum_{i=1}^{n}(z(x_i) - z(x_i + h))^2 \quad (13.16)$$

where h is the distance between known points, often referred to as the lag; n is the number of pairs of sample points separated by h; and z is the attribute value. The semivariance is expected to increase as h increases.

After semivariance values are computed at different distances, they are plotted in a **semivariogram,** with the semivariance along the y-axis and the distance between known points along the x-axis (Figure 13.12). A semivariogram can be dissected into three possible elements: nugget, range, and sill. The nugget is the semivariance at the distance of 0, representing the spatially uncorrelated noise. The range is the spatially correlated portion of the semivariogram that shows increases in the semivariance with the distance. Beyond the range, the semivariance levels off to a relatively constant value. The

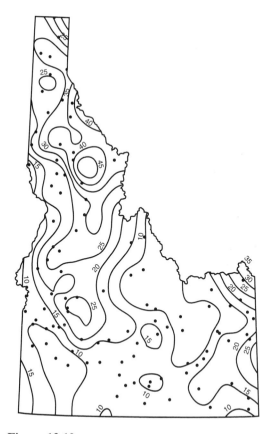

Figure 13.10
An isohyet map created by the regularized splines method.

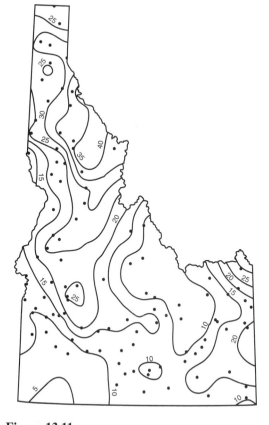

Figure 13.11
An isohyet map created by the splines with tension method.

semivariance at which the leveling takes place is called the sill.

A semivariogram relates the semivariance to the distance. It may be used alone as a measure of spatial correlation, similar to spatial autocorrelation (Chapter 11). But to be used as an interpolator in kriging, the semivariogram must be fitted with a mathematical function or a model. ARC/INFO, for example, offers five models: spherical, circular, exponential, linear, and Gaussian (Figure 13.13). The fitted semivariogram can then be used for estimating the semivariance at any given distance.

Ordinary kriging uses the fitted semivariogram directly in spatial interpolation. The general equation for estimating the z value at a point is

$$z_0 = \sum_{i=1}^{s} z_x W_x \qquad (13.17)$$

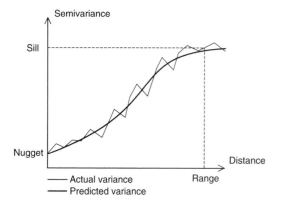

Figure 13.12
A semivariogram shows semivariances along the y-axis and distances along the x-axis.

where z_0 is the estimated value, z_x are values at known points, W_x are weights associated with each known point, and s is the number of known points used in estimation. The weights can be derived from solving a set of simultaneous equations. For example, the following simultaneous equations are needed for estimating the value at a point to be estimated from three known points:

$$W_1\gamma(h_{11}) + W_2\gamma(h_{12}) + W_3\gamma(h_{13}) + \lambda = \gamma(h_{10})$$
$$W_1\gamma(h_{21}) + W_2\gamma(h_{22}) + W_3\gamma(h_{23}) + \lambda = \gamma(h_{20})$$
$$W_1\gamma(h_{31}) + W_2\gamma(h_{32}) + W_3\gamma(h_{33}) + \lambda = \gamma(h_{30})$$
$$W_1 + W_2 + W_3 + 0 = 1.0 \text{ (13.18)}$$

where $\gamma(h_{ij})$ is the semivariance between known points i and j, $\gamma(h_{io})$ is the semivariance between the ith known point and the point to be estimated, and λ is a Lagrange multiplier, which is added to ensure the minimum possible estimation error. The above equations can be rewritten in matrix form

$$\begin{bmatrix} \gamma(h_{11}) & \gamma(h_{12}) & \gamma(h_{13}) & 1 \\ \gamma(h_{21}) & \gamma(h_{22}) & \gamma(h_{23}) & 1 \\ \gamma(h_{31}) & \gamma(h_{32}) & \gamma(h_{33}) & 1 \\ 1 & 1 & 1 & 0 \end{bmatrix} \cdot \begin{bmatrix} W_1 \\ W_2 \\ W_3 \\ \lambda \end{bmatrix} = \begin{bmatrix} \gamma(h_{10}) \\ \gamma(h_{20}) \\ \gamma(h_{30}) \\ 1 \end{bmatrix}$$

The vector of the weights can be solved by multiplying the inverse of the matrix on the left by the matrix on the right. Once the weights are known, Equation 13.17 can be used to estimate z_0

$$z_0 = z_1 W_1 + z_2 W_2 + z_3 W_3$$

The above example shows that weights used in kriging involve not only semivariances between the point to be estimated and the known points but also those between the known points. This is different

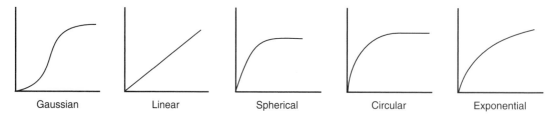

Figure 13.13
Five mathematical models for fitting semivariograms: Gaussian, linear, spherical, circular, and exponential.

from the inverse distance weighted method, which uses only weights applicable to known points and the point to be estimated. Another important difference between kriging and other local methods is that kriging produces a variance measure for each estimated point to indicate the reliability of the estimation. For the above example, the variance estimation can be calculated by

$$s^2 = W_1\gamma(h_{10}) + W_2\gamma(h_{20}) + W_3\gamma(h_{30}) + \lambda \quad (13.19)$$

Figure 13.14 shows an annual precipitation surface created by ordinary kriging with the linear model (Box 13.7). Figure 13.15 shows the distribution of the standard deviation of the kriged surface. As expected, the standard deviation is highest in data-poor areas, such as the southwest corner of Idaho.

13.4.5.2 Universal Kriging

Universal kriging assumes that the spatial variation in z values has a drift or a structural component in addition to the spatial correlation between sampled known points. Typically, universal kriging incorporates a trend such as a trend surface equation in the kriging process. ARC/INFO, for example, offer two types of universal kriging. Universal 1 uses a plane surface, which is defined by a first-order polynomial:

$$M = b_1x_i + b_2y_i \quad (13.20)$$

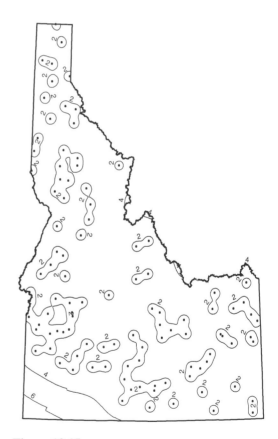

Figure 13.14
An isohyet map created by ordinary kriging with the linear model.

Figure 13.15
The map shows the standard deviation of the annual precipitation surface created by ordinary kriging with the linear model.

Box 13.7 — A Worked Example of Using Ordinary Kriging for Estimation

This worked example uses ordinary kriging for spatial interpolation. The first step is to construct a semivariogram by using point data of average annual precipitation values at 105 weather stations in Idaho (Figure 13.16). The distance interval in the semivariogram is measured in meters. The next step is to fit the semivariogram with a mathematical function. This example chooses the linear model, which is defined by

$$\gamma(h) = C_0 + C\,(h/a),\ 0 < h <= a$$
$$\gamma(h) = C_0 + C,\ h > a$$
$$\gamma(0) = 0$$

where $\gamma(h)$ is the semivariance at distance h, C_0 is the semivariance at distance 0, a is the range, and C is the sill, or the semivariance at a. The output from ARC/INFO shows: $C_0 = 0$, $C = 112.475$, and $a = 458000$.

Now we use the model for spatial interpolation. The scenario is the same as in Box 13.2: using five points with known values to estimate an unknown value. This step begins with the computation of distances between points and the semivariances at those distances based on the linear model. The results are shown in the following table, with distances in thousands of meters:

Points ij	0,1	0,2	0,3	0,4	0,5
h_{ij}	18.000	20.880	32.310	36.056	47.202
$\gamma(h_{ij})$	4.420	5.128	7.935	8.855	11.592
Points ij	1,2	1,3	1,4	1,5	2,3
h_{ij}	31.240	49.476	34.526	59.666	40.000
$\gamma(h_{ij})$	7.672	12.150	8.479	14.653	9.823
Points ij	2,4	2,5	3,4	3,5	4,5
h_{ij}	56.920	62.032	47.412	26.076	40.200
$\gamma(h_{ij})$	13.978	15.234	11.643	6.404	9.872

Using the semivariances, we can write the simultaneous equations for solving the weights in matrix form

$$\begin{bmatrix} 0 & 7.672 & 12.150 & 8.479 & 14.653 & 1 \\ 7.672 & 0 & 9.823 & 13.978 & 15.234 & 1 \\ 12.150 & 9.823 & 0 & 11.643 & 6.404 & 1 \\ 8.479 & 13.978 & 11.643 & 0 & 9.872 & 1 \\ 14.653 & 15.234 & 6.404 & 9.872 & 0 & 1 \\ 1 & 1 & 1 & 1 & 1 & 0 \end{bmatrix} \cdot \begin{bmatrix} W_1 \\ W_2 \\ W_3 \\ W_4 \\ W_5 \\ \lambda \end{bmatrix} = \begin{bmatrix} 4.420 \\ 5.128 \\ 7.935 \\ 8.855 \\ 11.592 \\ 1 \end{bmatrix}$$

The matrix solutions are

$W_1 = 0.397$ $W_2 = 0.318$ $W_3 = 0.182$
$W_4 = 0.094$ $W_5 = 0.009$ $\lambda = -1.161$

Using Equation 13.17, we can estimate the unknown value at point 0 by

$z_0 = (0.397)(20.820) + (0.318)(10.910) +$
$(0.182)(10.380) + (0.094)(14.600) +$
$(0.009)(10.560) = 15.091$

We can also calculate the estimate variance at point 0 by

$s^2 = (4.420)(0.397) + (5.128)(0.318) +$
$(7.935)(0.182) + (8.855)(0.094) +$
$(11.592)(0.009) - 1.161 = 4.605$

In other words, the estimate standard deviation (s) at point 0 is 2.146.

---

where M is the drift, x_i and y_i are the x- and y-coordinates of the known point i, and b_1 and b_2 are the drift coefficients to be estimated. Universal 2 uses a quadratic surface, which is defined by a second-order polynomial:

$$M = b_1 x_i + b_2 y_i + b_3 x_i^2 + b_4 x_i y_i + b_5 y_i^2 \quad (13.21)$$

The b_i coefficients in the polynomial equation must be estimated along with the weights. This means that universal kriging requires a larger set of simultaneous equations than ordinary kriging for estimating an unknown value.

Figure 13.17 shows an annual precipitation surface created by universal kriging with the linear

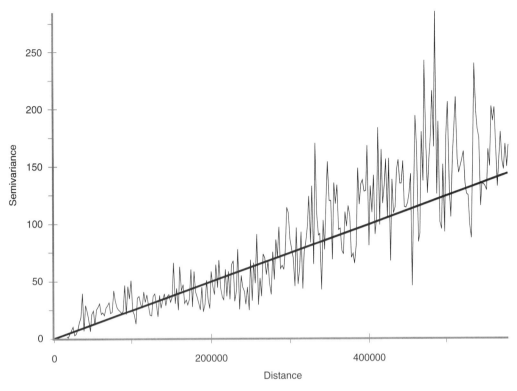

Figure 13.16.
A semivariogram constructed from annual precipitation values at 105 weather stations in Idaho. The linear model provides the trend line.

drift, and Figure 13.18 shows the distribution of the standard deviation of the kriged surface (Box 13.8). Both maps are similar to those created by ordinary kriging based on the linear model. The main difference is in the southwest corner of Idaho with the presence of negative values, a result of extrapolation by the linear drift of the universal kriging model.

13.4.5.3 Other Kriging Methods
Besides ordinary kriging and universal kriging, other kriging methods, such as block kriging and co-kriging (Bailey and Gatrell 1995; Burrough and McDonnell, 1998) have been proposed in the literature. Block kriging estimates the average value of a variable over some small area or block rather than at a point. Co-kriging uses one or more secondary

variables, which are correlated with the primary variable of interest, in interpolation. The idea is that the correlation between the variables can be used to improve the prediction of the value of the primary variable. For example, better results in precipitation interpolation have been reported by including elevation as an additional variable in co-kriging (Martinez-Cob 1996).

13.4.6 Comparison of Local Methods
Using the same data set, different local methods are expected to yield different results. Figure 13.19 shows the difference of annual precipitation estimates between the regularized splines method and the inverse distance squared method. Figure 13.20 shows the difference between

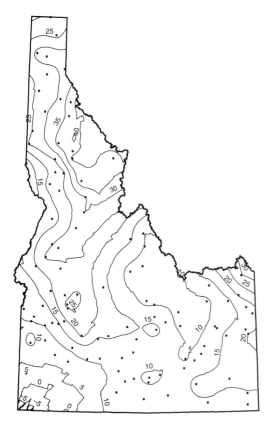

Figure 13.17
An isohyet map created by universal kriging with the linear drift.

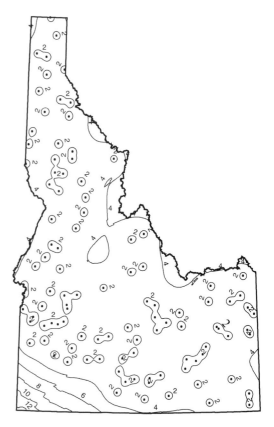

Figure 13.18
The map shows the standard deviation of the annual precipitation surface created by universal kriging with the linear drift.

ordinary kriging with the linear model and the regularized splines method. The two maps have similar general patterns, although the difference between the ordinary kriging and regularized splines methods is smaller than that between the regularized splines and inverse distance squared methods. On both maps, the largest differences, that is, more than 3 inches either positively or negatively, are all located in data-poor areas. This suggests that spatial interpolation—no matter which method is used—can never substitute for observed data.

Mapping the difference of estimates cannot tell which interpolation method is more accurate. Cross-validation analysis is commonly used for assessing the accuracy of estimates from different interpolation methods. For each interpolation method, a cross-validation analysis repeats the procedure of removing a known point from the data set, estimating the value at the point by using the remaining known points, and calculating the error of the estimation by comparing the estimated with the known values. Error statistics can then be calculated to assess the accuracy of the interpolation method. Cross-validation analysis has shown, for example, the use of elevation as an additional variable to the x, y location in elevation-detrended kriging or co-kriging yields better results than kriging (Phillips et al. 1992; Garen et al. 1994; Carroll and Cressie 1996).

Box 13.8 A Worked Example of Using Universal Kriging for Estimation

This example uses universal kriging to estimate the unknown value at point 0 (Box 13.2) and assumes that the drift is linear and the semivariogram is fitted with a linear model. Because of the additional drift component, this example uses eight simultaneous equations:

$$W_1\gamma(h_{11})+W_2\gamma(h_{12})+W_3\gamma(h_{13})+W_4\gamma(h_{14})+$$
$$W_5\gamma(h_{15})+\lambda+b_1x_1+b_2y_1=\gamma(h_{10})$$

$$W_1\gamma(h_{21})+W_2\gamma(h_{22})+W_3\gamma(h_{23})+W_4\gamma(h_{24})+$$
$$W_5\gamma(h_{25})+\lambda+b_1x_2+b_2y_2=\gamma(h_{20})$$

$$W_1\gamma(h_{31})+W_2\gamma(h_{32})+W_3\gamma(h_{33})+W_4\gamma(h_{34})+$$
$$W_5\gamma(h_{35})+\lambda+b_1x_3+b_2y_3=\gamma(h_{30})$$

$$W_1\gamma(h_{41})+W_2\gamma(h_{42})+W_3\gamma(h_{43})+W_4\gamma(h_{44})+$$
$$W_5\gamma(h_{45})+\lambda+b_1x_4+b_2y_4=\gamma(h_{40})$$

$$W_1\gamma(h_{51})+W_2\gamma(h_{52})+W_3\gamma(h_{53})+W_4\gamma(h_{54})+$$
$$W_5\gamma(h_{55})+\lambda+b_1x_5+b_2y_5=\gamma(h_{50})$$

$$W_1+W_2+W_3+W_4+W_5+0+0+0=1$$

$$W_1x_1+W_2x_2+W_3x_3+W_4x_4+W_5x_5+0+0+0=x_0$$

$$W_1y_1+W_2y_2+W_3y_3+W_4y_4+W_5y_5+0+0+0=y_0$$

where x_0 and y_0 are the x-, y-coordinates of the point to be estimated, and x_i and y_i are the x-, y-coordinates of the known point i; otherwise, the notations are the same as in Box 13.7. The x-, y-coordinates are actually rows and columns in the output grid using a cell size of 2000 meters.

Similar to ordinary kriging, semivariance values for the equations can be derived from the semivariogram and the linear model. The next step is to rewrite the equations in matrix form:

$$\begin{bmatrix} 0 & 7.672 & 12.150 & 8.479 & 14.653 & 1 & 69 & 76 \\ 7.672 & 0 & 9.823 & 13.978 & 15.234 & 1 & 59 & 64 \\ 12.150 & 9.823 & 0 & 11.643 & 6.404 & 1 & 75 & 52 \\ 8.479 & 13.978 & 11.643 & 0 & 9.872 & 1 & 86 & 73 \\ 14.635 & 15.234 & 6.404 & 9.872 & 0 & 1 & 88 & 53 \\ 1 & 1 & 1 & 1 & 1 & 0 & 0 & 0 \\ 69 & 59 & 75 & 86 & 88 & 0 & 0 & 0 \\ 76 & 64 & 52 & 73 & 53 & 0 & 0 & 0 \end{bmatrix} \cdot \begin{bmatrix} W_1 \\ W_2 \\ W_3 \\ W_4 \\ W_5 \\ \lambda \\ b_1 \\ b_2 \end{bmatrix} = \begin{bmatrix} 4.420 \\ 5.128 \\ 7.935 \\ 8.855 \\ 11.592 \\ 1 \\ 69 \\ 67 \end{bmatrix}$$

The solutions are

$$W_1 = 0.387 \quad W_2 = 0.311 \quad W_3 = 0.188 \quad W_4 = 0.093$$
$$W_5 = 0.021 \quad \lambda = -1.154 \quad b_1 = 0.009 \quad b_2 = -0.010$$

The estimate value at point 0 is

$$z_0 = (0.387)(20.820) + (0.311)(10.910) +$$
$$(0.188)(10.380) + (0.093)(14.600) +$$
$$(0.021)(10.560) = 14.981$$

And, the estimate variance at point 0 is

$$s^2 = (4.420)(0.387) + (5.128)(0.311) +$$
$$(7.935)(0.188) + (8.855)(0.093) +$$
$$(11.592)(0.021) - 1.154 = 4.710$$

The estimate standard deviation (s) at point 0 is 2.170. The results from universal kriging are very similar to those from ordinary kriging.

Another method for assessing the accuracy of estimates from different interpolation methods is to divide the known points into two samples: one sample for developing the model from each interpolation method and the other sample for testing the accuracy of the models (Chang and Li 2000). Sample splitting may not be feasible if the number of known points is small.

13.4.7 Local Methods in ARC/INFO and ArcView

ARC/INFO has commands to operate local methods covered in this chapter: THIESSEN for Thiessen polygons, POINTDENSITY for density estimation (simple and kernel), IDW for the inverse distance weighted interpolation method, SPLINE for thin-plate splines with tension and regularized splines, and KRIGING for ordinary kriging and universal kriging. These commands will probably be combined and expanded in Geostatistical Analyst, a menu-driven extension to ArcInfo 8.1.

The Spatial Analyst extension to ArcView has menu access to density estimation, inverse distance weighted interpolation, and splines (splines with tension and regularized splines). Kriging can be performed in ArcView using requests and Avenue scripts. Task 5 in the Applications section uses an Avenue script to run ordinary kriging.

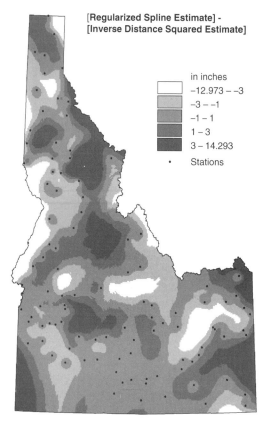

Figure 13.19
The map shows the difference between surfaces generated from the regularized splines method and the inverse distance squared method. A local operation, in which one surface grid was subtracted from the other, created the map.

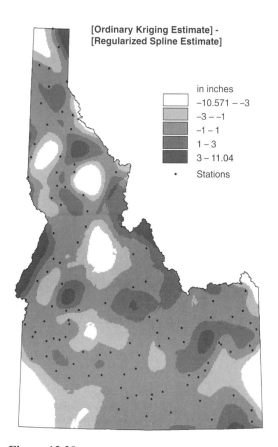

Figure 13.20
The map shows the difference between surfaces generated from the ordinary kriging with linear model method and the regularized splines method.

KEY CONCEPTS AND TERMS

Control points: Points with known values in spatial interpolation.

Density estimation: A local interpolation method, which measures densities in a grid based on a distribution of points and point values.

Global interpolation method: The interpolation method that uses every control point available in estimating an unknown value.

Inverse distance weighted interpolation: A local interpolation method, which assumes that the unknown value of a point is influenced more by nearby control points than those farther away.

Kernel estimation: A local interpolation method, which associates each known point with a kernel function in the form of a bivariate probability density function.

Kriging: A local interpolation method, which assumes that the spatial variation of an attribute includes a spatially correlated component, representing the variation of the regionalized variable.

Local interpolation method: The interpolation method that uses a sample of control points in estimating an unknown value.

Ordinary kriging: A kriging method, which assumes the absence of a drift or trend and focuses on the spatially correlated component.

Regression model: A global interpolation method that uses a number of independent variables to estimate a dependent variable.

Regularized splines: A variation of thin-plate splines for spatial interpolation.

Semivariance: A measure of the degree of spatial dependence among known points used in kriging.

Semivariogram: A diagram relating the semi-variance to the distance between known points used in kriging.

Spatial interpolation: The process of using points with known values to estimate unknown values at other points.

Thiessen polygons: A local interpolation method, which ensures that every unsampled point within a polygon is closer to the polygon's known point than any other known points.

Thin-plate splines: A local interpolation method, which creates a surface passing through control points with the least possible change in slope at all points.

Thin-plate spline with tension: A variation of thin-plate splines for spatial interpolation.

Trend surface analysis: A global interpolation method that uses points with known values and a polynomial equation to approximate a surface.

Universal kriging: A kriging method, which assumes that the spatial variation of an attribute has a drift or a structural component in addition to the spatial correlation between sampled known points.

APPLICATIONS: SPATIAL INTERPOLATION

This applications section covers five tasks. Task 1 uses an Avenue script to run trend surface analysis. Tasks 2 to 5 deal with the local interpolation methods of kernel density estimation, inverse distance weighted, thin-plate splines, and ordinary kriging. An Avenue script is used to perform ordinary kriging.

Task 1. Trend Surface Analysis Using an Avenue Script

What you need: *stations.shp*, a shapefile containing 105 weather stations in Idaho; *idoutl.shp*, an Idaho outline shapefile; *trend.ave*, an Avenue script to run trend surface analysis.

Task 1 lets you use an Avenue script to run trend surface analysis using the attribute called ann_prec in *stations.shp*. Ann_prec is the average annual precipitation from 1961 to 1990.

1. Start ArcView and load Spatial Analyst. Open a new view and add *stations.shp* and *idoutl.shp* to view. Select Properties from the View menu and set the Map Units as meters.

2. Click on Scripts in the Project window and New to open Script 1. Click the Load Text File button. Navigate the path to *trend.ave* and double-click on it. This action copies *trend.ave* to Script 1.

3. To use an Avenue script, you need to first compile it by clicking the Compile button. Because *trend.ave* stipulates that the view document is the active document, you must activate the view document and then click the Run button in Script 1 to run the script.

4. *Trend.ave* uses the Analysis Properties dialog to set the analysis environment. In the dialog, set the Analysis Extent to be the same as *idoutl.shp*, set the Analysis Cell Size to be As

Specified Below, and enter 2000 (meters) for the Cell Size. Click OK.

5. *Grid 1* is the output grid from trend surface analysis based on a third-order polynomial equation. Contouring is probably the most common method for mapping a surface. Select Create Contours from the Surface menu. Specify 5 for the Contour Interval and 10 for the Base Contour in the Contour Parameter dialog. Click OK.

6. Add *Contour of Grid 1* to view and make it active. Select Auto-label from the Theme menu. Select Contour for the Label Field in the Auto-label dialog, and click OK. Now you can see the interpolated surface better with the isohyet values.

7. Both *Grid 1* and *Contour of Grid 1* include areas outside Idaho. To limit interpolation and contouring to inside Idaho, you need to use an analysis mask. Task 3 will show you how to set up an analysis mask.

Task 2. Kernel Density Estimation

What you need: *deer.shp*, a point shapefile showing deer locations.

Task 2 uses the kernel estimation method to compute the average number of deer sightings per hectare from *deer.shp*. Deer location data have a 50-meter minimum discernible distance; therefore, some locations have multiple sightings.

1. Start ArcView and load Spatial Analyst. Open a new view and add *deer.shp* to view. Select Properties from the View menu, and specify the Map Units as meters.

2. Make *deer.shp* active and open its theme table. The field count shows the number of sightings at a point location. You can display deer sightings using graduated symbols. Double-click on *deer.shp* in the Table of Contents to open its legend editor. Select Graduate Symbol for the Legend Type and Count for the Classification Field. Click Apply. The graduated symbols show the number of sightings from 1 to 15.

3. Select Calculate Density from the Analysis menu. In the Output Grid Specification dialog, select Same As *Deer.shp* for the Output Grid Extent and enter 100 (meters) for the CellSize. Click OK. In the next dialog for Calculate Density, select Count for the Population Field, enter 100 (meters) for the Search Radius, select Kernel for the Density Type, and select Hectares for the Area Units. Click OK. *Density from Deer* is now added to the Table of Contents. To simplify the look of the density map, open its legend editor, change the number of classes to 4, and enter the class intervals as 0, 0–10, 10–20, and 20–24.

Task 3. Spatial Interpolation Using IDW

What you need: *stations.shp* and *idoutl.shp*, the same shapefiles from Task 1.

This task involves three steps: first, it converts *idoutl.shp* to a grid, *idoutlgd*; second, it creates a precipitation grid using the inverse distance weighted method and *idoutlgd* as the mask grid; third, it prepares an isoline map from the precipitation grid.

1. Start ArcView and load Spatial Analyst. Open a new view and add *stations.shp* and *idoutl.shp* to view. Select Properties from the View menu and set the Map Units as meters.

2. Make *stations.shp* active and open its theme table. One of its attributes is ann_prec, which is the z value to be used in spatial interpolation.

3. Make *idoutl.shp* active. Select Convert to Grid from the Theme menu. Name the output grid *idoutlgd*. In the Conversion Extent dialog, select Same As *Idoutl.shp* for the Output Grid Extent and enter 2000 (meters) for the Output Grid Cell Size. Click OK. In the Conversion Field dialog, select Idoutl_id for cell values and click OK. You do not want to join feature attributes to *idoutlgd*, but you do want to add *idoutlgd* to view. *Idoutlgd* has only two cell values: 1 within the state border and no data outside the border.

4. Next use *idoutlgd* as the analysis mask. Select Properties from the Analysis menu. In the Analysis Properties dialog, select Same As *Idoutlgd* for the Analysis Extent and As Specified Below for the Analysis Cell Size. Enter 2000 (meters) as the Cell Size, select *idoutlgd* for the Analysis Mask, and click OK.

5. Now you are ready to interpolate a surface from *stations.shp*. Make *stations.shp* active, and select Interpolate Grid from the Surface menu. In the Interpolate Surface dialog, select IDW for the Method, and ann_prec for the Z Value Field. Nearest Neighbors and Fixed Radius are two options for selecting control points. The Nearest Neighbor option uses a specified number of control points closest to a cell to be estimated. The Fixed Radius option uses a specified radius to select control points. For this task, choose Nearest Neighbor and the default number of 12 neighbors. Use the default power of 2, that is, the inverse distance squared method. Barriers are linear features that limit the selection of control points from the side of the cell to be estimated. Use the default of no barriers. Click OK to run the interpolation.

6. The output grid called *Surface from Stations* is added to the Table of Contents. Add the output grid to view. The grid is a rectangle defined by the map extent of *idoutlgd*. Cells within the state border contain values, whereas cells outside the border have no data. Notice that the z values of the surface range from 7.149 to 41.449 (inches).

7. Activate the interpolated surface. Select Create Contours from the Surface menu. In the Contour Parameters dialog, enter 5 for the contour interval and 10 for the base contour. Click OK.

8. Add the newly created contours theme to view. The term contours normally applies to an elevation surface. The proper term for isolines depicting precipitation is isohyets. To see the isohyet values, you can select Auto-label from the Theme menu.

Task 4. Comparing Two Methods of Thin-plate Splines

What you need: *stations.shp* and *idoutlgd*, from Task 3.

Task 4 compares the results from the two thin-plate splines methods available through menu access in ArcView. The task has three parts: one, create an interpolated grid using the regularized splines method; two, create an interpolated grid using the thin-plate splines with tension method; and three, use a local operation to compare the two grids. The result can show you the difference between the two interpolation methods.

1. Start ArcView, and load Spatial Analyst. Open a new view, and add *stations.shp* and *idoutlgd* to view. Select Properties from the View menu and set the Map Units as meters.

2. First set the analysis properties. Select Properties from the Analysis menu. In the Analysis Properties dialog, select Same As *Idoutlgd* for the Analysis Extent and As Specified Below for the Analysis Cell Size. Enter 2000 (meters) as the Cell Size, select *idoutlgd* for the Analysis Mask, and click OK.

3. Now create an interpolated grid using the regularized splines method. Activate *stations.shp* in the Table of Contents. Select Interpolate Grid from the Surface menu. In the Interpolate Surface dialog, choose Spline for the Method, ann_prec for the Z Value Field, and Regularized for the Type. Click OK to run the regularized thin-plate splines interpolation.

4. Check the box next to *Surface from Stations* in the Table of Contents to view the output grid. Make the output grid active, select Properties from the Theme menu, and rename the output grid *Regularized*.

5. Next create an interpolated grid using the thin-plate splines with tension method. Activate *stations.shp*. Select Interpolate Grid from the Surface menu. In the Interpolate Surface dialog, choose Spline for the Method, ann_prec for the Z Value Field, and Tension for the Type. Click OK to run the thin-plate splines with tension interpolation.

6. View the output grid and rename it *Tension*.

7. The final part of Task 4 is to compare the two grids, *Regularized* and *Tension*. Select the Map Calculator from the Analysis menu. Prepare the following statement in the Map Calculator's expression box: ([Regularized] - [Tension]). Click Evaluate.

8. The output called *Map Calculation 1* shows the difference in cell values between *Regularized* and *Tension*. To better compare the two grids, use a new legend. Activate *Map Calculation 1* and open its legend editor. In the Legend Editor dialog, click on Classify and change the number of classes to 4. Now change the Values of the four classes to -20 - -3, -3 - 0, 0 - 3, and 3 - 29. Highlight the cells that have difference values of greater than 3 in either direction. Also change the Color Ramp to Blues to Reds dichromatic. Click Apply.

9. Cells that have difference values of greater than 3 in *Map Calculation 1* are all in the data-poor areas within Idaho. You can also inspect the cell values in *Regularized* and *Tension* in detail. Make *stations.shp*, *Regularized*, and *Tension* all active. Zoom in a small area in Idaho. Press the Identify tool and click a point in the map. The Identify Results dialog includes all three themes. By highlighting each of the themes, you can read the estimated values (cell values) and the known value of the closest weather station. The three values should be close to each other if the point you click is near a weather station.

Task 5. Ordinary Kriging Using an Avenue Script

What you need: *stations.shp* and *idoutlgd* from Task 3; *kriging.ave*, an Avenue script to run kriging.

Task 5 lets you use *kriging.ave* to run ordinary kriging with the linear model. Kriging is a complex topic and requires expert knowledge in selecting the proper model for the data to be interpolated. This task is designed to only show you the steps to go through in running kriging in ArcView.

1. Start ArcView and load Spatial Analyst. Open a new view and add *stations.shp* and *idoutlgd* to view. Select Properties from the View menu and set the Map Units as meters.

2. Click on Scripts and New to open Script 1. Click the Load Text File button. Navigate the path to *kriging.ave* and double-click on it. This action copies *kriging.ave* to Script 1. Read the information at the top of *kriging.ave*. If you need to change the path for the estimate variance grid or remove it, do so. When you are ready, click the Compile button, activate the view document (the active document in *kriging.ave*), and click the Run button in Script 1.

3. *Kriging.ave* uses the Analysis Properties dialog to set the analysis environment. In the dialog, set both the Analysis Extent and the Analysis Cell Size to be the same as *idoutlgd*, and select *idoutlgd* for the Analysis Mask. Click OK.

4. *Kriging.ave* creates two output grids, *grid 1* and *vargrid*, and places them in the Table of Contents. Check the boxes next to the grids to view them. *Vargrid* is the estimated variance grid. To change it to an estimated standard deviation grid, select Map Calculator from the Analysis menu and prepare the following statement in the expression box: ([vargrid].Sqrt). This statement is to take the square root of *vargrid* to create a standard deviation grid. Sqrt is a Power function in the Map Calculator dialog. Click Evaluate. *Map Calculation 1* is the estimated standard deviation grid.

5. Again, you want to use contouring to map the kriged surface and the estimated standard deviation surface. Select Create Contours from the Surface menu. Use a contour interval of 5 and a base contour of 10 to map *grid 1*, and use a contour interval of 2 and a base contour of 0 to map *Map Calculation 1*. The contour maps should look the same as Figure 13.14 and Figure 13.15, respectively, in the chapter.

REFERENCES

Abramowitz, M., and I.A. Stegun. 1964. *Handbook of Mathematical Functions.* New York: Dover.

Bailey, T.C., and A.C. Gatrell. 1995. *Interactive Spatial Data Analysis.* Harlow, England: Longman Scientific & Technical.

Burrough, P.A., and R. A. McDonnell. 1998. *Principles of Geographical Information Systems.* Oxford: Oxford University Press.

Carroll, S.S., and N. Cressie. 1996. A Comparison of Geostatistical Methodologies Used to Estimate Snow Water Equivalent. *Water Resources Bulletin* 32: 267–78.

Chang, K., and Z. Li. 2000. Modeling Snow Accumulation with a Geographic Information System. *International Journal of Geographical Information Science* 14: 693–707.

Cressie, N. 1991. *Statistics for Spatial Data.* Chichester, NY: John Wiley & Sons.

Davis, J.C. 1986. *Statistics and Data Analysis in Geology.* 2^d ed. New York: John Wiley & Sons.

Franke, R. 1982. Smooth Interpolation of Scattered Data by Local Thin Plate Splines. *Computers and Mathematics with Applications* 8: 273–81.

Franke, R. 1985. Thin Plate Spline with Tension. *Computer Aided Geometrical Design* 2: 87–95.

Garen, D.C., G.L. Johnson, and C.J. Hanson. 1994. Mean Areal Precipitation for Daily Hydrologic Modeling in Mountainous Regions. *Water Resources Bulletin* 30: 481–91.

Hutchinson, M.F. 1995. Interpolating Mean Rainfall Using Thin Plate Smoothing Splines. *International Journal of Geographical Information Systems* 9: 385–403.

Issaks. E.H., and R.M. Srivastava. 1989. *An Introduction to Applied Geostatistics.* Oxford: Oxford University Press.

Martinez-Cob, A. 1996. Multivariate geostatistical analysis of evapotranspiration and precipitation in mountainous terrain. *Journal of Hydrology* 174: 19–35.

Middleton, G.V. 2000. *Data Analysis in the Earth Sciences Using Matlab.* Upper Saddle River, NJ: Prentice Hall.

Mitas, L., and H. Mitasova. 1988. General Variational Approach to the Interpolation Problem. *Computers and Mathematics with Applications* 16: 983–92.

Mitasova, H., and L. Mitas. 1993. Interpolation by Regularized Spline with Tension: I. Theory and Implementation. *Mathematical Geology* 25: 641–55.

Monmonier, M.S. 1982. *Computer-Assisted Cartography: Principles and Prospects.* Englewood Cliffs, NJ: Prentice Hall.

Phillips, D.L., J. Dolph, and D. Marks. 1992. A Comparison of Geostatistical Procedures for Spatial Analysis of Precipitation in Mountainous Terrain. *Agricultural and Forest Meteorology* 58: 119–41.

Robinson, A.H., J.L. Morrison, P.C. Muehrcke, A.J. Kimerling, and S.C. Guptill. 1995. *Elements of Cartography.* 6th ed. New York: John Wiley & Sons.

Scott, D.W. 1992. *Multivariate Density Estimation: Theory, Practice, and Visualization.* New York: John Wiley & Sons.

Silverman, B.W. 1986. *Density Estimation.* London: Chapman and Hall.

Tabios, G.Q., III, and J.D. Salas. 1985. A Comparative Analysis of Techniques for Spatial Interpolation of Precipitation. *Water Resources Bulletin* 21: 365–80.

Watson, D.F. 1992. *Contouring: A Guide to the Analysis and Display of Spatial Data.* Oxford: Pergamon Press.

Webster, R., and M.A. Oliver. 1990. *Statistical Methods in Soil and Land Resource Survey.* Oxford: Oxford University Press.

14

GIS MODELS AND MODELING

14.1 INTRODUCTION

The previous chapters presented basic tools for exploring, manipulating, and analyzing vector data and raster data. One of many uses of these tools is to build models. What is a model? A **model** is a simplified representation of a phenomenon or a system. Several types of models have been covered in this book. A map such as a USGS quad is a model. So are the vector and raster data models for representing spatial features and the relational database model for representing a database system. A model helps us better understand a phenomenon or a system by retaining the significant features or relationships of reality.

This chapter discusses GIS modeling. **GIS modeling** is defined here as the use of GIS in the process of building models with spatial data. Two points are important in this definition. First, the emphasis is models with spatial data, or spatially explicit models, because a GIS uses geographically referenced data. Second, the emphasis is the use of GIS in modeling rather than the models. A basic requirement in modeling is the modeler's interest and knowledge of the system be modeled (Hardisty et al. 1993). This is why many models are discipline specific models. Models of the environment, for example, are usually subdivided into atmospheric, hydrologic, land surface/subsurface, and ecological models. It would be impossible for a GIS book to address discipline specific models.

This chapter is divided into five sections. Section 1 discusses the characteristics of GIS modeling. Sections 2 through 5 cover binary models, index models, regression models, and process models respectively. Although they differ in the degree of complexity, these four types of models have two common elements: a set of selected spatial variables, and the functional or mathematical relationship between variables.

14.2 GIS MODELING

GIS can assist the modeling process in several ways. First, a GIS is a tool that can integrate different data sources including maps, DEMs, GPS data, images, and tables. These data sources can be displayed together and dynamically linked. A GIS is therefore useful for modeling related tasks such as exploratory data analysis, data visualization, and database management.

Second, models built with a GIS can be vector-based or raster-based. The choice depends on the nature of the model, data sources, and the computing algorithm. A raster-based model is preferred if the spatial phenomenon to be modeled varies continuously over the space such as soil erosion and snow accumulation. A raster-based model is also preferred if satellite images and DEMs constitute a major portion of the input data, or if the modeling involves intense and complex computation. Raster-based models, however, are not recommended for studies of travel demand because travel demand modeling requires the use of a topology-based road network. Vector-based models are generally recommended for spatial phenomena that involve well-defined locations and shapes.

Third, the distinction between raster-based and vector-based models does not preclude GIS users from integrating both types of data in the modeling process. Algorithms for conversion between vector and raster data are easily available in most GIS packages. The decision as to which data format to use in analysis should be based on the efficiency and the expected result, rather than restricted to the format of the original data. For example, if a vector-based model requires a precipitation map as the input, it would be easier to prepare a precipitation grid through spatial interpolation from known point data and then to derive a map showing precipitation zones from the grid.

Fourth, GIS modeling may take place in a GIS or require the linking of a GIS to other computer programs. Many GIS packages including ARC/INFO, ArcView, GRASS, Idrisi, SPANS, ILWIS, Vertical Mapper, and MFworks have extensive analytical functions for modeling. ArcView even has an extension called ModelBuilder to assist users in building raster-based models. But a GIS package cannot accommodate statistical analysis as well as a statistical analysis package can, or perform dynamic simulation efficiently.

Box **14.1** **Query or Map Overlay**

Often GIS users can choose between spatial query and map overlay to solve simple analysis tasks, especially those involving point or line features. Suppose an environmental enforcement agency has found a couple of its monitored wells with high nitrate readings and needs to notify private well owners who live within 10 miles of the contaminated wells. There are at least two methods to accomplish the task. The first method consists of two steps: buffer the contaminated wells with a buffer distance of 10 miles, and then overlay the coverage with private wells with the buffer zone coverage. The second method also has two steps: draw circles with a 10-mile radius around the contaminated wells, and then query which private wells fall within the circles. The second method uses spatial query and entirely skips buffering and map overlay.

There are three scenarios for linking a GIS to other computer programs (Corwin et al. 1997). GIS users may encounter all three scenarios in the modeling process, depending on the tasks to be accomplished.

A **loose coupling** involves transfer of data files between the GIS and other programs: data to be run in a statistical analysis package are exported from the GIS and results from statistical analysis are then imported to the GIS for data visualization or display. Under this scenario, GIS users must create and manipulate data files to be exported or imported unless the interface has already been established between the GIS and the targeted computer program. S-PLUS for ArcView is an example of such an interface; it allows GIS users to move tabular data from ArcView to S-PLUS and then returns graphics and analytical results to ArcView. A **tight coupling** gives the GIS and other programs a common user interface. For example, the GIS can have a menu selection to run a program using thin-plate splines for spatial interpolation, which resides outside the GIS. An **embedded system** bundles the GIS and other programs with shared memory and a common interface. For example, the GIS can have menu selections to run various programs for spatial interpolation, which are built as components of the GIS package.

14.3 BINARY MODELS

A **binary model** uses logical expressions to select map features from a composite map or multiple grids. The output of a binary model is in binary format: 1 (True) for map features that satisfy the logical expression and 0 (False) for map features that do not. Binary models can be considered as an expansion of data query or map query.

Prior to building a vector-based binary model, map overlay operations must be performed to combine attributes (variables) to be used in the logical expression into the same attribute table (Figure 14.1). Spatial query cannot replace map overlay in this case (Box 14.1). A raster-based binary model, on the other hand, can be derived directly from querying multiple grids, with each grid representing a selected variable (Figure 14.2).

A common application of binary models is siting analysis, with each logical expression corresponding to a siting criterion. Suppose a county government wants to select potential industrial sites by the following criteria:

- at least five acres in size;
- commercial zones;
- vacant or for sale;
- not subject to flooding;
- not more than one mile from a heavy duty road; and
- less than 10% slope.

To proceed with the task, the county government can do the following:

- Gather all digital maps relevant to the criteria and create a 1-mile buffer zone map by buffering heavy duty roads.
- Follow a series of map overlay operations to combine the road buffer zone map and other maps.
- Query the composite map involving all the selection criteria to reveal which parcels are potential industrial sites.

Another example of siting analysis is the Conservation Reserve Program (CRP) administered by the Farm Service Agency (FSA) of the U.S. Department of Agriculture (http://www.usda.gov/services.html/). Land eligible to be placed in the CRP includes cropland that is planted to an agricultural commodity during two of the five most recent crop years. Additionally, cropland must meet the following requirements:

- Have an Erosion Index of 8 or higher or be considered highly erodible land
- Be considered a cropped wetland
- Be devoted to any of a number of highly beneficial environmental practices, such as filter strips, riparian buffers, grass waterways,

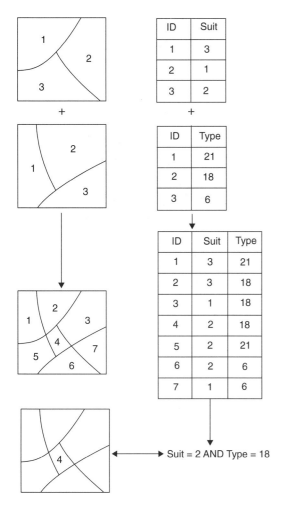

Figure 14.1

An illustration of a vector-based logical model. The two maps at the top are overlaid so that their spatial features and their attributes of Suit and Type are combined. A logical expression, Suite = 2 AND Type = 18, results in the selection of polygon 4 in the output.

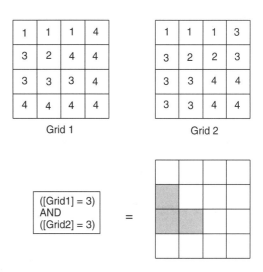

Figure 14.2

An illustration of a raster-based logical model. A query statement, ([Grid1] = 3) AND ([Grid2] = 3), results in the selection of 3 cells in the output.

Similar to the industrial site analysis example, the FSA can compile a map for each criterion, overlay the maps, and query the attributes of the composite map for agricultural lands that qualify for the CRP. Both the industrial site and CRP examples are likely to be vector-based because of the need for accurate area measurements.

Binary models can also be used for change detection. A change detection model can be vector- or raster-based, depending on the data sources and its intended use. The method for deriving a vector-based model is to overlay maps representing the spatial distribution of a variable of interest at two different points in time and to work with the attribute data of the composite map (Schlagel and Newton 1996). For example, a national forest is interested in change of vegetation covers within a ranger district from 1980 to 1990. This task would involve the following steps:

- Compile two vegetation cover maps, one for 1980 and the other for 1990.
- Overlay the two maps so that the boundaries and attributes of vegetation covers are combined to form the output.

shelter belts, wellhead protection areas, and other similar practices
- Be subject to scour erosion
- Be located in a national or state CRP conservation priority area, or be cropland associated with or surrounding non-cropped wetlands

- Query attribute data in the composite map to reveal which polygons have changed and which have not.

The same model can be prepared in raster format by replacing the vegetation cover maps with grids. The grids can be queried for specific types of vegetation cover change such as from white pine-dominant to Douglas fir-dominant. Another option is to first use the local operation of Combine on the two grids. Combine assigns a unique output value to each type of vegetation cover change. The output grid can then be queried for vegetation cover change of interest.

Sometimes the role of a binary model is to help build a more refined model. A binary model is often used at the beginning stage of the model building process when data needed for the analysis are incomplete and the relationship between spatial features is still not clear. The candystick is a rare plant species that has been recorded in central Idaho and western Montana. The candystick's favorable habitat cannot be defined by only a few candystick observations, although some believe that the species mainly inhabits the gentle to moderate slopes of high-elevation lodgepole pine forests (Lichthardt 1995). A binary model based on the preliminary findings of the candystick's habitat conditions can narrow the areas for field survey, which may lead to more observations of the candystick and a better defined habitat model for the species.

14.4 INDEX MODELS

An **index model** uses the index value calculated from a composite map or multiple grids to produce a ranked map. Selected variables in an index model are evaluated at two levels (Figure 14.3). First, the relative importance of each variable is evaluated against other variables. This evaluation leads to assigning a weight to each variable. For example, a wildlife habitat model may assign weights of 3, 2, and 1 to elevation, slope, and aspect, respectively.

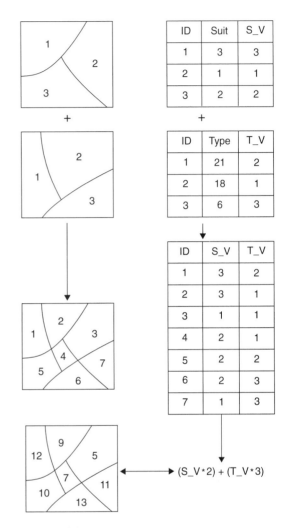

Figure 14.3

An illustration of a vector-based index model. First, the observed values of each map are given the numeric scores from 1 to 3. The Suit values of 1, 2, and 3 are given the scores of 1, 2, and 3 respectively. The Type values of 6, 18, and 21 are given the scores of 3, 1, and 2 respectively. Second, the two maps are overlaid. Third, a weight of 2 is assigned to the map with Suit and a weight of 3 to the map with Type. Finally, the index values are calculated for each polygon in the output. For example, Polygon 4 has an index value of 7 (2x2 + 1x3).

Second, the observed values of each variable are evaluated and given numeric scores. Typically, the observed values are grouped into classes and each class is given a score. For example, elevation values may be grouped into three classes such as < 1000, 1000–1200, and > 1200 meters, and the classes are given the numeric scores of 1, 2, and 3 respectively. The range of numeric scores is user-defined and should apply to all selected variables. The range may be from 0.0 to 1.0, 1 to 5, or 1 to 9.

After the selected variables and the numeric scores of each variable are evaluated, an index model can be expressed as a linear equation, with the index variable on the left and the selected variables and their weights on the right. The index value can be calculated by summing the weighted numeric scores from each variable. Figure 14.3 shows a vector-based index model, which requires a map overlay operation to combine attributes into a single table and attribute data computation to derive the index values. Figure 14.4 shows a raster-based index model, which uses a series of local operations to compute the index values.

The final step in producing the ranked map for an index model is often to normalize the index values so that they are scaled from 0 to 1. The formula for such data normalization is

$$(X \text{ - lowest index value}) / (\text{range of index values}) \tag{14.1}$$

where X is an index value. In Figure 14.3, for example, the index values range from 5 to 13, the index value of 9 would have a normalized score of 0.5 (4 / 8).

Building an index model is not difficult but it does require GIS users to document numeric scores and weights in detail. User interface in a GIS to simplify the process of building an index model is therefore welcome. ArcView's Model-Builder is intended for that purpose (Box 14.2). One of its operations is grid overlay, which may be Arithmetic Overlay and Weighted Overlay. Arithmetic Overlay uses an equal weight for the selected variables to build an index model, whereas Weighted Overlay uses different weights. GIS users can assign numeric scores and weights to an index model through menu dialogs.

There are variations to the above method for creating an index model, especially if the model is raster-based. For example, instead of summing the weighted numeric scores from each variable, one can assign the lowest score, the highest score, or the most frequent score among the variables to the index value (Tomlin 1990; Chrisman 1997). These variations can be incorporated easily through local operations for raster data analysis.

Box **14.2**   **ModelBuilder in ArcView**

Introduced as a new document in the Spatial Analyst extension version 2.0, ModelBuilder allows users to build a raster-based index model by using a model diagram or flow chart. ModelBuilder provides operations under five categories: data conversion, reclassification, terrain analysis, distance measure, and grid overlay. The data conversion operations include vector to grid, DEM to grid, and point interpolation. The terrain analysis operations include contour, slope, aspect, and hillshade. The grid overlay operations include arithmetic overlay and weighted overlay. Both types of grid overlay can be used to build index models, as shown in Task 5 of the Applications section.

Using ModelBuilder, an ArcView user can build a model diagram by stringing together a series of the input, the spatial function to be operated on the input, and the output. ModelBuilder executes the series of operations in sequence to create the final map. Once a model is built, it can be re-run by simply changing any of the inputs or the functions in the model diagram. For example, to determine the sensitivity of an index model, one can re-run the model with different schemes of numeric scores and weights.

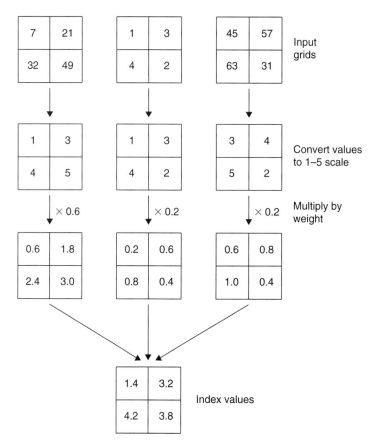

Figure 14.4
An illustration of a raster-based index model. First, the cell values of each input grid are assigned the numeric scores from 1 to 5. Second, the index values in the output grid are calculated by summing the products of each grid multiplied by its assigned weight. For example, the index value of 1.4 is calculated from: 1x0.6 + 1x0.2 + 3x0.2, or 0.6 + 0.2 + 0.6.

Index models are commonly used in suitability analysis and vulnerability analysis. The literature has many examples, but here we will look at five. The first example is a study of prioritizing lands for conservation protection in Sterling Forest on the New York-New Jersey border (Lathrop and Bognar 1998). The study selected the following five parameters as inputs for the assessment:

■ Development limitations due to soil conditions/steep slopes/flooding.

■ Non-point source pollution potential due to proximity to water/wetlands.
■ Habitat fragmentation potential due to distance from existing roads and development.
■ Sensitive wildlife habitat areas.
■ Aesthetic impact (visibility from the Appalachian and Sterling Ridge Trails).

The study applied the following procedure:

■ Ranked the environmental cost/development constraints for each parameter from 1 to 5, with 1 being very slight and 5 being very severe.

- Made a grid for each parameter and overlaid the input grids.
- Assigned the maximum value of any of the five input parameters to each cell in the output grid.
- Selected lands suitable for conservation protection from cells with low values in the output grid.

The second example is the DRASTIC model developed by the Environmental Protection Agency (EPA) for evaluating groundwater pollution potential (Aller et al. 1987). The acronym DRASTIC stands for the seven parameters used in the model: Depth to water, net Recharge, Aquifer media, Soil media, Topography, Impact of the vadose zone, and hydraulic Conductivity. The use of DRASTIC involves rating each parameter, multiplying the rating by a weight, and summing the total score by

$$\text{TotalScore} = \sum_{i=1}^{7} W_i P_i \qquad (14.2)$$

where P_i is the input parameter i and W_i is the weight applied to P_i.

The third example is a habitat suitability index (HSI) model. HSI models evaluate habitat quality by using attributes considered to be important to the wildlife species. The process of developing a HSI model involves interpreting habitat variables from the life requisites of a wildlife species and converting these habitat variables into spatial variables that can be rated and used in a HSI model. Kliskey et al. (1999), for example, used the following equation to calculate habitat suitability for pine marten:

$$\text{HSI} = \text{sqrt}\,(((3\text{SR}_{\text{BSZ}} + \text{SR}_{\text{SC}} + \text{SR}_{\text{DS}}) / 6)\\ ((\text{SR}_{\text{CC}} + \text{SR}_{\text{SS}}) / 2)) \qquad (14.3)$$

where SR_{BSZ}, SR_{SC}, SR_{DS}, SR_{CC}, and SR_{SS} are the ratings for biogeoclimatic zone, site class, dominant species, canopy closure, and seral stage, respectively. The model was scaled so that the HSI values ranged between 0 for unsuitable habitat to 1 for optimal habitat.

The fourth example is a forest fire hazard index model. Chuvieco and Congalton (1989) constructed their model for a study area in the Mediterranean coast of Spain by using the following five factors:

- vegetation species, classified according to fuel class, stand conditions, and site (v);
- elevation (e);
- slope (s);
- aspect (a); and
- proximity to roads and trails, campsites, or housing (r).

Each of the factors was divided into different levels, which were assigned standardized scores of 0, 1, and 2 for high, medium, and low fire hazard respectively. Then, each factor was weighted according to its impact on increasing the fire hazard in the model:

$$H = 1 + 100v + 30s + 10a + 5r + 2e \qquad (14.4)$$

where H is the hazard index. To make it more operational, this forest fire hazard index model was later modified by including new variables such as weather data, more objective criteria for weighting variables, and a new scheme for integrating the variables into the synthetic fire danger index (Chuvieco and Salas 1996).

The last example is a model of human vulnerability to chemical accident. Finco and Hepner (1999) built their model by including the following demographic parameters:

- total population;
- number of people younger than 18 years of age and older than 65 years;
- economic status as measured by household income; and
- proximity to sensitive institutions such as schools, hospitals, and health clinics.

Each of the parameters was rated from 0 to 10, with 10 being most vulnerable. The total population and sensitive population parameters were positively related to vulnerability, whereas the economic status parameter was inversely related to vulnerability. The study assumed that the economically disadvantaged had less access to information about chemical accidents and were therefore more vulnerable than others. The study also assumed 100 meters as the zone of influence for sensitive institutions. Finally,

Finco and Hepner (1999) combined the four parameters into a single measure of vulnerability.

The usefulness of an index model depends on the selection of variables and the interpretation of numeric scores and weights (Robel et al. 1993; Merchant 1994). To minimize subjective biases, the consensus of a panel of experts should determine numeric scores and weights. Additionally, it may be helpful to assess the sensitivity of the vulnerability measure to the weighting scheme by using different weight combinations in a sensitivity analysis.

14.5 REGRESSION MODELS

A **regression model** relates a dependent variable to a number of independent variables in an equation, which can be used for prediction or estimation. Like an index model, a regression model can use map overlay operations in a GIS to combine all the independent variables needed for the analysis. There are two types of regression models: linear regression and logistic regression. Linear regression is used when the dependent and independent variables are all numeric variables. Logistic regression is used when the dependent variable is a binary phenomenon (e.g., presence or absence) and the independent variables are categorical or numeric variables.

Chang and Li (2000) used linear regression models to model snow accumulation. In their models, snow water equivalent (SWE) was the dependent variable and location and topographic variables derived from a DEM were the independent variables. For example, one of their regional models was expressed as

$$SWE = a + b_1 \text{ EASTING} + b_2 \text{ SOUTHING} + b_3 \text{ ELEV} \quad (14.5)$$

where a, b_1, b_2, and b_3 are regression coefficients, EASTING is the column number of a grid cell, SOUTHING is the row number of a cell, and ELEV is the elevation value of a cell.

A habitat suitability model for red squirrel developed by Pereira and Itami (1991) from a logistic regression analysis is as follows:

$$Y = 0.002 \text{ elevation} - 0.228 \text{ slope} + 0.685 \text{ canopy1} + 0.443 \text{ canopy2} + 0.481 \text{ canopy3} + 0.009 \text{ aspectE-W} \quad (14.6)$$

where canopy1, canopy2, and canopy3 represent three categories of canopy. Using the above equation, the researchers were able to calculate the probability (p) of squirrel presence for each grid cell by

$$p = 1/(1 + \exp(-Y)) \quad (14.7)$$

Mladenoff et al. (1995) also built a logistic regression model to estimate the amount and spatial distribution of favorable gray wolf habitat. The study used vector data and required a fair amount of data processing. It began by creating wolf pack areas and nonpack areas in Wisconsin in polygon coverages. Wolf pack areas were home ranges derived from telemetry data of wolf location points. Nonpack areas were randomly located in the study area at least 10 kilometers from known pack territories. Both pack and nonpack areas were then overlaid with the landscape coverages of human population density, prey density, road density, land cover, and land ownership. Using the composite coverages, stepwise logistic regression analysis converged on the following model:

$$\text{Logit}(p) = -6.5988 + 14.6189\, R \quad (14.8)$$

where p is the probability of occurrence of a wolf pack and R is road density. Probability values for occurrence of wolf presence can be calculated by

$$p = 1 / (1 + e^{\text{logit}(p)}) \quad (14.9)$$

where e is the natural exponent.

The logistic regression model, which was based on data from Wisconsin, was then applied to a three state region (Wisconsin, Minnesota, and Michigan) to map the amount and distribution of favorable wolf habitat. The same model was used in a subsequent study (Mladenoff and Sickley 1998) to predict habitat suitable for wolves in the Northeast from New York to Maine. The later study was based on raster data with a cell resolution of 5 kilometers.

Some GIS packages are capable of performing linear and logistic regression analysis. Both ARC/INFO and IDRISI have commands to build raster-based linear or logistic models. GRASS has a command to build linear regression models. Unlike statistical analysis packages, these GIS commands do not offer choices of methods for running linear or logistic regression analysis.

14.6 PROCESS MODELS

A **process model** integrates existing knowledge about the environmental processes in the real world into a set of relationships and equations for quantifying the processes (Beck et al. 1993). Unlike the previous models, which are either descriptive or statistical, process models offer both a predicative capability and an explanation that is inherent in the proposed processes (Hardisty et al. 1993). Some process models use generalized equations similar to those of index models (Coroza et al. 1997). Others use sophisticated equations to describe the interaction among large amounts of environmental data. The output of a process model is usually a set of equations that can be used for prediction.

A well-known example of a generalized process model, the Universal Soil Loss Equation (USLE) uses the product of six factors to predict soil losses for agricultural land (Wischmeier and Smith 1978):

$$A = R\,K\,L\,S\,C\,P \qquad (14.10)$$

where A is the average soil loss in tons, R is the rainfall intensity, K is the erodibility of the soil, L is the slope length, S is the slope gradient, C is the cultivation factor, and P is the supporting practice factor. Of the six factors, the R, K, C, and P factors can usually be derived from data on precipitation, soils, and land use. The L and S factors, which can be estimated from field measurements, pose the major challenge to USLE users when the topography is complex and irregular. One approach is to combine L and S into a single topographic factor, which can then be calculated from the upslope contributing area of each cell and the slope of the cell (Moore et al.1993; Desmet and Govers 1996).

The AGNPS (Agricultural Nonpoint Source) model analyzes nonpoint source pollution and estimate runoff water quality from agricultural watersheds (Young et al. 1987). AGNPS is event-based and operates on a cell basis. Using various types of input data, the model simulates runoff, sediment, and nutrient transport. For example, AGNPS uses a modified form of USLE to estimate upland erosion for single storms (Young et al. 1989):

$$SL = (EI)\,K\,LS\,C\,P\,(SSF) \qquad (14.11)$$

where SL is the soil loss, EI is the product of the storm total kinetic energy and maximum 30-minute intensity, K is the soil erodibility, LS is the topographic factor, C is the cultivation factor, P is the supporting practice factor, and SSF is a factor to adjust for slope shape within the cell. Detached sediment calculated from the equation is routed through the cells according to yet another equation based on the characteristics of the watershed.

The SWAT (Soil and Water Assessment Tool) model predicts the impact of land management practices on water quality and quantity, sediment, and agricultural chemical yields in large complex watersheds (Srinivasan and Arnold 1994). SWAT is a process-based continuous simulation model. Inputs to SWAT include land management practices such as crop rotation, irrigation, fertilizer use, and pesticide application rates, as well as the physical characteristics of the basin and subbasins such as precipitation, temperature, soils, vegetation, and topography. The creation of the input data files requires substantial knowledge at the subbasin level. Model outputs include simulated values of surface water flow, groundwater flow, crop growth, sediment, and chemical yields.

USLE, AGNPS, and SWAT are three examples of process models. The literature offers other process models on such topics as nonpoint source pollutants in the vadose zone (Corwin et al. 1997), landslides (Montgomery et al. 1998), and groundwater contamination (Loague and Corwin 1998).

Process models are typically raster-based. The role of a GIS in building a process model depends on the complexity of the model. A simple process model may be prepared and run entirely within a GIS. But more often a GIS is delegated to the role of performing modeling-related tasks such as data visualization, database management, and exploratory data analysis. The GIS is then linked to other computer programs for complex and dynamic analysis.

Commercial GIS packages do not offer commands for building process models. GRASS has commands for preparing input variables to AG-NPS, running the model, and viewing the model. GRASS also has a command that can produce the *LS* and *S* factors of USLE. The Natural Resources Conservation Service (NRCS) has a demonstration project using Soil Survey Geographic (SSURGO) Database and the ArcView interface to develop SWAT models of five small watersheds in Iowa (http://waterhome.tamu.edu/NRCSdata/SWAT_SSURGO).

KEY CONCEPTS AND TERMS

Binary model: A GIS model that uses logical expressions to select map features from a composite map.

Embedded system: GIS is bundled with other computer programs in a system with shared memory and a common interface.

GIS modeling: Use of GIS in the process of building spatially explict models.

Index model: A GIS model that uses the index value calculated from a composite map or multiple grids to produce a ranked map.

Loose coupling: The process for linking GIS and other computer programs through transfer of data files.

Model: A simplified representation of a phenomenon or a system.

Process model: A GIS model that integrates existing knowledge into a set of relationships and equations for quantifying physical processes.

Regression model: A GIS model that involves a dependent variable and a number of independent variables in developing a regression equation for prediction or estimation.

Tight coupling: The process for linking GIS and other computer programs through a common user interface.

APPLICATIONS: GIS MODELS AND MODELING

This applications section covers five tasks. Tasks 1 and 2 let you build binary models using vector data and raster data respectively. Tasks 3 and 4 let you build index models using vector data and raster data respectively. Task 5 introduces the ModelBuilder extension to ArcView as a tool for building a weighted overlay model.

Task 1: Build a Vector-based Binary Model

What you need: *elevzone.shp*, an elevation zone map; *stream.shp*, a stream map.

Task 1 asks you to locate the potential habitats of a plant species. Both *elevzone.shp* and *stream.shp* are measured in meters and spatially registered. The field zone in *elevzone.shp* shows three elevation zones. The potential habitats must meet the following criteria: (1) in elevation zone 2 and (2) within 200 meters of streams.

1. Start ArcView and check the GeoProcessing extension. Open a new view and add *elevzone.shp* and *stream.shp* to view.

2. Select Properties from the View menu. In the View Properties dialog, set the map units and distance units to meters.

3. The first step is to buffer streams with a distance of 200 meters. Make *stream.shp* active, and select Create Buffers from the Theme menu. In the Create Buffers dialog, make sure you want to buffer the features of *stream.shp*. Click Next. Set a specified distance of 200 meters. Click Next. Click Yes to dissolve barriers between buffers. Save the buffers as a new theme and call the theme *strmbuf.shp*. Click Finish. *Strmbuf.shp* appears in the Table of Contents as *Buffer1 of stream.shp*. Select Properties from the Theme menu and rename *Buffer1 of stream.shp strmbuf.shp*.

4. The next step is to overlay *elevzone.shp* and *strmbuf.shp*. Choose the geoprocessing operation of Intersect Two Themes. Select *strmbuf.shp* as the input theme to intersect, select *elevzone.shp* as the overlay theme, and specify the intersect output as *pothab.shp*. Click Finish.

5. Now you want to query *pothab.shp* and select areas in elevation zone 2. Open the theme table of *pothab.shp*. Click the Query Builder button. Set the logical expression in the Query Builder dialog as, zone = 2, and click New Set. You should see the potential habitats highlighted in *pothab.shp*.

6. Often you want to save the results from a binary model into a new theme so that you can use it for reference or further analysis. To convert the selected polygons in *pothab.shp* into a new theme, do the following. Make sure *pothab.shp* is active. Select Convert to Shapefile from the Theme menu. In the dialog, specify the path and name (e.g., *finalhab.shp*) for the new theme. Click OK.

Task 2: Build a Raster-based Binary Model

What you need: *elevzone_gd*, an elevation zone grid; *stream_gd*, a stream grid.

Task 2 tackles the same problem as Task 1 except that Task 2 uses raster data. Both *elevzone_gd* and *stream_gd* have the cell resolution of 30 meters. The cell value in *elevzone_gd* corresponds to the elevation zone. The cell value in *stream_gd* corresponds to the stream ID.

1. Start ArcView and load Spatial Analyst. Open a new view and add *elevzone_gd* and *stream_gd* to view. Select Properties from the View menu and set the map units and distance units to meters.

2. The first step in data processing is to create continuous distance measures from *stream_gd*. Activate *stream_gd*. Select Find Distance from the Analysis menu. The output called *Distance to Stream_gd* is added to the Table of Contents.

3. Now you can query *elevzone_gd* and *Distance to Stream_gd* to locate the potential habitats for the plant species. Select Map Query from the Analysis menu. In the Map Query dialog, set up the query statement as: ([Distance to stream_gd] <= 200) AND ([elevzone_gd] = 2). Click Evaluate.

4. The output called *Map Query 1* shows the potential habitats as True.

Task 3: Build a Vector-based Index Model

What you need: *soil.shp*, a soil theme; *landuse.shp*, a land use theme; *depwater.shp*, a depth to water theme.

Task 3 simulates a project on mapping groundwater vulnerability. The project assumes that groundwater vulnerability is related to three variables: soil characteristics, depth to water, and land use. Each variable has been rated on a scoring system from 0 to 50. For example, scores of 50, 35, 20, and 10 have been assigned to the depth to water classes of 1–25, 26–50, 51–100, and 101–250 feet, respectively. Soilrate shows the scores in *soil.shp*, dwrate in *depwater.shp*, and lurate in *landuse.shp*. The score value of 99 is assigned to areas such as urban and built-up areas in *landuse.shp*, which should not be included in the model. The project also assumes that the soil factor is more important than the other two factors and is assigned a weight of 3, compared to 1 for the other two factors. The index model can therefore be expressed as

Index value = 3 • soilrate + lurate + dwrate

1. Start ArcView and check the GeoProcessing extension.
2. Add *soil.shp*, *landuse.shp*, and *depwater.shp* to view.
3. The main part of Task 3 is to overlay all three themes. You can only overlay two themes at a time. Therefore, you need to perform overlay twice for three themes. Use INTERSECT as the overlay method. Select GeoProcessing Wizard from the View menu. Choose the geoprocessing operation of Intersect Two Themes. Select *landuse.shp* as the input theme to intersect, select *soil.shp* as the overlay theme, and specify the output as *landsoil.shp*. Click Finish.
4. Next, overlay *landsoil.shp* with *depwater.shp*. Repeat Step 3 but use *landsoil.shp* and *depwater.shp* as the two themes to be overlaid. Specify the output as *vulner.shp*. Click Finish.
5. *Vulner.shp* has all three rates in its theme table. Now calculate the index value from the three factors. But you have to go through a couple of steps before computation. Add a new field to the *vulner.shp* theme table for the index value. Then exclude areas with scores of 99 from computation.
6. This step is to add a new field called total to the *vulner.shp* theme table. Open the *vulner.shp* theme table, and select Start Editing from the Table menu. Then select Add Field from the Edit menu. In the Field Definition dialog, specify total for the field name, numeric for the type, 6 for the width, and 0 for the decimal places. Click OK.
7. Next, exclude areas with scores of 99 from computation. Click the Query Builder button. Set the query expression as follows: ([Lurate] <> 99), and click New Set. You are ready to compute the index value. Select Calculate from the Field menu. In the Field Calculator dialog, put the following expression in the

lower left box: 3 * [Soilrate] + [Lurate] + [Dwrate]. Then click OK. You should see the field total is populated with the index values. Select Stop Editing from the Table menu and opt for Save Edits.

8. Make the View window active because you probably want to display the index value. Double-click on *vulner.shp* to open its legend editor. Specify in the Legend Editor Graduated Color as the Legend Type and total as the Classification Field. Click Apply.
9. Once the index value map is made, you can modify the classification so that the grouping of index values may represent a rank order such as very severe, severe, moderate, slight, and very slight. You can then convert the index value map into a ranked map by doing the following: save the rank of each class under a new field called rank, and then use the dissolve operation to remove boundaries of polygons that fall within the same rank. The ranked map should look much simpler and cleaner than the index value map.

Task 4: Build a Raster-based Index Model

What you need: *soil_gd*, a soils grid; *landuse_gd*, a land use grid; *depwater_gd*, a depth to water grid.

Task 4 performs the same analysis as Task 3 but uses raster data. All three grids have the cell resolution of 90 meters. The cell value in *soil_gd* corresponds to soilrate, the cell value in *landuse _gd* corresponds to lurate, and the cell value in *depwater_gd* corresponds to dwrate.

1. Start ArcView and load Spatial Analyst. Open a new view and add *soil_gd*, *landuse_gd*, and *depwater_gd* to view.
2. First reclassify 99 in *landuse_gd* as nodata. The cell value of 99 represents urban and built-up areas that are not included in the model. Make *landuse_gd* active. Select Reclassify from the Analysis menu. In the Reclassify Values dialog, click the first cell under New Value and enter 20. Enter 40, 45, and 50 in the next three cells. Then enter No Data in the next two cells. Click OK. A new

grid called *Reclass of landuse_gd* is added to view. *Reclass of Landuse_gd* should have the cell values of 20, 40, 45, 50, and No Data.

3. To calculate the index value from *soil_gd*, *reclass of landuse_gd*, and *depwater_gd*, select Map Calculator from the Analysis menu. Prepare the following expression in the window area of the Map Calculation 1 dialog: ([Soil_gd] * 3 + [Reclass of Landuse_gd] + [Depwater_gd]). Click Evaluate. The evaluation adds to view a new grid called *Map Calculation 1*, which is the grid representation of the index model. You can compare the index grid to the index model from Task 3. (Use the same data classification to compare the maps.) Although one is in raster format and the other in vector format, the spatial distribution of the index values should look similar.

4. *Map Calculation 1* has an index value range from 145.2 to 250. Many GIS users prefer to have a value range from 0 to 1. To convert the index value range, select Map Calculator from the Analysis menu. In the next dialog, prepare the calculation statement as: ((([Map Calculation 1] – 145.2) / (250 – 145.2)). The output called *Map Calculation 2* shows the value range from 0 to 1.

Task 5. Use ModelBuilder to Build a Raster-based Index Model

What you need: *soilint_gd*, a soils grid; *landuse_gd*, a land use grid; *depwater_gd*, a depth to water grid.

Task 5 uses ArcView's ModelBuilder to build an index model similar to that in Task 4. Using ModelBuilder, however, requires two modifications. The first is the use of *soilint_gd*, an integer copy of *soil_gd*, as an input theme. The second is the use of percentages, instead of whole numbers, as weights.

1. Start ArcView and load the ModelBuilder and Spatial Analyst extensions. Open a new view and add *soilint_gd*, *landuse_gd*, and *depwater_gd* to view.

2. Select Start ModelBuilder from the Model menu. After the ModelBuilder splash screen

is briefly displayed, an empty window opens. This window is independent of the ArcView application window.

3. Begin with the input data. Click Add Data and then click a point on the left side of the ModelBuilder window for the start of the model diagram. Right-click the Data icon and choose Theme. Right-click the Theme icon and choose Properties. In the Project Theme dialog, enter Soils as the project data name, choose *soilint_gd* as the input theme, and choose value as the input field. Click OK. The input data icon now turns blue and has the label of Soils.

4. Repeat the same procedure as in (3) to enter Land use and Depth to Water as the project data names and *landuse_gd* and *depwater_gd* as the input themes in the model diagram.

5. Add an operation in the ModelBuilder window. Click Add Function and then click a point to the right of the input themes. The icons of Function and Derived Data appear in the model diagram. Right-click the Function icon and select Overlay and then Weighted Overlay. The two icons are now labeled Weighted Overlay and Weighted Overlay Map.

6. Next, link the function to the input themes. Click Add Connection and drag the cursor from Soils to the Weighted Overlay function to add a connection line between them. Do the same to add the connection lines between the function and Land use and Depth to Water.

7. This step is to define the process of building an index model from the three input themes. Click Add Function. Right-click the Weighted Overlay Function, and select Properties to open the Weighted Overlay dialog. The dialog has two tabs at the top. First select Evaluation Scale. The Evaluation Scale dialog determines the scoring systems for the selected variables in the index model. Because the highest value is 50 for the three input grids, a custom evaluation scale must be defined. After choosing to define a custom evaluation scale, set the scale from 1 to 50 by 1.

8. Select Overlay Table. The table has five fields: Input Theme, % Inf, Input Field, Input

Label, and Scale Value. You need to work with % Inf and Scale Value. % Inf allows you to assign weights to each selected variable in percentage. Enter 60 for *soilint_gd*, 20 for *landuse_gd*, and 20 for *depwater_gd*. The sum of the weights must equal 100%.

9. This step is to fill in the Scale Value column. Click on the first cell and enter the Input Field Value as its Scale Value. For example, if the first record belongs to *soilint_gd* and has an Input Field Value of 38, enter 38 as its Scale Value. Do the same for the rest of the table. The exceptions are No Data and the record of *landuse_gd* with an Input Field Value of 99, for which the Scale Value of Restricted is entered.

10. After the Weighted Overlay dialog is completed, click Run. In the next dialog, specify the path to save the model and its output. View the Weighted Overlay map, which has been added to the Table of Contents. The Weighted Overlay map shows the index value range of 1 to 50 in the legend, but its theme table shows that the range is actually from 29 to 50.

11. Compare the Weighted Overlay map with *Map Calculation 1* in Task 4. They should look the same. Because ModelBuilder uses percentages as weights, the highest index value in the Weighted Overlay map is 50, which is one-fifth of the highest value (250) in *Map Calculation 1*.

12. ModelBuilder saves the model diagram in a file with the extension of xmd. The .xmd file can be reopened in ModelBuilder, and % Inf and Scale Value can be altered before rerunning the model.

REFERENCES

Aller, L., T. Bennett, J. H. Lehr, R. J. Petty, and G. Hackett, 1987. *DRASTIC: A Standardized System for Evaluating Groundwater Pollution Potential Using Hydrogeologic Settings*. U.S. Environmental Protection Agency, EPA/600/2-87/035, 622 pp.

Beck, M.B., A.J. Jakeman, and M.J. McAleer. 1993. Construction and Evaluation of Models of Environmental Systems. In A.J. Jakeman, M.B. Beck, and M.J. McAleer (eds.). *Modelling Change in Environmental Systems*. Chichester, NY: John Wiley & Sons, pp. 3–35.

Chang, K., and Z. Li. 2000. Modeling Snow Accumulation with a Geographic Information System. *International Journal of Geographical Information Science* 14: 693–707.

Chrisman, N. 1997. *Exploring Geographic Information Systems*. New York: John Wiley & Sons.

Chuvieco, E., and R.G. Congalton. 1989. Application of Remote Sensing and Geographic Information Systems to Forest Fire Hazard Mapping. *Remote Sensing of the Environment* 29: 147–59.

Chuvieco, E., and J. Salas. 1996. Mapping the Spatial Distribution of Forest Fire Danger Using GIS. *International Journal of Geographical Information Systems* 10: 333–45.

Coroza, O., D. Evans, and I. Bishop. 1997. Enhancing Runoff Modeling with GIS. *Landscape and Urban Planning* 38: 13–23.

Corwin, D.L., P.J. Vaughan, and K. Loague. 1997. Modeling Nonpoint Source Pollutants in the Vadose Zone with GIS. *Environmental Science & Technology* 31: 2157–75.

Desmet, P.J.J., and G. Govers. 1996. A GIS Procedure for Automatically Calculating the USLE LS Factor on Topographically Complex Landscape Units. *Journal of Soil and Water Conservation* 51: 427–33.

Desmet, P.J.J., and G. Govers. 1996. Comparison of Routing Systems for DEMs and Their Implications for Predicting Ephemeral Gullies. *International Journal of Geographical Information Systems* 10: 311–31.

Finco, M.V., and G.F. Hepner. 1999. Investigating US–Mexico Border Community Vulnerability to

Industrial Hazards: A Simulation Study in Ambos Nogales. *Cartography and Geographic Information Science* 26:243–52.

Hardisty, J., D.M. Taylor, and S.E. Metcalfe. 1993. *Computerized Environmental Modelling*. Chichester, NY: John Wiley & Sons.

Kliskey, A.D., E.C. Lofroth, W.A. Thompson, S. Brown, and H. Schreier. 1999. Simulating and Evaluating Alternative Resource-Use Strategies using GIS-based Habitat Suitability Indices. *Landscape and Urban Planning* 45: 163–75.

Lathrop, R.G. Jr., and J.A. Bognar. 1998. Applying GIS and Landscape Ecologic Principles to Evaluate Land Conservation Alternatives. *Landscape and Urban Planning* 41: 27–41.

Lichthardt, J. 1995. *Conservation Strategy for ALLOTROPA VIRGATA (CANDYSTICK), U.S. Forest Service, Northern and Intermountain Regions.* Report of a cooperative challenge cost-share project, Nez Perce National Forest and the Idaho Department of Fish and Game.

Loague, K., and D.L. Corwin. 1998. Regional-scale Assessment of Non-point Source Groundwater Contamination. *Hydrologic Processes* 12: 957–65.

Merchant, J.W. 1994. GIS-Based Groundwater Pollution Hazard Assessment: A Critical Review of the DRASTIC Model. *Photogrammetric Engineering & Remote Sensing* 60: 1117–27.

Mladenoff, D.J., T.A. Sickley, R.G. Haight, and A.P. Wydeven. 1995. A Regional Landscape Analysis and Prediction of Favorable Gray Wolf Habitat in the Northern Great Lakes Regions. *Conservation Biology* 9: 279–94.

Mladenoff, D.J., and T.A. Sickley. 1998. Assessing Potential Gray Wolf Restoration in the Northeastern United States: A Spatial Prediction of Favorable Habitat and Potential Population Levels. *Journal of Wildlife Management* 62: 1–10.

Montgomery, D.R., K. Sullivan, and H.M. Greenberg. 1998. Regional Test of A Model for Shallow Landsliding. *Hydrologic Processes* 12: 943–55.

Moore, I.D., A.K. Turner, J.P. Wilson, S.K. Jenson, and L.E. Band. 1993. GIS and Land-Surface-Subsurface Process Modelling. In M.F. Goodchild, B.O. Park, and L.T. Styaert (eds.). *Environmental Modelling with GIS*. Oxford, UK: Oxford University Press, pp. 213–30.

Pereira, J.M.C., and R.M. Itami. 1991. GIS-Based Habitat Modeling Using Logistic Multiple Regression: A Study of the Mt. Graham Red Squirrel. *Photogrammetric Engineering & Remote Sensing* 57: 1475–86.

Robel, R.J., L.B. Fox, and K.E. Kemp. 1993. Relationship between Habitat Suitability Index Values and Ground Counts of Beaver Colonies in Kansas. *Wildlife Society Bulletin* 21: 415–21.

Schlagel J.D., and C.M. Newton. 1996. A GIS-Based Statistical Method to Analyze Spatial Change. *Photogrammetric Survey and Remote Sensing* 62: 839–44.

Srinivasan, R., and J.G. Arnold. 1994. Integration of a Basin-scale Water Quality Model with GIS. *Water Resources Bulletin* 30: 453–62.

Tomlin, C.D. 1990. *Geographic Information Systems and Cartographic Modeling*. Englewood Cliffs, NJ: Prentice Hall.

Wischmeier, W.H., and D.D. Smith. 1978. Predicting Rainfall Erosion Losses: A Guide to Conservation Planning. *Agricultural Handbook 537*. Washington, DC: U.S. Department of Agriculture.

Young, R.A., C.A. Onstad, D.D. Bosch, and W.P. Anderson. 1987. AGNPS, Agricultural Non-Point-Source Pollution Model: A Large Watershed Analysis Tool. *Conservation Research Report 35.* Washington, DC: Agricultural Research Service, U.S. Department of Agriculture.

Young, R.A., C.A. Onstad, D.D. Bosch, and W.P. Anderson. 1989. AGNPS: A Nonpoint-Source Pollution Model for Evaluating Agricultural Watersheds. *Journal of Soil and Water Conservation* 44: 168–73.

15

REGIONS

CHAPTER OUTLINE

15.1 INTRODUCTION

Regions are higher-level objects built on top of simple arcs and polygons (Chapter 3). The regions data model organizes polygons into regions and groups regions into region layers or subclasses. A region consists of polygons that may or may not be connected spatially. Regions of different region subclasses may overlap or cover the same area. And each region subclass can have its own attributes and can function independently for data display, query, and analysis. Some of these characteristics of the regions data model are similar to those of polygon shapefiles and polygon features based on the object-oriented data model.

The regions data model corresponds to the concept of geographic regions in such disciplines as geography, landscape ecology, and forestry (Berry 1968; Bailey 1976; Forman and Godron 1986). A geographic region defines an area that has certain attributes to give the area a particular character. Examples of geographic regions include ethnic regions, linguistic regions, hydrologic units, and ecological units. Like the regions data model, these geographic regions may overlap and each region may include spatially joint or disjoint areas.

This chapter is divided into seven sections. Section 1 covers two types of geographic regions. Section 2 discusses applications of the regions data model. Section 3 reviews methods for creating regions from new or existing arcs and polygons. Section 4 discusses dropping and converting regions. Section 5 describes attribute data management with regions. Sections 6 and 7 cover regions-based overlay and query and other tools.

Both ArcView and ARC/INFO can display, query, and analyze regions by treating region subclasses as separate layers in a polygon coverage. But ArcView cannot create regions or perform regions-based analyses. Although regions are similar to polygon features of the geodatabase model, the two models remain separate in ArcInfo 8. Regions are only accessible in ArcInfo Workstation or ARC/INFO.

15.2 GEOGRAPHIC REGIONS

15.2.1 Uniform Regions

A **uniform region** can be broadly defined as a geographic area with similar characteristics. Uniform regions may be related to a single theme. For example, *The National Atlas of the United States of America* of 1970 contains a map at a scale of 1:7,500,000 that divides the country into uniform

landform regions by using a classification scheme of slope, local relief, and profile type. The same atlas also includes uniform regions of potential natural vegetation, soils, and major land uses. Traffic analysis zones are another example. Defined as geographic areas with similar land use and economic activities, traffic analysis zones represent the origins and destinations of trips in travel demand modeling (Miller 1999).

Uniform regions may also be defined by multiple themes. An example of this type of uniform region is Omernik's ecoregions (1987). By combining land use, potential natural vegetation, land surface form, soils, and other supporting data through the overlay process, Omernik's map divides the conterminous United States into 76 ecoregions at a scale of 1:7,500,000. Some of Omernik's ecoregions have disjoint areas. For example, the ecoregion called Arizona/New Mexico Mountains consists of several disjoint areas in Arizona and New Mexico.

15.2.2 Hierarchical Regions

Hierarchical regions represent different spatial scales of a hierarchy. A hierarchical structure divides the Earth's surface into progressively smaller regions of increasingly uniform characteristics. Two well-known examples of hierarchical regions are census units and the USGS hydrologic units.

The U.S. Bureau of the Census compiles and distributes census data by state, county, census tract, block group, and block. These area units form a nested hierarchy: blocks are contained within a block group, block groups within a census tract, census tracts within a county, and counties within a state (Figure 15.1). Census units in a hierarchy represent different levels of spatial aggregation.

USGS hydrologic units are also organized in a nested hierarchy. The United States is divided into progressively smaller hydrologic units at four levels: regions, sub-regions, accounting units, and cataloging units (http://water.usgs.gov/GIS/huc.htm). A two digit code is given at each level. Each hydrologic unit in the United States therefore has a unique code of two to eight digits, depending on the unit's classification within the hierarchy.

Ecological units are the more complex examples of hierarchical regions. In 1993 the U.S. Forest Service adopted the National Hierarchical Framework of Ecological Units, primarily based on the work of Bailey (1976, 1983) (http://svinet2.fs.fed .us/land/pubs/ecoregions/). This hierarchical framework provides a systematic method for classifying and mapping areas of the Earth's surface by the associations of ecological factors at the scales of ecoregion, subregion, landscape, and land unit (Cleland et al. 1997). Climate and landform define ecoregion and subregion boundaries. Relief, geologic parent materials, and potential natural communities are added to climate and landform in defining landscape boundaries. Topography, soil characteristics, and plant associations subdivide landscape units into land units. The Canadian system of ecological land classification has a similar framework by having four hierarchical levels of

land region, land district, land system, and land type (Wiken and Ironside 1977) (http://sis.agr.gc.ca/ cansis/nsdb/ecostrat/intro.html/).

According to Cleland et al. (1997), ecological units are useful for delineating ecosystems, assessing resources, conducting environmental analyses, and managing and monitoring natural resources.

15.3 APPLICATIONS OF THE REGIONS DATA MODEL

15.3.1 Building Geographic Regions

Geographic regions can be built using either polygons or regions. One method for mapping ecological units, for example, is to compile necessary base maps such as topography, geology, soils, climate, and potential natural communities; overlay these polygon maps; and to query the associations

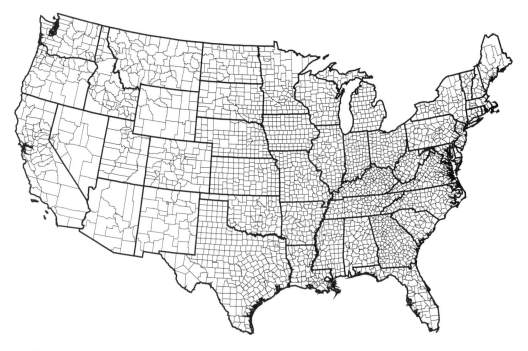

Figure 15.1
A nested hierarchy of counties and states in the conterminous United States.

of the input factors. This method produces a map series of ecological units, one map for each spatial scale. The regions method uses the same input but integrates each spatial scale as a region subclass in a polygon coverage. Therefore, the regions method produces an **integrated coverage**, combining a polygon coverage with built-in region subclasses.

An integrated coverage has several advantages over a series of polygon coverages. First, the regions data model allows nested, overlapped region subclasses and regions that may have disjoint areas. In contrast, a polygon coverage cannot have overlapped polygons or polygons with spatially disjoint components. Second, an integrated coverage is simpler than a map series for data management. This is particularly true for a long map series such as forest fire records for the past 100 years. Third, regions-based query and analysis commands, which can only be used on an integrated coverage, are more efficient than polygon-based commands in many instances. These regions-based commands are covered later in the chapter.

An integrated coverage has its disadvantages as well. First, an integrated coverage with its region subclasses requires large computer memory and extra processing time. Second, an integrated coverage is more difficult than a polygon coverage to revise or update. If one of the input layers is revised, the integrated coverage must be rebuilt, no matter how extensive the revision is.

Third, an integrated coverage is also more difficult than a polygon coverage to manipulate, such as in the operations of reselect, clip, and erase (Chapter 10). Altering the hierarchical data structure of arc, polygon, and region can corrupt an integrated coverage. Fourth, use of an integrated coverage is primarily limited to ARC/INFO and ArcView users, although region subclasses can be easily converted to shapefiles, a data format that can be used by many other GIS packages.

15.3.2 Incorporating Spatial Scales in GIS Analysis

Hierarchical regions represent different spatial scales of a hierarchy and are therefore ideal for including spatial scale in GIS analysis. Here we will examine two areas of active research dealing with spatial scale.

15.3.2.1 Studies of Ecosystems

Ecosystems are places where life forms and environment interact. The structure and function of ecosystems are affected by environmental and biological factors including climate, geology, soils, and flora, and these factors vary at different spatial scales (Box 15.1). Hierarchy theory in ecology has emphasized that ecosystems and their underlying biophysical environments occur in a spatial hierarchy, with smaller ecosystems embedded in larger ones (Allen and Starr 1982; O'Neill et al. 1986).

Box **15.1**   **Map Scale and Spatial Scale**

Scale is a confusing term. Cartographers define the map scale as the ratio between a map distance and the ground distance it represents. A map scale of 1:24,000 means that an inch on the map represents 24,000 inches (2000 feet) on the ground. At a scale of 1:100,000, a one-inch map distance represents 100,000 inches (about 1.58 mile) on the ground. To cartographers, 1:24,000 is a larger scale than 1:100,000, and a large-scale map covers a smaller area than a small-scale map.

Ecologists use the term spatial scale. Unlike map scale, spatial scale is not rigidly defined. Spatial scale refers to the size of area or extent. A large spatial scale therefore covers a larger area than a small spatial scale. To put it another way, a large spatial scale to an ecologist is a small map scale to a cartographer.

The literature has shown that ecological patterns and processes are scale dependent and scale is an important variable in ecological studies (Meentemeyer and Box 1987; Urban et al. 1987; Wiens 1989; Costanza and Maxwell 1994).

According to hierarchy theory, the conditions and processes of larger ecosystems can affect or constrain those of smaller ecosystems. For example, logging operations on upper slopes of an ecological unit will likely affect streams and riparian habitats in downslope smaller units. At the same time, local conditions and processes can have cumulative effects on the next higher level. For example, pines in a snow-pine forest landscape convert solar radiation into sensible heat, which moves to the snow cover and melts it faster than would happen in either a wholly snow covered or a wholly forested basin (Bailey 1983).

Scaling in ecology is a complex topic. A GIS using the regions data model can assist research in this area in two ways. First, it can build an integrated coverage by incorporating spatial scales into region subclasses. In other words, the regions data model provides a means to build a framework similar to a hierarchical GIS proposed by Walker and Walker (1991). Second, the GIS can provide the analytical functions for manipulating and analyzing attributes of different region subclasses.

15.3.2.2 Modifiable Area Unit Problem

The **Modifiable Area Unit Problem (MAUP)** addresses the 'modifiable' nature of area units used in spatial analysis and the influence it has on the analysis and modeling results. The modifiable nature of area units appears in two forms: the level of spatial data aggregation, and the definition or partitioning of area units for which data are collected. The former is referred to as the scale effect and the latter as the zoning effect in the MAUP literature (Openshaw and Taylor 1979; Wong and Amrhein 1996). Users of census data, for example, must be aware of the scale effect because they can choose block groups, census tracts, or counties for analysis. The zoning effect does not exist if the boundaries of area units such as census units are predefined. The zoning

effect exists, for example, when a random procedure is used to aggregate area units (Fotheringham and Wong 1991).

An early study of the MAUP found that the correlation strengthened with spatial aggregation (Gehlke and Biehl 1934). Robinson (1950) reported a similar relationship between estimated correlation coefficients and the spatial scale of census data used in the computation. Other studies have reported MAUP effects in bivariate regression (Clark and Avery 1976; Amrhein, 1995), multivariate regression (Fotheringham and Wong 1991), spatial interaction (Openshaw 1977; Batty and Sikdar 1982a, 1982b), and location-allocation (Goodchild 1979). The MAUP has also been studied in landscape ecology (Jelinski and Wu 1996) and transportation planning (Khatib et al. 2001).

MAUP effects appear to be consistent for bivariate regression: as the level of spatial aggregation increases, one can expect a general increase in the correlation coefficient. This is likely resulted from the smoothing effect of data aggregation. But MAUP effects are essentially unpredictable in multivariate analysis (both linear regression and logistic regression) according to Fotheringham and Wong (1991). Their regression models relating mean family income to demographic variables in Buffalo, New York initially showed a coefficient of determination of 0.81 at the census tract level and 0.37 at the block group level. Subsequent experimentation with scaling and zoning, however, showed a wide range of results in terms of estimated regression parameters and the coefficient of determination.

Given the uncertainty of MAUP effects, it is wise to incorporate spatial scale as a variable in studies that are based on modifiable area units. Fotheringham and Wong (1991), for example, recommended the reporting of results at different levels of aggregation and with different zoning systems at the same scale. The regions data model can be useful for studying the MAUP by including each level of spatial aggregation to be studied as a region subclass.

15.4 CREATE REGIONS

This section discusses methods for creating regions in ARC/INFO. As mentioned earlier, regions are similar to polygon shapefiles (Box 15.2) and polygon features based on the geodatabase model. The object-oriented data model may add new options for creating polygons/regions in the future.

Regions may be created in ARC/INFO in several ways: from existing arcs or polygons, from data conversion methods, from a related table, or from regions-based commands. No matter which method is used, the creation of regions must follow the regions data model and its data structure. For example, whenever the choice for making regions is between arcs and polygons, polygons should be used because regions are built directly on polygons. If arcs are chosen, the process of creating regions must go through the hierarchical data structure of arcs, polygons, and regions.

One option in creating regions is the choice between noncontiguous and contiguous regions. A **noncontiguous region** may have spatially disjoint components, as allowed in the regions data model. A **contiguous region**, however, must be spatially connected. Although regions created from both options function the same, the difference

is the number of records in the region subclass' attribute table. The noncontiguous option typically has a smaller number of records than the contiguous option.

15.4.1 Create Regions Interactively

Regions can be digitized directly but the process is slow because of overlapping regions and their spatially disjoint components. The preferred method is to select arcs or polygons that have been already digitized to make up regions. Figure 15.2 shows a soil suitability map in three class values. Suppose we want to build a region subclass from the coverage with three different regions corresponding to the three suitability values respectively. The following shows how the region subclass can be created from existing polygons or arcs.

It is relatively simple to create regions from the existing polygons. To make up the region with the suitability value of 1, for example, requires only to select polygons that have the suitability value of 1 and to use the ARC/INFO command MAKERE-GION to make a region from the selected polygons (Figure 15.3). We can build regions with the suitability values of 2 and 3 in the same way. In this case, the output uses the noncontiguous option. Alternatively, the contiguous option can make a

---

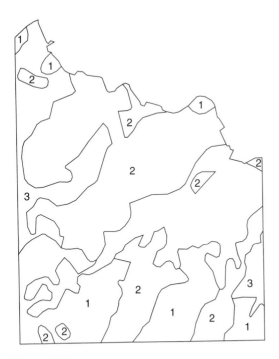

Figure 15.2
A soil map with the suitability values of 1, 2, and 3.

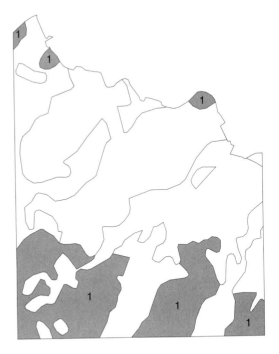

Figure 15.3
MAKEREGION creates a region subclass containing polygons that have the suitability value of 1. The region subclass is noncontiguous.

region from each polygon and code the region with the polygon's suitability value (Figure 15.4).

Creating regions from existing arcs is more tedious than from polygons. For two adjacent polygons that belong to the same region, the arcs making up each polygon must be selected and processed separately to support the hierarchical data structure of arcs, polygons, and regions (Figure 15.5). Otherwise, a region may not form a closed loop. Using the previous example of suitability regions, we must select all arcs that define polygons with the suitability value of 1 to make the first region, and repeat the same process for the second and third regions (Figure 15.6). The output consists of noncontiguous regions. Again, the contiguous alternative is to make a region from arcs that define each polygon on the coverage.

15.4.2 Create Regions by Data Conversion

Regions can be converted from polygons or arcs. Several ARC/INFO commands can perform the conversion. POLYREGION converts a polygon coverage to a region subclass and each polygon on the coverage to a region. In other words, POLYREGION creates a region subclass using the contiguous option and converts polygons to regions using a one-to-one relationship.

REGIONCLASS converts arcs to regions. The process is more complex because it must support the hierarchical data structure of arcs, polygons, and regions and work with the PAL (polygon arc list) file and the RXP (region subclass cross-reference table) file (Box 15.3). The PAL file contains a list of arcs for each region, whereas the RXP file relates regions to polygons for a region subclass. The method is best used if the data are in x-, y-coordinates and are already stored in a text file.

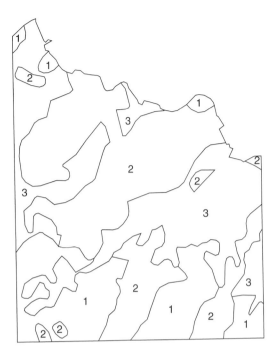

Figure 15.4
The contiguous option of MAKEREGION converts each polygon to a region.

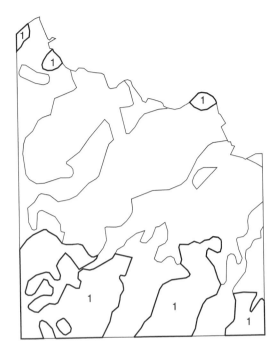

Figure 15.6
To create a region from arcs that make up the polygons with the value of 1, select the arcs (thicker lines) first and use the MAKEREGION command.

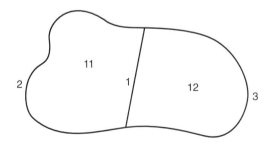

Figure 15.5
Polygon 11 shares arc 1 with polygon 12. Arc 1 must be selected and processed twice to complete the hierarchical data structure of arcs, polygons, and regions used by the regions data model.

Vector Product Format (VPF) is a standard format used by the National Imagery and Mapping Agency (NIMA) for large geographic databases

(Chapter 4). VPFIMPORT is an ARC/INFO command that can convert VPF files into ARC/INFO coverages. If a VPF file contains multiple features, such as a boundary file containing political boundaries, ocean boundaries, and inland water boundaries, VPFIMPORT uses region subclass attribute tables to separate them.

15.4.3 Create Regions by Using a Related Table

Using a related table to create regions is unique. The method is only possible with the regions data model because the related table can direct ARC/INFO in how to construct the RXF file. The command REGIONJOIN requires that (1) a region subclass must be present and (2) a relate must be established between the region subclass and a

---

Box 15.3   **Convert Existing Arcs to Regions**

The following shows the steps of using REGION-CLASS in ARC/INFO to convert existing arcs to regions.

1. Code all arcs that make up a region component with the same ID value. For example, all arcs that make up the first region in the above suitability example must have the same ID value of 1, and so on (Figure 15.7).
2. Convert the data file into a line coverage using the GENERATE command.
3. Build the line coverage's topology, and make sure that the ID item in the attribute table

separates arcs by region so that all arcs that make up region 1 have the ID value of 1, and so on.
4. Use the ID item as the input item and the REGIONCLASS command to create preliminary regions from the arcs in the line coverage.
5. Use the CLEAN command to build the topology of preliminary regions, that is, to construct the RXP file, and to complete the conversion process.

If the arcs that make up a region do not close properly and the resulting overshoot or undershoot is not taken care of by cleaning the coverage, the region will have topological errors.

```
             1
    4111.058105    7188.290039
    4125.195801    7126.065918
    4121.023926    7073.250000
    4075.225098    7039.095215
    4051.218994    6993.986816
    4013.513916    6957.342773
 END
             1
    4111.058105    7188.290039
    4011.874023    7100.367188
    4013.513916    6957.342773
 END
             1
    4404.659180    6859.535156
    4382.346191    6902.346191
    4329.016113    6954.854980
    4322.033203    6974.992188
 END
    .
    .
    .
```

Figure 15.7
The text file to be used for creating regions must have the same ID value (1 in this example) for all arcs that make up a region.

related table (Chapter 6). The related table may reside in the same workspace as the existing coverage or in an external database.

Figure 15.8 shows how REGIONJOIN works. Figure 15.8a shows the attribute table of an existing region subclass called *soils1*. To establish a relate between *soils1* and a suitability table, we can use soils1-id as the key or relate item (Figure 15.8b). A one-to-one relationship exists between *soils1* 's attribute table and the suitability table. But the suitability table contains only three suit values of 1, 2, and 3. When *soils1* and the related table are used to create a new region subclass called *soils2*, REGIONJOIN, with the noncontiguous option, groups regions with the same suit value in *soils1* into a single region in *soils2*. Having only three values, *soils2* has three regions, each with spatially disjoint components (Figure 15.8c).

If a one-to-many relationship exists between a region and the related table, REGIONJOIN stacks duplicates of the region in the output and lists a different attribute value with each duplicate. Figure 15.9 is the same as Figure 15.8 except that (1) the related table is a crop table and (2) one of the regions in *soils1* (soils1-id = 10) has three values (2, 3, and 1). *Soils3*, a new region subclass created by REGIONJOIN, has five regions (Figure 15.9c).

The first two are duplicates of soils1-id = 10 and have the soils1_crop values of 3 and 1 respectively. The other duplicate of soils1-id = 10 with the soils1_crop value of 2 is merged with other region components in *soils3*.

15.4.4 Create Regions by Using Regions-based Commands

Regions-based commands can create regions by querying and analyzing existing coverages or region subclasses. These commands are covered later in the chapter.

15.5 DROP OR CONVERT REGIONS

Region subclasses in an integrated coverage can be dropped or converted into polygon coverages or shapefiles. The ARC/INFO command DROPFEATURES can drop both the geometry and attributes of a region subclass from an integrated coverage. REGIONPOLY can convert a region subclass into a polygon coverage.

A region subclass, after it is added as a theme and activated in ArcView, can be converted into a polygon shapefile. The ArcView MIFSHAPE is a stand-alone utility program that can convert a MapInfo Interchange Format (MIF) file into an ArcView shapefile. MIF files use regions instead of polygons. MIFSHAPE converts region features created in MapInfo into polygons.

15.6 ATTRIBUTE DATA MANAGEMENT WITH REGIONS

An integrated coverage includes a polygon coverage and region subclasses. The polygon coverage has its own attribute table and so does each region subclass. The naming convention for a region subclass' attribute table is to have the coverage name as the prefix and the concatenated string of PAT and the region subclass name as the extension. For example, the file name *all.patveg* refers to the attribute table for a region subclass named *veg* in a polygon coverage called *all*. The coverage *all* may have other region subclasses such as *soil* and *elev*, each having its own PAT file (Figure 15.10).

The PAT for a region subclass is maintained and managed separately from the polygon coverage and other region subclasses. Each region subclass' PAT can have its own attributes that are different from other PATs and can function as a relation in a relational database. In short, each region subclass can be treated as a separate layer in GIS operations.

15.7 REGIONS-BASED QUERY AND OVERLAY

Logical expressions can be used to query a region subclass' attribute table. The procedure is the same as querying an attribute table of a polygon coverage. Because region subclasses are integrated in a polygon coverage, this section discusses a different kind of regions-based query. The query actually overlays polygon coverages, selects a subset of polygons or regions, and saves the result of the selection into a new region subclass. ARC/INFO has two commands, AREAQUERY and REGION-QUERY, which can perform overlay, query, and creation of region subclasses.

AREAQUERY uses a set of polygon coverages as the input and processes the input in two steps. First, it converts each coverage into a region layer and stores the layer as a region subclass in the integrated polygon coverage. Second, AREAQUERY evaluates a logical expression operated on the existing region subclasses and saves the result into a new region subclass. The second step of AREAQUERY is the same as REGIONQUERY.

REGIONQUERY performs query and creates regions from a polygon coverage or an integrated coverage with region subclasses. Suppose a polygon coverage records past fires with three items: fire1 lists the year of fire for areas burned once, fire2 lists the year of the second fire for areas burned twice, and fire3 lists the year of the third fire for areas burned three times. The logical expression, fire1 > 0 AND fire2 > 0, used with

(a)

Area	Perimeter	Soils1#	Soils1-ID
508308.531	8336.992	1	2
14021.991	553.177	2	3
19000.115	531.830	3	4
23931.438	673.881	4	5
1283917.250	8736.511	5	6
58753.410	1519.348	6	7
22764.352	577.357	7	8
833005.563	9419.269	8	9
8438.227	412.599	9	10
18385.139	606.922	10	11
188282.297	3157.877	11	12
303357.188	3476.662	12	13
86360.703	1849.829	13	14
516075.688	6647.514	14	16
301924.563	3017.327	15	17
90327.867	1566.401	16	18
11328.362	395.495	17	19
13428.370	470.184	18	20

(b)

Soils1-ID	Suit
2	3
3	1
4	1
5	2
6	2
7	3
8	1
9	3
10	2
11	2
12	2
13	2
14	3
16	1
17	1
18	1
19	2
20	2

(c)

Area	Perimeter	Soils2#	Soils2-ID	Suit
1486428.250	21125.438	1	1	3
964114.563	12893.606	2	2	1
1851068.250	17930.133	3	3	2

Figure 15.8

REGIONJOIN creates regions from a related table. Soils1-ID relates the region subclass' attribute table (a) to the suitability table (b). Table (c) is the attribute table of the new region subclass.

(a)

Area	Perimeter	Soils1#	Soils1-ID
508308.531	8336.992	1	2
14021.991	553.177	2	3
19000.115	531.830	3	4
23931.438	673.881	4	5
1283917.250	8736.511	5	6
58753.410	1519.348	6	7
22764.352	577.357	7	8
833005.563	9419.269	8	9
8438.227	412.599	9	10
18385.139	606.922	10	11
188282.297	3157.877	11	12
303357.188	3476.662	12	13
86360.703	1849.829	13	14
516075.688	6647.514	14	16
301924.563	3017.327	15	17
90327.867	1566.401	16	18
11328.362	395.495	17	19
13428.370	470.184	18	20

(b)

Soils1-ID	Crop
2	3
3	1
4	1
5	2
6	2
7	3
8	1
9	3
10	2
10	3
10	1
11	2
12	2
13	2
14	3
16	1
17	1
18	1
19	2
20	2

(c)

Area	Perimeter	Soils3#	Soils3-ID	Crop	Soils1_crop
8438.227	412.599	1	1	2	3
8438.227	412.599	2	2	2	1
1486428.250	21125.438	3	3	3	3
964114.563	12893.606	4	4	1	1
1851068.250	17930.133	5	5	2	2

Figure 15.9

A one-to-many relationship exists between the region subclass' attribute table (a) and the suitability table (b) for soils1-ID = 10. REGIONJOIN creates a stack of 3 regions for soils1-ID = 10: 2 are shown in the first 2 records of table (c), and the third is merged with the region for crop = 2.

all.pat   — attribute table for the polygon coverage *all*

all.patveg — attribute table for the region subclass *veg*
             that resides in the polygon coverage *all*

all.patsoil — attribute table for the region subclass *soil*
              that resides in the polygon coverage *all*

all.patelev — attribute table for the region subclass *elev*
             that resides in the polygon coverage *all*

Figure 15.10
The PATs of the integrated polygon coverage *all* and its region subclasses, *veg*, *soil*, and *elev*.

TABLE 15.1	Attribute items and item descriptions of region subclasses, elev, strmbuf, and slope.

Subclass	Attribute Item	Item Value Description
elev	zone	1 - less than 3000 ft
		2 - 3000 to 4000 ft
		3 - greater than 4000 ft
strmbuf	inside	100 - inside buffer zone
		1 - outside buffer zone
slope	slope-code	1 - less than 20%
		2 - 20–40%
		3 - 40–60%
		4 - greater than 60%

REGIONQUERY can create region subclasses for areas burned twice.

Suppose an integrated coverage called *allcov*, which may have been created by AREAQUERY, has three region subclasses of *elev* (elevation), *strmbuf* (stream buffer), and *slope*. Attribute data in each region subclass attribute table are listed in Table 15.1. The following shows two examples of using REGIONQUERY to create new region subclasses. The first example creates a new region subclass containing areas that are in elevation zone 2 and are within the stream buffer:

elev.zone = 2 AND strmbuf.inside = 100

The above expression looks the same as any logical expression except for the operands of elev.zone and strmbuf.inside. Each operand consists of two parts: the first part refers to the region subclass, and the second part refers to the item in the region subclass.

The second example creates a new region subclass containing areas that are in elevation zone 2, have the slope code of 1, and are outside the stream buffer:

elev.zone = 2 AND slope.slope-code = 1
NOT $strmbuf

$strmbuf in the expression means that all region features in the *strmbuf* subclass are used in the selection. The Boolean connector NOT negates $strmbuf, meaning that the selection excludes areas covered by the *strmbuf* subclass.

The above examples show that AREAQUERY combines map overlay and attribute data query into a single command, thus providing an alternative to traditional vector-based overlay analysis, which separates map overlay operations from attribute data query (Chapter 10). AREAQUERY has two other advantages over traditional overlay analysis. First, AREAQUERY can overlay up to 32 coverages in a single operation. Second, REGIONQUERY, when used on the integrated coverage created by AREAQUERY, can create additional region subclasses based on different selection criteria. All residing in the same polygon coverage, these region subclasses can be used as scenarios in a decision making process.

Regions-based operations, however, differ from traditional vector-based overlay analysis in three areas. First, regions-based operations are limited to polygon coverages. Traditional map overlay analysis can work with point, line, and polygon coverages. Second, regions-based operations use logical expressions and the proper Boolean connectors of AND, OR, XOR, and NOT to duplicate the overlay methods of UNION, INTERSECT, and IDENTITY. Box 15.4 shows additional examples of using the Boolean connectors

---

Box 15.4   Use Logical Expressions to Select Regions

Figure 15.11a shows two region subclasses: elevation zones (*elev*) in thin lines and stream buffer zones (*strmbuf*) in thick lines. Notice that the area extent of the stream buffer goes slightly beyond that of the elevation zone. The logical expression, $elev OR $strmbuf, duplicates a UNION operation, with the result shown in Figure 15.11b. Figure 15.11c shows the output if the expression is changed to $elev AND $strmbuf, or an INTERSECT operation. And Figure 15.11d shows the evaluation of the expression, ($elev AND $strmbuf) OR $elev, or an IDENTITY operation using *elev* as the input coverage.

XOR and NOT add more varieties to regions-based operations. Figure 15.12 shows the output if the Boolean operation, $elev XOR $strmbuf, is carried out. The output retains map features of the input region subclasses except those in the overlapped area of the two subclasses. Figure 15.13 shows the output from the operation, $elev NOT $strmbuf. Region features of the subclass *elev* that fall within *strmbuf* are removed from the output.

---

in selecting regions. Third, AREAQUERY allows the user to assign a relative weight to an input coverage to influence its role in line snapping. Line snapping takes place when arcs in the overlay output are within a specified fuzzy tolerance. Lines on a coverage with a greater weight are less likely to be snapped to lines on a coverage with a smaller weight.

15.8 REGIONS-BASED TOOLS IN ARC/INFO

Besides overlay and query, ARC/INFO has other regions-based tools for dissolve, buffering, and area tabulation between region subclasses. REGIONDISSOLVE aggregates regions and polygons with the same value for a specified item and saves the output to a new region subclass.

REGIONBUFFER creates buffers around points, lines, polygons, or regions, and saves the output into a region subclass. Like the BUFFER command (Chapter 10), REGIONBUFFER allows either constant buffer distance or different buffer distances. But REGIONBUFFER differs from BUFFER in two aspects. First, buffer zones created around individual map features by REGIONBUFFER may be noncontiguous or contiguous. Noncontiguous means

that each buffer zone is maintained separately. Contiguous means that overlapped buffer zones are dissolved to create a spatially connected buffer zone. Second, REGIONBUFFER combines map features from both the input and output (Figure 15.14). To retain the input coverage, the user must specify a new coverage for the output.

REGIONXAREA tabulates all overlap combinations of region subclasses in an integrated coverage and, for each combination, lists the area of overlap and the percent of the first region subclass that is overlapping with the second. Table 15.2 shows the output from REGIONXAREA for a small coverage with three region subclasses. The region subclass of *fire1* covers the entire area. The region subclass of *fire12* has two regions. And the region subclass of *fire123* has one region. The first record in the table shows that 43.15% of *fire1* is overlapping with the first region of *fire12*. The second record shows that 18.48% of *fire1* is overlapping with the second region of *fire12*, and so on.

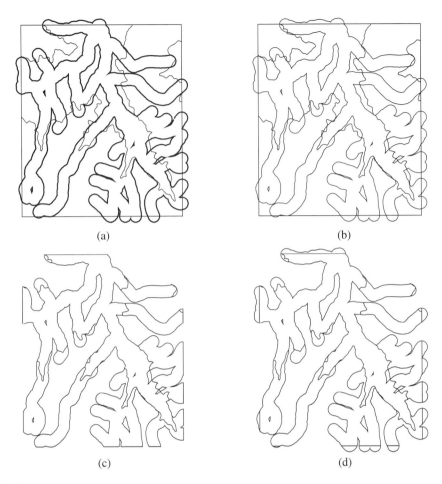

(a)   (b)

(c)   (d)

Figure 15.11

Logical expressions are used on the two region subclasses as shown by different line widths in (a) to simulate the overlay methods of UNION (b), INTERSECT (c), and IDENTITY (d).

TABLE 15.2   **A sample output from REGIONXAREA**

Subclass1	Subclass1#	Subclass2	Subclass2#	Area	Percent
fire1	1	fire12	1	6325234.000	43.15
fire1	1	fire12	2	2709613.500	18.48
fire1	1	fire123	1	963893.500	6.58
fire12	1	fire1	1	6325234.000	100.00
fire12	2	fire1	1	2709613.500	100.00
fire12	2	fire123	1	963893.500	35.57
fire123	1	fire1	1	963893.500	100.00
fire123	1	fire12	2	963893.500	100.00

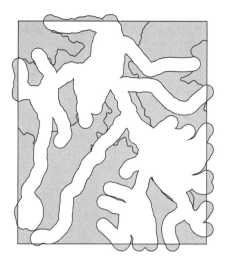

Figure 15.12
The XOR connector retains map features of the two region subclasses (shaded) in Figure 15.11a except features in the overlapped area.

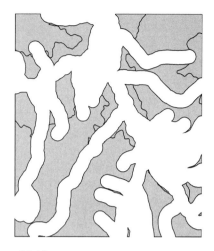

Figure 15.13
The NOT connector removes map features that fall within *strmbuf* in Figure 15.11a from the output.

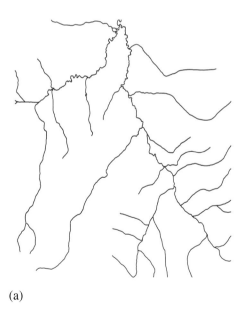

(a)

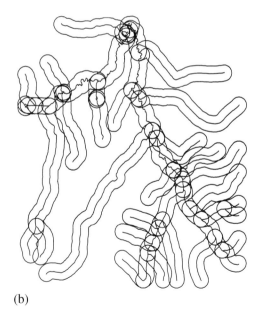

(b)

Figure 15.14
REGIONBUFFER combines map features from the input coverage (a) into the output polygon coverage (b).

KEY CONCEPTS AND TERMS

Contiguous region: A region with spatially joint components.

Hierarchical regions: Regions representing different spatial scales of a hierarchy.

Integrated coverage: A polygon coverage with built-in region layers.

Noncontiguous region: A region with spatially disjoint components.

Modifiable Area Unit Problem (MAUP): The problem related to the influence that the definition of units for which data are collected and the level of spatial data aggregation have on results of spatial analysis.

Uniform region: A geographic area with similar characteristics.

APPLICATIONS: REGIONS

This applications section covers five tasks. Task 1 shows how to display and query regions using ArcView. Although ArcView cannot be used for creating regions, it can handle region subclasses as themes. The remaining four tasks use ARC/INFO. Tasks 2 and 3 focus on creating regions, and Tasks 4 and 5 tabulate region overlaps and perform regions-based overlay respectively.

Task 1: Display and Query Regions in ArcView

What you need: a regions-based coverage called *intcov*.

The integrated polygon coverge *intcov* has six region subclasses: *land* from a land use coverage, *soil* from a soils coverage, *stream* from a stream buffer coverage, *sewer* from a sewer buffer coverage, and *potsites* and *finalsites* from results of regions-based queries. The purpose of this task is to display and query regions in ArcView.

1. Start ArcView and open a new view. Click the Add Theme button. In the Add Theme dialog, click on the icon next to *intcov* to open a list of themes included in *intcov*: *region.finalsites*, *region.land*, *region.potsites*, *region.sewer*, *region.soil*, *region.stream*, *polygon*, and *labelpoint*. The first six on the list are the region subclasses in the coverage

and the last two apply to the coverage itself. Add all themes except *labelpoint* to view.

2. Make *region.soil* active and open its theme table. The table contains soil attributes of soilcode and suit. Use the Query Builder to query: suit = 2. All polygons in *region.soil* that have the suit value of 2 are now highlighted. A region subclass behaves like a regular theme for data display and query.

3. Make *region.potsites* active and open its table. Click the Query Builder and set the query expression as area > 2000. The query selects 7 out of 10 polygons in *region.potsites*. If the result of this query is saved as a new shapefile, the shapefile will be the same as *region.finalsites*.

Task 2: Create Regions from Existing Polygons in ARC/INFO

What you need: A polygon coverage called *landuse*.

This task asks you to create a region from polygons that have the lucode value of 700 in the coverage *landuse*. Go to ArcEdit and set up the editing environment:

```
Arcedit: display 9999
Arcedit: mapextent landuse
Arcedit: editcov landuse
Arcedit: drawenv poly label
```

Arcedit: draw /* draw polygons and labels of *landuse*
Arcedit: editfeature label
Arcedit: select lucode = 700 /* select labels that have lucode = 700
Arcedit: drawselect /* highlight selected labels
Arcedit: editfeature polygon /* change the edit feature to polygons
Arcedit: select many /* click the labels previously highlighted and press 9 to exit
Arcedit: makeregion marsh /* create a region called *marsh* from the selected polygons
Arcedit: quit /* save the changes made in the session
Arc: describe landuse

The DESCRIBE command provides a description of the coverage *landuse* and its content. The feature class should include regions, and the subclass should list *marsh*. Type 'list landuse.patmarsh' to view the *marsh* subclass' PAT. Only one record is listed in the PAT because the three polygons are grouped into a single region using the noncontiguous option.

Task 3: Create Regions by Using Regions-based Commands

What you need: a polygon coverage called *fire.*

The coverage *fire* shows past fires in a national forest. Three attribute items (fire1, fire2, and fire3) record up to three fire events in a given polygon. For example, a polygon with recorded fires in 1909, 1923, and 1970 has the values of 1970, 1923, and 1909 under fire1, fire2, and fire3 respectively. The purpose of this task is to use the command REGIONDISSOLVE to create a region subclass.

Arc: regiondissolve
Usage: REGIONDISSOLVE <in_cover> {out_cover} <out_subclass> <dissolve_item | #ALL> {POLY | REGION.Subclass}
Arc: regiondissolve fire # fire2 fire2

REGIONDISSOLVE uses the dissolve item fire2 to create a region subclass called *fire2* in the

coverage *fire.* You can use ArcTools, a graphical user interface to ARC/INFO, to display the region subclass *fire2*:

1. Type arctools at the Arc prompt. Select Map Tools from the ArcTools menu. Then select New from the View menu in the Map Tools dialog.
2. In the Add New Theme dialog, select Coverage for Categories and Region for Theme Classes.
3. In the Region Theme Properties dialog, enter *fire2* as the Identifier, *fire* as the Data Source, and *fire2* as the Region. Select red as the symbol to draw regions. Click Preview to view the region subclass of *fire2.*

Now, go to Tables. Use the dir command in Tables to list all the INFO files in your workspace. You should see fire.pat and fire.patfire2.

Tables: select fire.patfire2 /* select the PAT of *fire2*
Tables: list /* list records in the PAT

You can remove the *fire2* region subclass from the *fire* coverage by using the command DROPFEATURES:

Arc: dropfeatures
Usage: DROPFEATURES <cover> <feature_class> {ATTRIBUTE | GEOMETRY}
Arc: dropfeatures fire region.fire2

REGIONQUERY is another regions-based command that can create regions. REGIONQUERY uses logical expressions to select polygons to be grouped into regions. In the following, REGIONQUERY creates nested regions of *fire1*, *fire12*, and *fire123* from the coverage *fire.* The *fire1* region layer represents areas burned once, *fire12* burned twice, and *fire123* burned three times. The three layers are superimposed on one another, forming a set of nested regions.

Arc: regionquery
Usage: REGIONQUERY <in_cover> {out_cover} <out_subclass> {selection_file}

{NONCONTIGUOUS | CONTIGUOUS}
{out_item…out_item}
Arc: regionquery fire # fire1 # # fire1
>: reselect fire1 > 0
>:
Arc: regionquery fire # fire12 # # fire1 fire2
>: reselect fire1 > 0 and fire2 > 0
>:
Arc: regionquery fire # fire123 # # fire1 fire2 fire3
>: reselect fire1 > 0 and fire2 > 0 and fire3 > 0
>:

Now display the nested regions of *fire1*, *fire12*, and *fire123* using ArcTools:

1. Type arctools at the Arc prompt. Select Map Tools from the ArcTools menu. Then select New from the View menu in the Map Tools dialog.
2. Select Coverage for Categories and Region for Theme Classes in the Add New Theme dialog.
3. First, set up the draw environment for *fire1*. In the Region Theme Properties dialog, enter *fire1* as the Identifier, *fire* as the Data Source, and *fire1* as the Region. Select red as the symbol to draw regions, and click OK.
4. Repeat Steps 2 and 3 to set up the draw environment for *fire12*. Enter *fire12* as the Identifier, *fire* as the Data Source, and *fire12* as the Region in the Region Theme Properties dialog. Select green for the Symbol, and click OK.
5. Repeat Steps 2 and 3 to set up the draw environment for *fire123*. Enter *fire123* as the Identifier, *fire* as the Data Source, and *fire123* as the Region in the Region Theme Properties dialog. Select blue for the Symbol, and click OK.
6. *Fire1, fire12,* and *fire123* are now listed under Themes in the Theme Manager dialog. Use the arrow to move the three region layers to the Draw list. Then click on the Draw button to draw the region layers in three different colors.

You can list the attributes contained in the PAT of each region subclass in either Arc or Tables.

   Arc: list fire.patfire12
   Arc: list fire.patfire123

Task 4: Tabulate Region Overlaps

What you need: a polygon coverage called *sfire*.

   The ARC command REGIONXAREA runs an AML macro to tabulate all possible region overlaps and their areas and percentage areas. The result is saved into an INFO file. Depending on the number of region subclasses and the number of regions in each subclass, the INFO file can be massive. For this task, you will use a small coverage called *sfire* so that you can better understand how REGIONXAREA works.

1. Use ArcTools to display the nested region subclasses in *sfire*.
2. Now, run REGIONXAREA in Arc. Ignore messages from the execution of REGIONXAREA.

   Arc: regionxarea
   Usage: REGIONXAREA <in_cover>
   <out_info_file> {$ALL |
   subclass…subclass}
   Arc: regionxarea sfire sfire.dat

3. List the INFO file created by regionxarea.

   Arc: list sfire.dat

Each record in *sfire.dat* represents an overlap of regions. Subclass1 and subclass2 are the two region subclasses in the overlap. Subclass1# and subclass2# list the regions in the overlap. Area is the size of the overlap, and percent shows the area percentage of subclass1# overlapping with subclass2#. As examples, the first record in *sfire.dat* shows that 43.15% of *fire1* is overlapping with the first region of *fire12* and the second record shows that 18.48% of *fire1* is overlapping with the second region of *fire12*.

Task 5: Perform Regions-based Overlay

What you need: coverages of *landuse, soils, strmbuf,* and *sewerbuf.*

As described in this chapter, AREAQUERY can perform both map overlay and data query in a single operation. AREAQUERY is an alternative to the traditional overlay method for GIS analysis. This task simulates a suitability analysis with the following selection criteria:

- Preferred land use is brushland (lucode = 300 in *landuse*)
- Choose soil types suitable for development (suit >= 2 in *soils*)
- Site must be within 300 meters of existing sewer lines (*sewerbuf*)
- Site must be beyond 20 meters of existing streams (*strmbuf*)
- Site must contain an area at least 2000 square meters

Arc: areaquery
Usage: AREAQUERY <out_cover> <out_subclass> {tolerance} {NONCONTIGUOUS|CONTIGUOUS} {out_item…out_item}
Arc: areaquery allcov potsites # contiguous /* initiate AREAQUERY

A dialog command, AREAQUERY first asks for input coverages and their region subclass names.

[Enter the 1st coverage:] landuse land /* land is the subclass name for *landuse*

[Enter the 2nd coverage:] soils soil /* soil is the subclass name for *soils*
[Enter the 3rd coverage:] sewerbuf sewer /*sewer is the subclass name for *sewerbuf*
[Enter the 4th coverage:] strmbuf stream /*stream is the subclass name for *strmbuf*
[Enter the 5th coverage:] end

AREAQUERY then asks for a logical expression. Each operand in the logical expression consists of two parts: the first part refers to the region subclass, and the second part refers to the item in the region subclass. The logical expression can include the Boolean connectors of AND, OR, XOR, and NOT.

>: resel land.lucode = 300 and soil.suit >= 2 and $sewer and not $stream
>: [hit return]
Arc: list allcov.patpotsites

The output from AREAQUERY is saved into the region subclass *potsites*, which includes sites that satisfy the first four criteria. To complete the task, use REGIONQUERY to select sites that are larger than 2000 m^2.

Arc: regionquery allcov # finalsites # contiguous
>: resel potsites.area > 2000
>: [hit return]

REFERENCES

Allen, T.H.F., and T.B. Starr. 1982. *Hierarchy: Perspectives for Ecological Complexity*. Chicago: The University of Chicago Press.

Amrhein C.G. 1995. Searching for The Elusive Aggregation Effect: Evidence from Statistical Simulations. *Environment and Planning A* 27: 105–19.

Bailey, R.G. 1976. *Ecoregions of the United States*. Map (scale 1:7,500,000). Ogden, Utah: U.S. Department of Agriculture, Forest Service, Intermountain Region.

Bailey, R.G. 1983. Delineation of Ecosystem Regions. *Environmental Management* 7: 365–73.

Batty, M., and P.K. Sikdar. 1982a. Spatial Aggregation in Gravity Models. 1. An Information-Theoretic Framework. *Environment and Planning A* 14: 377–405.

Batty, M., and P.K. Sikdar. 1982b. Spatial Aggregation in Spatial Interaction Models. 2. One-dimensional Population Density Models. *Environment and Planning A* 14: 524–53.

Berry, B.J.L. 1968. Approaches to Regional Analysis: A Synthesis. In B.J.L. Berry and D.F. Marble (eds.). *Spatial Analysis: A Reader in Statistical Geography*. Englewood Cliffs, NJ: Prentice-Hall, pp. 24–34.

Clark, W.A.V., and K. L. Avery. 1976. The Effect of Data Aggregation in Statistical Analysis. *Geographical Analysis* 8: 428–37.

Cleland, D.T., R.E. Avers, W.H. McNab, M.E. Jensen, R.G. Bailey, T. King, and W.E. Russell. 1997. National Hierarchical Framework of Ecological Units. In M.S. Boyce and A. Haney (eds.). *Ecosystem Management Applications for Sustainable Forest and Wildlife Resources.* New Haven, CT: Yale University Press, pp. 181–200.

Costanza, R., and T. Maxwell. 1994. Resolution and Predictability: An Approach to the Scaling Problem. *Landscape Ecology* 9: 47–57.

Forman, R.T.T., and M. Godron. 1986. *Landscape Ecology.* New York: John Wiley & Sons.

Fotheringham, A.S., and D.W.S. Wong. 1991. The Modifiable Areal Unit Problem in Multivariate Statistical Analysis. *Environment and Planning A* 23: 1025–44.

Gehlke, C.E., and K. Biehl. 1934. Certain Effects of Grouping Upon the Size of the Correlation Coefficient in Census Tract Material. *Journal of the American Statistical Association* 24:169–70.

Goodchild, M.F. 1979. The Aggregation Problem in Location-Allocation. *Geographical Analysis* 11: 240–55.

Jelinski, D.E., and J. Wu. 1996. The Modifiable Areal Unit Problem and Implications for Landscape Ecology. *Landscape Ecology* 11:129–40.

Khatib, Z., K. Chang, and Y. Ou. 2001. Impacts of Analysis Zone Structures on Modeled Statewide Traffic. *ASCE Journal of Transportation Engineering* 127: 31–38.

Meentemeyer, V., and E.O. Box. 1987. Scale Effects in Landscape Studies. In M.G. Turner (ed.). *Landscape Heterogeneity and Disturbance.* New York: Springer-Verlag, pp. 15–34.

Miller, H.J. 1999. Potential Contributions of Spatial Analysis to Geographic Information Systems for Transportation. *Geographical Analysis* 31: 373–99.

Omernik, J.M., 1987. Ecoregions of the Conterminous United States. *Annals of the Association of American Geographers* 77: 118–25.

O'Neill, R.V., D.L. DeAngelis, J.B. Waide, and T.F.H. Allen. 1986. *A Hierarchical Concept of Ecosystems.* Princeton, NJ: Princeton University Press.

Openshaw, S. 1977. Optimal Zoning Systems for Spatial Interaction Models. *Environment and Planning A* 9:169–84.

Openshaw, S., and P.J. Taylor. 1979. A Million or So Correlation Coefficients: Three Experiments on the Modifiable Areal Unit Problem. In N. Wrigley (ed.), *Statistical Applications in the Spatial Sciences.* London: Pion. pp. 127–44

Robinson, W.S. 1950. Ecological Correlation and The Behavior of Individuals. *American Sociological Review* 15: 351–57.

The National Atlas of the United States of America. 1970. Washington, D.C.: U.S. Geological Survey.

Urban, D.L., D. O'Neill, and H.H. Shugart. 1987. Landscape Ecology. *BioScience* 37: 119–27.

Walker, D.A., and M. D. Walker. 1991. History and Pattern of Disturbance in Alaskan Arctic Terrestrial Ecosystems: A Hierarchical Approach to Analysing Landscape Change. *Journal of Applied Ecology* 28: 244–76.

Wiens, J.A. 1989. Spatial Scaling in Ecology. *Functional Ecology* 3: 385–97.

Wiken, E.B., and G. Ironside. 1977. The Development of Ecological (Biophysical) Land Classification in Canada. *Landscape Planning* 4: 273–75.

Wong, D., and C. Amrhein. 1996. Research on the MAUP: Old Wine in a New Bottle or Real Breakthrough. *Geographical Systems* 3: 73–76.

NETWORK AND DYNAMIC SEGMENTATION

16.1 INTRODUCTION

A network consists of connected linear features. A familiar network is a road system. Other networks include railways, public transit, bicycle paths, streams, and shorelines. Dynamic segmentation is a data model that is built upon lines of a network and allows the use of real-world coordinates with linear measures such as mileposts (Chapter 3). By using the dynamic segmentation model, one can link linearly referenced data such as accidents and pavement conditions to a geographically referenced road network.

Both network and dynamic segmentation models must have the appropriate attributes for real-world applications. To use a road network for path finding or distribution of resources, attributes such as travel time, impedance at turn, and one-way streets must be added to the arcs and nodes of the network. Likewise, linear measures must be added to arcs in a road network so that data that are measured in mileposts, for example, can be directly linked to road segments.

This chapter is divided into five sections. Section 1 covers the preparation of a road network with an example from Moscow, Idaho. Section 2 considers network analysis and applications. Section 3 discusses the building of a dynamic segmentation model, and Section 4 discusses the linking of linearly referenced data to the model. Section 5 reviews applications of dynamic segmentation.

This chapter uses ARC/INFO and ArcView as examples. ARC/INFO can work with both the network and dynamic segmentation models. ARC/INFO also has specific commands for building dynamic segmentation. The Network Analyst to ArcView covers network applications but has limited capability in preparing dynamic segmentation. Other topology-based GIS packages also have built-in network and segmentation functions. MGE Network, PAMAP NETWORKER, and TransCAD can perform network analysis included in this chapter. MGE Segment Manager and TransCAD have dynamic segmentation functions.

16.2 NETWORK

A **network** is a line coverage, which is topology-based and has the appropriate attributes for the flow of objects such as traffic. The geometry of a network can be digitized or imported from existing data sources. A network, however, must have the appropriate attributes for real-world applications. Because most network applications involve road systems, the following discussion concentrates on feature and attribute data for road networks, including impedance values assigned to network links, turns, one-way streets, and overpasses and underpasses. Later an example from Moscow, Idaho shows how these data can be put together to form a street network.

16.2.1 Link Impedance

A **link** refers to a segment separated by two nodes in a road network. **Link impedance** is the cost of traversing a link. A simple measure of the cost is the physical length of the link. But the length may not be a reliable measure of cost, especially in cities where speed limits and traffic conditions vary significantly along different streets. A better measure of link impedance is the travel time estimated from the length and the speed limit of a link. For example, if the speed limit is 30 miles per hour and the length is 2 miles, then the link travel time is 4 minutes (2/30 x 60 minutes).

There are variations in measuring the link travel time. The travel time may be directional: the travel time in one direction may be different from the other direction. In that case, the travel time can be entered separately as from-to and to-from, depending on the arc direction. The travel time may also vary by the time of the day and by the day of the week, thus requiring the set-up of different network attribute data for different applications.

16.2.2 Turn Impedance

A turn is a transition from one arc to another in the network. **Turn impedance** is the time it takes to complete a turn, which is significant in a congested street network (Ziliaskopoulos and Mahmassani 1996). Because a network typically has many turns with different conditions, a **turn table** is used to assign the turn impedance values. A turn table has three items: the node number for the intersection, the arc numbers involved in a turn, and the turn impedance values.

A turn table must list all possible turns by arc numbers at each intersection. A driver approaching a street intersection usually has three options: go straight, turn right, or turn left. In some cases the driver has the fourth option of making a U turn. Assuming the intersection involves four street segments, as most intersections do, this means at least 12 possible turns at the intersection excluding U turns (Figure 16.1).

Depending on the level of detail for a study, GIS users may not have to include every intersection and every possible turn in a turn table. A partial turn table listing only turns at intersections with stop lights may be sufficient for a network application.

Turn impedance is usually directional. For example, it may take 5 seconds to go straight, 10 seconds to make a right turn, and 30 seconds to make a left turn at a stop light. A negative impedance value for a turn means a prohibited turn, such as turning the wrong way onto a one-way street.

16.2.3 One-way or Closed Streets

One-way or closed streets can be denoted in a network's attribute table by a designated field. The field value can show the traffic direction of a one-way street. FT, for example, means that travel is permitted in the direction from the from-node to the to-node of the arc. TF means that travel is permitted from the to-node to the from-node of the arc. And N means that travel is not allowed in either direction.

16.2.4 Overpasses and Underpasses

There are two methods for including overpasses and underpasses in a network. The first method is to use non-planar features: an overpass and the street underneath it are both represented as continuous lines without a node at their intersection (Figure 16.2). The second method treats overpasses and underpasses as planar features: the two arcs representing the overpass meet at one node, and the two arcs representing the street underneath the overpass meet at another (Figure 16.3). Elevation items are assigned to the arcs: T-Elev is the elevation of the arc entering the crossing and F-Elev is the elevation of the arc leaving the crossing. To differentiate the overpass from the street, one can give the node used by the overpass a higher elevation value (1) than the node used by the street (0).

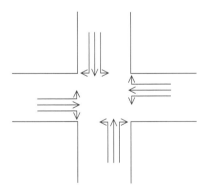

Figure 16.1
Possible turns at an intersection with four street segments. No U turns are allowed in this example.

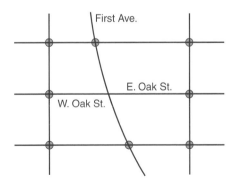

Figure 16.2
First Ave. crosses Oak St. with an overpass. A non-planar representation with no nodes is used at the intersection of Oak St. and First Ave.

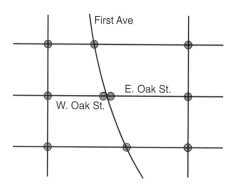

Street Name	F-elev	T-elev
First Ave	0	1
First Ave	1	0
W. Oak St.	0	0
E. Oak St.	0	0

Figure 16.3
First Ave crosses Oak St with an overpass. A planar representation with two nodes is used at the intersection: one for First Ave, and the other for Oak St. The elevation value of 1 shows that the overpass is along First Ave.

16.2.5 Putting Together a Street Network

Putting together a road network with the appropriate attributes for real-world applications is not an easy task. The example of Moscow, Idaho shows how to accomplish the task. A university town of 20,000 people, Moscow has more or less the same network features as other larger cities except for overpasses or underpasses.

TIGER/Line files from the U.S. Bureau of the Census are a common data source for street networks. Step 1 in this example is to make a preliminary street coverage from the TIGER/Line file. TIGER/Line files use longitude and latitude values as measurement units and NAD83 as the datum. Therefore, the Moscow street coverage must be converted to real-world coordinates.

Step 2 is to edit and update the preliminary street coverage. TIGER/Line files include a long list of attribute data, many of which are not relevant to network applications and can be deleted. Mistakes on the street coverage, which are transferred

from the TIGER/Line file, must be corrected, and new streets built in the 1990s must be added. Also, pseudo nodes, that is, nodes that are not required topologically, must be removed so that a street segment between two intersections is not unnecessarily broken up. But a pseudo node is needed at the location where one street changes into another along a continuous arc. Without the pseudo node, the continuous arc will be treated as the same street.

Step 3 is to attribute the street network with link impedance values. Speed limits can be assigned to streets by functional classification. Moscow, Idaho has three speed limits: 35 miles/hour for principal arterials, 30 miles/hour for minor arterials, and 25 miles/hour for all other city streets. With speed limits in place, the travel time for each street segment can be computed from the segment's length and speed limit.

Step 4 is to attribute one-way streets. Moscow has two one-way streets serving as the northbound and the southbound lanes of a state highway. One-way streets are denoted by the value of 1 for an item called direction. The arc direction of all street segments that make up a one-way street must be consistent and pointing in the right direction. Street segments that are incorrectly oriented must be flipped.

Step 5 is to prepare a turn table. The ARC command TURNTABLE can generate a turn table with all intersections and possible turns in a network. One can then add the turn impedance values to the turn table using an item, such as minutes or seconds. Attributing turn impedance requires selecting each intersection or node, determining different types of turns at the intersection, and assigning an estimated impedance value to each turn. In this example, a partial turn table is made for street intersections with stop lights only in Moscow, Idaho.

Figure 16.4 shows a street intersection at node# 341with no restrictions for all possible turns except U turns. An angle value of 90 means a left turn, for example, turning left from arc# 503 to arc# 467 (Box 16.1). An angle value of −90 means a right turn. And an angle value of 0 means

The turn angle is the angle between the 'from' arc and the 'to' arc of a turn. In reality, the angle is often 90° for a right or left turn and 0° for going straight. As measured from the coverage prepared from the TIGER/Line file, the angle is likely to deviate from 90° or 0°. Because the purpose of the angle is simply to determine the type of turn, we can ignore the deviation.

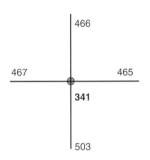

Node#	Arc1#	Arc2#	Angle	Minutes
341	503	467	90	0.500
341	503	466	0	0.250
341	503	465	−90	0.250
341	467	503	−90	0.250
341	467	466	90	0.500
341	467	465	0	0.250
341	466	503	0	0.250
341	466	467	−90	0.250
341	466	465	90	0.500
341	465	503	90	0.500
341	465	467	0	0.250
341	465	466	−90	0.250

Figure 16.4
Possible turns at node 341.

Node#	Arc1#	Arc2#	Angle	Minutes
265	339	342	−87.412	0.000
265	339	340	92.065	0.000
265	339	385	7.899	0.000
265	342	339	87.412	0.500
265	342	340	−0.523	0.250
265	342	385	−84.689	0.250
265	340	339	−92.065	0.250
265	340	342	0.523	0.250
265	340	385	95.834	0.500
265	385	339	−7.899	0.000
265	385	342	84.689	0.000
265	385	340	−95.834	0.000

Figure 16.5
Turn impedance values for turns at node 265, which has stop signs for the east-west traffic.

going straight. This example uses two turn impedance values: 30 seconds or 0.5 minute for a left turn, and 15 seconds or 0.25 minute for either a right turn or going straight.

Some street intersections do not allow certain types of turns. Figure 16.5 shows a street intersection at node# 265 with stop signs posted only for the east-west traffic. Therefore, the turn impedance values are assigned only to turns to be made from arc# 342 and arc# 340. Figure 16.6 shows a street intersection with stop lights between a southbound one-way street and an east-west two-way street. An impedance value of −1 means a prohibited turn, such as a right turn from arc# 467 to arc# 461 or a left turn from arc# 462 to arc# 461.

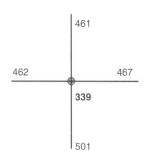

Node#	Arc1#	Arc2#	Angle	Minutes
339	467	501	90.190	0.500
339	467	462	1.152	0.250
339	467	461	−92.197	−1.000
339	462	501	−90.962	0.250
339	462	467	−1.152	0.250
339	462	461	86.651	−1.000
339	461	501	2.386	0.250
339	461	467	92.197	0.500
339	461	462	−86.651	0.250

Figure 16.6
Turn impedance values for turns at node 339. Node 339 is an intersection between a southbound one-way street and an east-west two-way street.

16.3 NETWORK APPLICATIONS

This section discusses applications that use a network directly for path finding and accessibility measure as well as applications that link a network to a broader topic, such as location-allocation and transportation planning modeling. Direct network applications are usually accessible through commands or menu choices in GIS packages. Applications that involve a network as input often link a GIS package with a discipline specific software package, such as a transportation planning software.

16.3.1 Shortest Path Analysis

Shortest path analysis finds the path with the minimum cumulative impedance between nodes on a network. The path may connect just two nodes—the origin and destination—or have specific stops between the nodes. Shortest path analysis can help a traveler plan a trip, a van driver to set up a schedule for dozens of deliveries, or an emergency service to connect a dispatch station, accident location, and hospital.

The idea behind shortest path analysis is similar to that of cost distance measure operations using raster data (Chapter 11). The main difference is the data model. Shortest path analysis is vector-based and uses an existing network. The cost distance measure operation is raster-based and uses an input grid and a cost grid to search for the least accumulative cost path for proposed facilities such as water lines and pipelines.

The shortest path problem has been studied extensively in operations research, computer science, spatial analysis, and transportation engineering. A new application of shortest path analysis is in the development of in-vehicle route guidance systems (RGS). Important to any RGS is an algorithm that can find the optimal (shortest) route from the origin to the destination (Fu and Rilett 1998).

Shortest path analysis typically begins with an impedance matrix, in which a value represents the impedance of a direct link between two nodes on a network and a 0 or infinity means no direct connection. The analysis then follows an iterative process of finding the shortest distance from node 1 to all other nodes (Dijkstra 1959). Different computing algorithms for solving the shortest path have been proposed in the literature (Dreyfus 1969).

Here we will look at an example of six cities on a road network (Figure 16.7). Table 16.1 shows the impedance matrix of travel time measured in minutes. The value of 0 above and below the principal diagonal in the impedance matrix means no direct path between two nodes. Suppose we want to find the shortest path from node 1 to all other nodes in Figure 16.7. We could solve the problem by using an iterative procedure (Lowe and Moryadas 1975). At each step, we choose the shortest path from a list of candidate paths and place the node of the shortest path in the solution list.

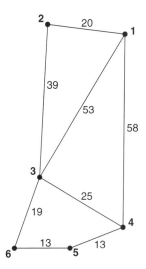

Figure 16.7
Six cities on a road network for shortest path analysis.

TABLE 16.1	The impedance matrix among 6 nodes in Figure 16.7

	(1)	(2)	(3)	(4)	(5)	(6)
(1)	0	20	53	58	0	0
(2)	20	0	39	0	0	0
(3)	53	39	0	25	0	19
(4)	58	0	25	0	13	0
(5)	0	0	0	13	0	13
(6)	0	0	19	0	13	0

TABLE 16.2	Shortest paths from node 1 to all other nodes in Figure 16.7

From-node	To-node	Shortest Path	Minimum Cumulative Impedance
1	2	p_{12}	20
1	3	p_{13}	53
1	4	p_{14}	58
1	5	$p_{14} + p_{45}$	71
1	6	$p_{13} + p_{36}$	72

The first step is to choose the minimum among the three paths from node 1 to nodes 2, 3, and 4 respectively:

$$\min (p_{12}, p_{13}, p_{14}) = \min (20, 53, 58)$$

We choose p_{12} because it has the minimum impedance value among the three candidate paths. We then place node 2 in the solution list with node 1.

The second step is to prepare a new candidate list of paths that are directly or indirectly connected to nodes in the solution list (nodes 1 and 2):

$$\min (p_{13}, p_{14}, p_{12} + p_{23}) = \min (53, 58, 59)$$

We choose p_{13} and add node 3 to the solution list. To complete the solution list with other nodes on the network we need to go through the following steps:

$$\min (p_{14}, p_{13} + p_{34}, p_{13} + p_{36}) = \min (58, 78, 72)$$
$$\min (p_{13} + p_{36}, p_{14} + p_{45}) = \min (72, 71)$$
$$\min (p_{13} + p_{36}, p_{14} + p_{45} + p_{56}) = \min (72, 84)$$

Table 16.2 summarizes the solution to the shortest path problem from node 1 to all other nodes.

The **traveling salesman problem** is a more complex form of shortest path analysis because it has two constraints: (1) the salesman must visit each of the select stops only once and (2) the salesman may start from any stop but must return to the original stop. The objective is to determine which route, or tour, the salesman can take to minimize the total impedance value.

One solution to this problem is to use a heuristic method (Lin 1965). Beginning with an initial random tour, the method runs a series of locally optimal solutions by swapping stops that yield reductions in the cumulative impedance. The iterative process ends when no improvement can be found by swapping stops. This heuristic approach can usually create a tour with a minimum, or near minimum, cumulative impedance.

ARC/INFO has the commands, PATH and TOUR, for solving the shortest path problem. PATH uses a user-defined order of stops to solve the problem, whereas TOUR determines the order in which stops are visited to minimize the cumulative impedance. The Network Analyst extension to Arc-View has a menu selection called Find Best Route, which combines the functionalities of PATH and TOUR. The shortest path solution from ARC/INFO or ArcView is saved as a route—a higher-level object based on the dynamic segmentation model. Therefore, the solution can be displayed on top of a network. Both ARC/INFO and ArcView offer the option of providing directions for the shortest path with street names, turns, and distances.

16.3.2 Closest Facility

One type of shortest path analysis is to find the closest facility, such as a hospital, fire station, or ATM, to any location on a network. The **closest facility** algorithm first computes the shortest paths from the select location to all candidate facilities, and then chooses the closest facility

among the candidates. ArcView's Network Analyst has a menu selection called Find Closest Facility for this type of application. Figure 16.8 shows the closest fire station to a street address in Moscow, Idaho.

16.3.3 Allocation

Allocation is the study of the spatial distribution of resources through a network. Resources in allocation studies often refer to public facilities, such as fire stations or schools, whereas the distribution of resources defines service zones. The main objective of spatial allocation analysis is to measure the efficiency of public facilities.

A common measure of efficiency in the case of emergency services is the response time—the time it takes for a fire truck or an ambulance to reach an incident. Figure 16.9, for example, shows

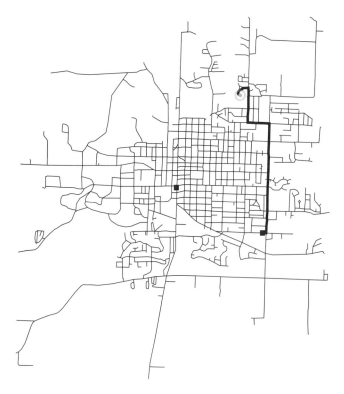

Figure 16.8
Shortest path from a street address in Moscow, Idaho to its closest fire station (shown by the square symbol).

areas in Moscow, Idaho covered by two existing fire stations within a two-minute response time. The map shows a large portion of the city is outside the two-minute response zone. The response time to Moscow's outer zone is about five minutes (Figure 16.10).

If residents of Moscow demand that the response time to any part of the city be two minutes or less, then the options are either to relocate the fire stations or, more likely, to build new fire stations. A new fire station should be located to cover the largest portion of the city unreachable in two minutes by the existing fire stations. The problem then becomes a location and allocation problem, which is covered in the next section.

ARC/INFO's command ALLOCATE works with the allocation problem. ArcView's Network Analyst has a menu selection called Find Service Area, which also deals with allocation.

16.3.4 Location-allocation

An important topic in operations research and spatial analysis, **location-allocation** solves problems of matching the supply and demand by using sets of objectives and constraints. Location-allocation algorithms used to be available in stand-alone computer programs, but in recent years GIS packages, such as ARC/INFO and MGE Network, have incorporated some of the algorithms as application tools.

The private sector offers many location-allocation examples. Suppose a company operates soft-drink distribution facilities to serve supermarkets. The objective in this example is to minimize the total distance traveled, and a constraint, such as a two-hour drive distance, may be imposed on the problem. A location-allocation analysis is to match the distribution facilities and the supermarkets while meeting both the objective and constraint.

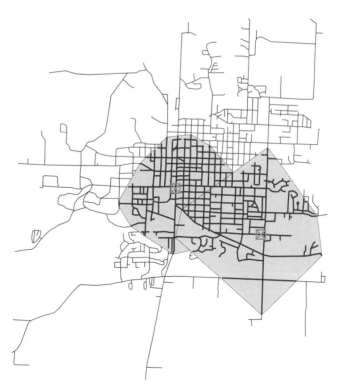

Figure 16.9
Service areas of two fire stations within a 2-minute response time in Moscow, Idaho.

Figure 16.10
Service areas of two fire stations within a 5-minute response time in Moscow, Idaho.

Location-allocation is also important in the public sector. For example, a local school board may decide that (1) all school-age children should be within 1 mile of their schools and (2) the total distance traveled by all children should be minimized. In this case, schools represent the supply, and school-age children represent the demand. The objective of this location-allocation analysis is to provide equitable service to a population, while maximizing efficiency in the total distance traveled.

The set up of a location-allocation problem requires inputs in supply, demand, and distance measures. The supply consists of facilities or centers at point locations. The demand may consist of points, lines, or polygons, depending on the data source and the level of spatial data aggregation. For example, the locations of school-age children may be represented as individual points, or aggregate points along streets, or aggregate points (centroids) in unit areas such as census block groups. Use of aggregate demand points is likely to introduce errors in location-allocation analysis (Francis et al. 1999). Distance measures between the supply and demand are often presented in a distance matrix or a distance list. Distances may be measured along the shortest path between two points on a road network or along the straight line connecting two points. In location-allocation analysis shortest path distances are likely to yield more accurate results than straight-line distances.

Perhaps the most important input to a location-allocation problem is the model for solving the problem. Two most common models are minimum distance and maximum covering (Ghosh et al. 1995; Church 1999). The **minimum distance model**, also called the *p*-median location model,

Figure 16.11
Demand points in Latah County, Idaho for a public library location problem.

minimizes the total distance traveled from all demand points to their nearest supply centers (Hakimi 1964). The p-median model has been applied to a variety of facility location problems, including food distribution facilities, public libraries, and health facilities. The **maximum covering model** maximizes the demand covered within a specified time or distance (Church and ReVelle 1974). Public sector location problems, such as the location of emergency medical and fire services, are ideally suited for the maximum covering model. The model is also useful for many convenience oriented retail facilities, such as movie theaters, banks, and fast food restaurants.

Both minimum distance and maximum covering models may have added constraints or options. A maximum distance constraint may be imposed on the minimum distance model so that the solution, while minimizing the total distance traveled,

ensures that no demand is beyond the specified maximum distance. A desired distance option may be used with the maximum covering model to cover all demand points within the desired distance.

Here we will use public libraries in Latah County, Idaho as an example to examine three location-allocation scenarios. The demand for public libraries in Latah County consists of 33 demand points, each representing the total number of households in a census block group (Figure 16.11). Distances are measured along straight lines between the demand points and public library locations.

The first scenario assumes that (1) every block group is a possible candidate for a public library and (2) the county government wants to set up three libraries using the maximum covering model with a desired distance of 10 miles. Figure 16.12 shows the three selected candidates. The solution leaves one demand point with 344

Figure 16.12
Three candidates selected by the maximum covering model and the demand points served by the candidates.

households uncovered. To cover every demand point, the county government must either increase the number of public libraries or relax the desired distance.

The second scenario is set up the same way as the first except that the county government wants to use the minimum distance model. Figure 16.13 shows the three selected candidates, one of which is not included in Figure 16.12 from the maximum covering model.

The third scenario assumes that one of the three library locations is fixed or has been previously chosen. Figure 16.14 shows the two selected new sites using the minimum distance model with the maximum distance constraint of 10 miles. The solution leaves three demand points with 1124 households uncovered.

ArcView does not have menu selections for solving location-allocation problems. ARC/INFO offers two location-allocation algorithms: GRIA (global regional interchange algorithm) (Densham and Rushton 1992), and TEITZBART (Teitz and Bart 1968). Both algorithms are heuristics. The algorithms can solve location-allocation problems reasonably fast but may not produce optimal solutions, especially with many supply and demand points.

The maximum covering model in ARC/INFO has a required parameter of maximum distance and an optional parameter of desired distance. Although the objective is maximum covering, the model can actually leave some demand points uncovered, as seen in Figure 16.12. This apparent discrepancy is a result of using a minimum distance solution technique to solve a maximum covering model (Church 1999).

Figure 16.13
Three candidates selected by the minimum distance model and the demand points served by the candidates.

16.3.5 Urban Transportation Planning Model

An urban transportation planning (UTP) model typically uses the four-step process of trip generation, trip distribution, modal choice, and trip assignment (Nyerges 1995). Trip generation estimates the number of trips to and from each traffic analysis zone (TAZ) in the model. Trip distribution uses the number of trips generated in each TAZ and the distance or travel time between TAZs to produce a trip interchange matrix between TAZs. Modal choice splits trips by travel mode. Trip assignment loads estimated traffic volumes from the model on the network.

UTP modeling is a complex process. Each step in the four-step process involves selecting algorithms or models and requires a good understanding of transportation planning from the software

developer and the user (Waters 1999). This is why specialized software packages are normally used for UTP modeling.

One of the necessary inputs to UTP modeling is a transportation network. The network is used to measure the travel impedance between TAZs. Both trip distribution and modal choice in the four-step process need the travel impedance data. The network is also used in trip assignment because at this step the estimated traffic volume is assigned to every link of the network.

To be used for UTP modeling a network must have the arc-node data structure. Nodes are road intersections and critical turns. Arcs or links are road segments between nodes. Link attributes needed for transportation planning usually include distance, speed limit, capacity, and functional classification.

Figure 16.14
One fixed and two selected candidates by the minimum distance model with the maximum distance constraint of 10 miles. The fixed candidate is the one to the west.

16.4 DYNAMIC SEGMENTATION

Built upon arcs of a network, the dynamic segmentation data model has the basic elements of routes, sections, and events. A route is a collection of sections and resides in a line coverage as a subclass. A section refers directly to arcs in the line coverage and provides measures to a route system. Events are occurrences, such as accidents or pavement condition, and are related to a route system by measures of their locations (Box 16.2). The process of developing a dynamic segmentation model therefore involves creating routes, computing measures for routes, and building event tables.

16.4.1 Create Routes on New Arcs

One can digitize arcs that make up a new route and compute the route measures by using the command REMEASURE. When the topology of the line

coverage including routes is built, ARC/INFO automatically creates a section table and a route table. The section table shows measures of each section by relating sections to arcs, and the route table uses the route ID to link to the section table.

16.4.2 Create Routes on Existing Arcs

Creating routes on the existing arcs of a line coverage is an efficient option. This option offers the interactive and data conversion methods. The interactive method consists of two steps: (1) select arcs that make up a route and (2) use the command MAKEROUTE command to convert the arcs to a route. MAKEROUTE automatically computes the measures for the route.

The data conversion method applies to a line coverage rather than a select group of arcs as in the interactive method and can create many routes at once by referencing an item in the line coverage's attribute table. For example, if the reference item is the highway number, the conversion algorithm creates a route system for each numbered highway.

The commands of ARCROUTE and MEASUREROUTE can perform the conversion. ARCROUTE works with topologically continuous routes, whereas MEASUREROUTE works with either continuous or discontinuous routes. A topologically discontinuous route refers to a route that is divided by a canal or river into two or more parts.

Figure 16.15 shows a line coverage with the following interstate highways in Idaho: 15, 84, 86, 90, and 184. Suppose we want to create a route system for each interstate highway. We can use ARCROUTE to perform the conversion. Table 16.3 is the output route table, which has a route for each interstate highway. Table 16.4 is the section table, which uses routelink# to link to the route system and arclink# to link to arcs in the line coverage. Each route system is measured separately as shown by the items of *f*-meas and *t*-meas in the section table.

16.4.3 Create Different Types of Routes

Routes may be grouped into the following four types.

- **simple route**: a route follows one direction and does not loop or branch.

Figure 16.15
Interstate highways in Idaho.

- **combined route**: a route is joined with another route.

- **split route**: a route subdivides into two routes.
- **looping route**: a route intersects itself.

Simple routes are easy to create, but the other three types require special handling. An example of combined routes is an interstate highway, which has different traffic conditions depending on the traffic direction. In this case, two routes can be built for the interstate highway, one for each direction. An example of split routes is highway ramps.

Like combined routes, split routes can be separated into two or more routes: one route for the continuous path of the highway and individual routes for split sections.

A looping route requires breaking up the route into sections. Unless the route is dissected into sections, the route measures will be off. To explain the procedure, we will use a bus route in Moscow, Idaho as an example (Figure 16.16). The bus route has two loops. Therefore, we can dissect the bus route into three sections: (1) from the origin on the west side of the city to the first crossing of the route, (2) between the first and the second crossing, and (3) from the second crossing back to the origin (Figure 16.17).

After selecting arcs that make up the first section, we can use MAKEROUTE to create the initial bus route. Next, we select arcs of the second section and append them to the route. We can then use the REMEASURE command to compute measures of the second section. The bus route is completed by appending arcs and re-measuring arcs of the third section.

TABLE 16.3	The route table for the interstates in Idaho

Inter#	Inter-Id	High_Number
1	1	15
2	2	84
3	3	86
4	4	90
5	5	184

TABLE 16.4   The section table for the interstates in Idaho

routelink#	arclink#	f-meas	t-meas	f-pos	t-pos	inter#	inter-id
1	5	0	44700	100	0	1	1
1	4	44700	123648	100	0	2	2
1	3	123648	199791	100	0	3	3
1	16	199791	239375	100	0	4	4
1	15	239375	315194	100	0	5	5
2	6	0	74024	0	100	6	6
2	7	74024	78964	0	100	7	7
2	9	78964	154873	0	100	8	8
2	10	154873	226153	0	100	9	9
2	11	226153	303050	0	100	10	10
2	12	303050	356992	0	100	11	11
2	17	356992	433769	0	100	12	12
2	18	433769	443570	0	100	13	13
3	13	0	78065	0	100	14	14
3	14	78065	101154	0	100	15	15
4	1	0	72033	0	100	16	16
4	2	72033	117974	0	100	17	17
5	8	0	6348	100	0	18	18

16.4.4 Create Routes with Measured PolyLine Shapefiles

ArcView can use routes prepared in ARC/INFO or from Measured PolyLine Shapefiles. Measured PolyLine Shapefiles are the same as regular Poly-Line shapefiles (that is, a set of one or more linear features) except that they contain an additional measured value.

16.5 EVENT TABLES

Events are attribute data measured on a linear reference system. Using the route ID and location measures of events, the dynamic segmentation model can link events to a geographically referenced route system. Events may be point, continuous, or linear events:

- **Point events**, such as accidents and stop signs, occur at point locations. To relate point events to a route system, a point event table must have the route ID, the location measures of the events, and attributes describing the events.
- **Continuous events**, such as speed limits, cover the entire route with no gaps. A continuous event table relates to a route system by the route ID and the TO measure of the event.
- **Linear events**, such as pavement conditions, are discontinuous events along a route system. To relate linear events to a route system, a linear event table must have the route ID and the FROM and TO measures. Because gaps may be present between linear events, both the FROM and

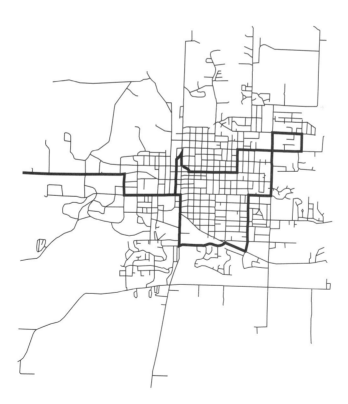

Figure 16.16
A bus route in Moscow, Idaho with two loops.

TO measures are required to mark the location of a linear event.

Except for the required route ID and measures, event tables are like regular attribute data tables in preparation and use.

16.5.1 Use Point or Polygon Coverages to Prepare Event Tables

An event table can be prepared from a point or polygon coverage so long as the coverage is based on the same coordinate system as the route system. The procedure is similar to map overlay except that the result is an event table. By placing a point coverage on a route system, the location of points in relation to the route system can be measured with the help of a fuzzy tolerance. If the distance between a point and the route system is within the fuzzy tolerance, then the point is a point event and its location is measured. By placing a polygon coverage on a route system, each segment of the route system is assigned with the attribute data of the polygon it crosses. The following shows two examples of creating event tables from coverages.

Figure 16.18 shows the bus stops along the bus route in Moscow, Idaho. To display these bus stops on the route system and the attributes at the stops, such as numbers of passengers getting on and off, we can treat the bus stops as point events. First, we prepare a point coverage containing bus stops. Next, we use the command ADDROUTEMEA-SURE to measure the locations of the bus stops along the bus route and to write the information to

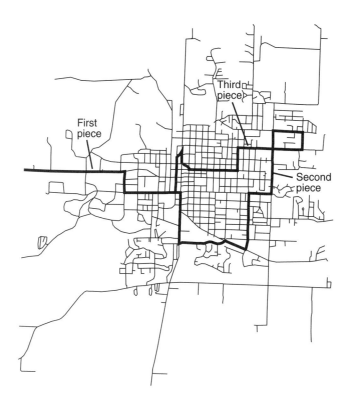

Figure 16.17
Because the bus route is a looping route, the route system is dissected into three sections.

a point event table (Table 16.5). Because the point event table is an INFO file, we can easily add to the file attribute data, such as numbers of passengers getting on and off.

For the second example we use a stream route system and a slope coverage with four slope classes to create a linear event table (Figure 16.19). After

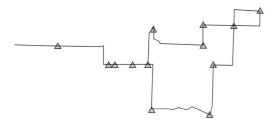

Figure 16.18
Bus stops along the bus route in Moscow, Idaho.

TABLE 16.5	A point event table showing bus stops along the bus route	
Busstations#	**Bus#**	**Measure**
1	1	899.930
2	1	2359.145
3	1	2476.239
4	1	2849.655
5	1	3163.485
6	1	4173.557
7	1	5446.844
8	1	6451.580
9	1	9368.944
10	1	8509.497
11	1	10002.686
12	1	10412.696
13	1	11728.987

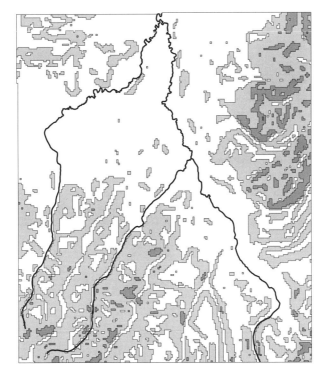

Figure 16.19
A stream route system and a slope coverage with four slope classes.

TABLE 16.6	A continuous event table showing the slope-code value for each segment of the stream route system		
Emida-Id	From	To	Slope-code
1	0	7638	1
1	7638	7798	2
1	7798	7823	1
1	7823	7832	2
1	7832	8487	1
1	8487	8561	1
1	8561	8586	2
1	8586	8639	1
1	8639	8643	2
2	0	2321	1
2	2321	2341	2
2	2341	2433	1
2	2433	2439	2
2	2439	2472	1
2	2472	2485	2
2	2485	2610	1
2	2610	2617	2
2	2617	2803	1
2	2803	2815	2
2	2815	3500	1
2	3500	3523	2
2	3523	3805	1
2	3805	3831	2
2	3831	3865	1
2	3865	3993	2
2	3993	4629	1
2	4629	4645	2
2	4645	4706	1
2	4706	5015	2
2	5015	5073	1
3	0	3	1
3	3	15	-9999
3	15	124	1
3	124	153	-9999
3	153	1911	1
3	1911	1937	2
3	1937	7473	1
3	7473	7478	2
3	7478	8367	1
3	8367	8380	2
3	8380	8435	1
3	8435	8438	2
3	8438	8667	1
3	8667	8761	2
3	8761	8822	1
3	8822	8884	2

TABLE 16.7	A linear event table showing year of pavement re-surfacing on the interstate highway route system		
Inter-Id	From	To	Year
1	44700	90000	1995
1	123648	180000	1989
1	239375	270000	1992
2	74024	78000	1991
2	154873	180000	1993
2	356992	400000	1987
3	78065	90000	1988
4	40000	72033	1986

computing the intersection between the route system and the slope coverage, the command POLYGONEVENTS assigns a slope-code value to each segment of the route system and writes the information to a linear event table (Table 16.6).

16.5.2 Prepare Event Tables as INFO Files

Although using point and polygon coverages to create event tables is relatively easy, the method cannot be applied to event data in tabular format. The alternative is to prepare event tables directly as INFO files or text files, as shown in the following two examples.

The first example is a linear event (pavement resurfacing) table for the interstate highway route system in Idaho. The event table or INFO file has four items: Inter-Id (interstate ID), From (from measure), To (to measure), and Year (year of pavement re-surfacing) (Table 16.7). The first three items relate the event table to the route system: (1) Inter-Id links the event table to the route table, and (2) From and To together relate the event table to the section table (Figure 16.20). Using the INFO file, we can query and display Year on the route system (Figure 16.21).

The second example also applies to the interstate highway route system in Idaho but uses a continuous event table. For this example we assume that each interstate highway must be divided into

(a)

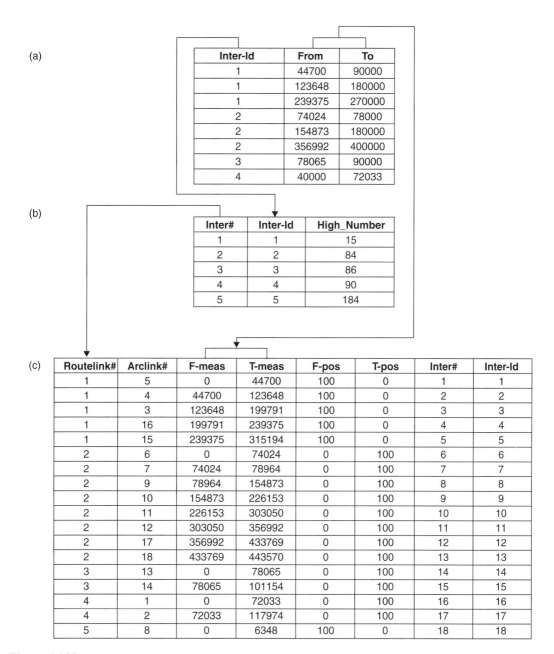

Figure 16.20
The linking of an event table (a), a route table (b), and a section table (c). See text for explanation.

50-mile segments (the last segment may be shorter than 50 miles) for the purpose of sampling. Table 16.8 is a continuous event table. Inter-Id, or the interstate highway number, links the event table to the route table. To, or the to measure, relates the event table to the section table for continuous

Figure 16.21
Display of the result from querying the event table
(Table 16.7) using the logical expression, Year <= 1990.

TABLE 16.8	**A continuous event table dividing each interstate highway into 50-mile segments**	

Inter-Id	To	Unit
1	80465	1
1	160930	2
1	241395	3
1	315194	4
2	80465	1
2	160930	2
2	241395	3
2	321860	4
2	402325	5
2	443570	6
3	80465	1
3	101155	2
4	80465	1
4	117974	2

events. (Because the highway coverage is measured in meters, the To values are also measured in meters rather than miles.) Unit is the unit number of 50-mile (80465-meter) segments, starting with 1 for each interstate highway. Figure 16.22 plots the first segment of each interstate in thicker line.

16.6 APPLICATIONS OF DYNAMIC SEGMENTATION

Applications of the dynamic segmentation model include the display, query, and analysis of routes, measures, and events. Similar to regions-based commands, ARC/INFO has specific commands to work with dynamic segmentation. For example, to link an event source and a route system as required by most applications, one can use EVENTSOURCE or EVENTMENU. EVENTSOURCE uses a command statement to set up the input arguments for the linkage, whereas EVENTMENU invokes a menu for the inputs.

16.6.1 Data Query with Events

Like attribute data tables, event tables can be queried by logical expressions. Suppose a linear event table has an item called Year for year of pavement re-surfacing. We can use the logical expression, Year <= 1990, for example, to select highway sections re-surfaced before 1990.

Querying event data spatially, such as querying past accidents within 10 miles of a recent accident, requires that we first convert event data to a coverage. The commands of EVENTARC and EVENTPOINT can perform the conversion. For example, EVENTPOINT can convert the point event table of accidents to a point coverage, which can then be queried graphically.

16.6.2 Data Analysis with Routes and Events

To buffer a route in ARC/INFO, the route must be converted into arcs first by using the command ROUTEARC or SECTIONARC. Conversion, however, is not required for buffering routes in Arc-View. Like buffering streams, buffering routes can be a useful management tool. For example, to study the efficiency or equity of public transit routes, we can buffer the routes with a distance of 1/2 of a mile and overlay the buffer zone with census blocks or block groups. Demographic data from the buffer zone (e.g., population density, income levels, number of vehicles owned per family, and commuting time to work) can provide useful information for the planning and operation of public transit.

The command OVERLAYEVENTS can overlay event data with the following options: line-on-line or point-on-line, and union or intersect. Suppose a point event table shows accidents, and a linear event table shows the pavement condition. To see if a relationship exists between accidents and the pavement condition, we can overlay the two event tables by using the point-on-line option and the intersect option. The result can determine if accidents are more likely to happen on stretches of the route with poor pavement conditions.

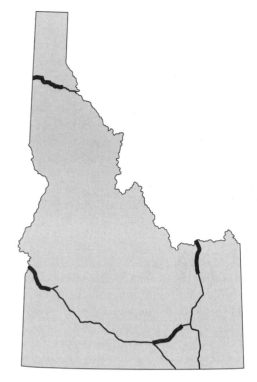

Figure 16.22
Display of the first 50-mile segment of Interstates 15, 84, 86, and 90.

KEY CONCEPTS AND TERMS

Allocation: A study of the spatial distribution of resources through a network.

Closest facility: A network analysis, which computes the shortest paths from the select location to all candidate facilities and then finds the closest facility among the candidates.

Combined route: A route that is joined with another route.

Continuous events: Events that cover the entire route with no gaps on a network, such as speed limits.

Linear events: Discontinuous events along a route system, such as pavement conditions.

Link: A segment separated by two nodes in a road network.

Link impedance: The cost of traversing a link, which may be measured by the physical length or the travel time.

Location-allocation: A spatial analysis, which matches the supply and demand by using sets of objectives and constraints.

Looping route: A route that intersects itself.

Maximum covering model: An algorithm for solving the location-allocation problem by maximizing the demand covered within a specified time or distance.

Minimum distance model: An algorithm for solving the location-allocation problem by minimizing the total distance traveled from all demand points to their nearest supply centers. The model is also called the *p-median location model.*

Network: A line coverage, which is based on the arc-node model and has the appropriate attributes for the flow of objects such as traffic.

Point events: Events that occur at point locations on a network, such as accidents and stop signs.

Shortest path analysis: A network analysis, which finds the path with the minimum cumulative impedance between nodes on a network.

Simple route: A route that follows one direction and does not loop or branch.

Split route: A route that subdivides into two routes.

Traveling salesman problem: A network analysis, which finds the best route with the conditions of visiting each stop only once, and returning to the original stop where the journey starts.

Turn impedance: The cost of completing a turn on a road network, which is usually measured by the time delay.

Turn table: A table that can be used to assign the turn impedance values on a road network.

APPLICATIONS: NETWORK AND DYNAMIC SEGMENTATION

This applications section has five tasks. The first three tasks cover the three types of network applications in ArcView: find shortest path, find closest facility, and find service area. Task 4 uses ArcView and the dynamic segmentation model to display a bus route, and point events along the route. Task 5 shows you how to create a bike route from the existing arcs of a street coverage in ARC/INFO.

Task 1. Find Best Route

What you need: a point theme called *uscities.shp*, a line theme called *interstates.shp*, and a polygon theme called *lower48.shp.*

   Lower48.shp shows the conterminous United States, *uscities.shp* contains cities, and *interstates.shp* includes the interstate highways. The objective of this task is to find the best (shortest) route between any two cities, measured in miles or in minutes. The speed limit for calculating the travel time is 65 miles/hour. The travel time considers only link impedance.

1. Start ArcView, and load the Network Analyst extension. Open a new view, and add *lower48.shp*, *uscities.shp*, and *interstates.shp* to view. Select Properties from the View menu. In the View Properties dialog, set the Map Units to meters and the Distance Units to miles. Make *uscities.shp* active.

2. Use the Query Builder button and the query expression, ([City_name] = "Helena") or ([City_name] = "Raleigh"), to select and highlight Helena, Montana and Raleigh, North Carolina in *uscities.shp*.

3. Make *interstates.shp* active. Select Find Best Route from the Network menu to add *Route1* to the Table of Contents and to display the *Route1* dialog (this is the Problem Definition dialog for the route).

4. Click Properties in the dialog. In the Properties menu, select Line Length as the Cost Field. Click OK to dismiss the menu.

5. Click Load Stops in the *Route1* dialog. Then choose *uscities.shp* to load stops. After you dismiss the Load Stops menu, you should see two stops are added to the *Route1* dialog. Another way to load stops is to use the Add Location tool and point to Helena and then to Raleigh. This option requires zooming in to more accurately locate the cities. Cities to be used as stops must be on or near the interstate highways.

6. Click the Solve Network Problem button either in the *Route1* dialog or in the View window. This step calculates the total

distance to travel the route. The best route is now displayed in the view. The *Route1* dialog also shows the total route cost in miles between Helena and Raleigh.

7. If you change the Cost Field from Line Length to Minutes in the *Route1* Properties menu, the total route cost will be measured in travel time using the speed limit of 65 miles per hour.

8. You can add intermediate stops between Helena and Raleigh. For example, to add Chicago as an intermediate stop along the route, change the query expression in Step 2 to read: ([City_name] = "Helena") or ([City_name] = "Chicago") or ([City_name] = "Raleigh").

Task 2: Find Closest Facility

What you need: a line coverage called *moscowst* and a point theme called *firestat.shp.*

Moscowst is a street coverage of Moscow, Idaho. Originally derived from the TIGER/Line file, the coverage has been edited and updated. *Firestat.shp* shows two fire stations in Moscow. The purpose of this task is to find the closest fire station and the quickest way to get to the station from any location in Moscow. The estimation of travel time considers link impedance, turn impedance, and one-way streets.

1. Start ArcView and load the Network Analyst extension. Open a new view, and add *moscowst* and *firestat.shp* to view. Because *moscowst* contains more than one map layer, the procedure to add *moscowst* is different. Click the Add Theme button. In the Add Theme dialog, click on the icon next to *moscowst*. Three separate layers are in the folder: *route.bus*, *arc*, and *node*. Click on *arc* to add the layer to view. The view shows the streets of Moscow. Select Properties in the View menu and set the Map Units as meters and the Distance Units as miles.

2. Make *moscowst* active.

3. Select Find Closest Facility in the Network menu. Notice the addition of *Fac1* (Facility 1) to the Table of Contents. Click Properties in the *Fac1* dialog and make sure the cost

field is defined as either miles or minutes. Click OK to dismiss the Property menu.

4. Click the Add Location tool and click a point on the map.

5. Click the Solve Network Problem button.

6. *Fac1* shows which of the two fire stations is closer to the point and the shortest route connecting the point and the closer fire station.

7. Make *Fac1* active. Select Table from the Theme menu to read the total cost.

Task 3: Find Service Area

What you need: *moscowst* and *firestat.shp*, the same data sets from Task 2.

Task 3 deals with an allocation problem. The objective is to measure the efficiency of two fire stations in Moscow.

1. Start ArcView and load Network Analyst. Open a new view and add *moscowst* (arc) and *firestat.shp* to view. Select Properties in the View menu and set the Map Units as meters and the Distance Units as miles.

2. Make *moscowst* active. Select Find Service Area from the Network menu. The Find Service Area function adds two themes, a service area theme (e.g., *Sarea1*) and a service network theme (*Snet1*), to the Table of Contents. The Find Service Area function also opens the Problem Definition dialog. You need to interact with the dialog in three ways. First, click Properties in the dialog to open the Property menu, in which you select minutes as the Cost Field. Second, select Travel from Site in the Problem Definition dialog and click Load Sites. Third, choose *firestat.shp* in the Load Sites dialog. The Problem Definition dialog now has Site #1 and Site #2 under Label and some default values under Minutes. Double-click the first cell under Minutes, and enter 3 as the value. Do the same for the second cell. The value 3 defines the service area as area within the 3-minute response time from the fire stations. Now, click the Solve Network Problem button.

3. The view shows the area within the 3-minute response time from the two fire stations. The

Problem Definition dialog shows the total area covered and the total distance on the network.

4. You can also use distance instead of travel time to define service areas.

Task 4: Display Point Events on a Route System in ArcView

What you need: *moscowst* and a point event table called *stations.txt*.

Moscowst is a street coverage of Moscow, Idaho. A bus route system has been built on the arcs of *moscowst*. The point event table *stations.txt* has the fields of busstations#, bus#, measure, and adp. The first three fields refer to bus station ID, route ID, and measure of bus station respectively. Adp refers to the average daily number of passengers. The purpose of this task is for you to become familiar with use of the dynamic segmentation data model in ArcView.

1. Start ArcView and click the Add Theme button. In the Add Theme dialog, click on the icon next to *moscowst*. Add *route.bus* and *arc* to view. The view shows the streets of Moscow and the bus route.
2. Next, add the point events, that is, bus stops, to view. Click on Tables and then Add in the Project's Table of Contents. In the next dialog, select Delimited Text as the File Type. Double-click *stations.txt* to add the table.
3. Select Add Event Theme from the View menu. In the Add Event Theme dialog, click on the route event icon at the top. Then select bus as the route theme, bus# as the route field, points as the event type, *stations.txt* as the table, bus# as the event field, and measure as the location field. Click OK.
4. Bus stations are now added to view. Make the *stations.txt* theme active and open its theme table. Click on Query Builder and use the expression, adp > 30, to select bus stations that have the average daily number of passengers greater than 30. Those selected stations are highlighted. You can also display the average daily number of passengers in graduated symbols.

Task 5: Create a Bike Route in ARC/INFO

What you need: a line coverage called *mosst*.

Same as *moscowst* in Task 4, *mosst* is a street coverage of Moscow, Idaho. In Task 5, you will build a bike route on the arcs and nodes of *mosst* using the ArcEdit command MAKEROUTE. The bike route is a simple route along Sixth St. from Perimeter Rd. to Washington St.

1. Go to ArcEdit and set up the draw and edit environment.
   Arcedit: display 9999
   Arcedit: mapex mosst
   Arcedit: editcov mosst
   Arcedit: drawenv arc
   Arcedit: draw
   Arcedit: editfeature arc
2. Next, select and display Sixth St, Perimeter Rd, and Washington St.
   Arcedit: select fename = 'Sixth' /* fename contains the street name
   Arcedit: drawselect /* to highlight Sixth St
   Arcedit: aselect fename = 'Perimeter' /* add Perimeter Rd to the subset
   Arcedit: drawselect
   Arcedit: aselect fename = 'Washington' /* add Washington St to the subset
   Arcedit: drawselect
3. The bike route follows Sixth St and extends from Perimeter Rd to Washington St. To create the bike route on the existing arcs of *mosst*, first select the arcs that make up the route and then run the command MAKEROUTE.
   Arcedit: select many /* select arcs of the bike route and press 9 to exit
   Arcedit: makeroute
   Usage: makeroute <subclass> {route-id} {measure_item} {UL | UR | LL | LR | * | xy} {START <start_measure>} {NOGAP | GAP} {CONNECT <connect_distance>}
   Arcedit: makeroute bike /* use selected arcs to make a route called *bike*
4. Exit ArcEdit and save edits. Go to Tables.

Enter Command: select mosst.ratbike /*
select the *bike* route table
Enter Command: list /* the route table shows
a single route
Enter Command: select mosst.secbike /*
select the *bike* section table
Enter Command: list

5. The section table has the following items:
ROUTELINK#, ARCLINK#, *F*-MEAS, *T*-MEAS, *F*-POS, *T*-POS, BIKE#, and BIKE-ID. ROUTELINK# links the sections to the *bike* route, while ARCLINK# links the sections to the arcs in *mosst*. *F*-MEAS and *T*-MEAS provide the cumulative measures of the sections. *F*-POS and *T*-POS measure the beginning and ending position of each section relative to the underlying arc. BIKE# and BIKE-ID are the machine ID and label ID of the sections.

6. You can use the Arc command DROPFEATURES to drop routes and sections from a coverage.
Arc: dropfeatures
Usage: DROPFEATURES <cover>
<feature_class> {ATTRIBUTES |
GEOMETRY}
Arc: dropfeatures mosst route.bike /* remove the *bike* route
Arc: dropfeatures mosst section.bike /* remove the *bike* sections

REFERENCES

Church, R.L. 1999. Location Modelling and GIS. In P.A. Longley, M.F. Goodchild, D.J. MaGuire, and D.W. Rhind (eds.). *Geographical Information Systems*. New York: John Wiley & Sons , 2d ed. pp. 293–303.

Church, R.L., and C.S. ReVelle. 1974. The Maximal Covering Location Problem. *Papers of the Regional Science Association* 32: 101–18.

Densham, P.J., and G. Rushton. 1992. A More Efficient Heuristic for Solving Large *P*-median Problems. *Papers in Regional Science* 71: 307–29.

Dijkstra, E.W. 1959. A Note on Two Problems in Connexion with Graphs. *Numerische Mathematik* 1: 269–71.

Dreyfus, S. 1969. An Appraisal of Some Shortest Path Algorithms. *Operations Research* 17: 395–412.

Francis, R.L., T.J. Lowe, G. Rushton, and M.B. Rayco. 1999. A Synthesis of Aggregation Methods for Multifacility Location Problems: Strategies for Containing Error. *Geographical Analysis* 31: 67–87.

Fu, L., and L.R. Rilett. 1998. Expected Shortest Paths in Dynamic and Stochastic Traffic Networks. *Transportation Research B* 32: 499–516.

Ghosh, A., S. McLafferty, and C.S. Craig. 1995. Multifacility Retail Networks. In Z. Drezner (ed.). *Facility Location: A Survey of Applications and Methods*. New York: Springer, pp. 301–30.

Hakimi, S.L. 1964. Optimum Locations of Switching Centers and the Absolute Centers and Medians of a Graph. *Operations Research* 12: 450–59.

Lin, S. 1965. Computer Solutions of the Travelling Salesman Problem. *Bell System Technical Journal* 44: 2245–69.

Lowe, J.C., and S. Moryadas. 1975. *The Geography of Movement*. Boston: Houghton Mifflin Company.

Nyerges, T.L. 1995. Geographical Information System Support for Urban/regional Transportation Analysis. In S. Hanson (ed.). *The Geography of Urban Transportation*. 2d ed. New York: Guilford Press, pp. 240–65.

Teitz, M.B., and P. Bart. 1968. Heuristic Methods for Estimating the Generalised Vertex Median of a Weighted Graph. *Operations Research* 16: 953–61.

Waters, N. 1999. Transportation GIS: GIS-T. In P.A. Longley, M.F. Goodchild, D.J. MaGuire, and D.W. Rhind (eds.). *Geographical Information Systems*. 2d ed. New York: John Wiley & Sons, pp. 827–44.

Ziliaskopoulos, A.K., and H.S. Mahmassani. 1996. A Note on Least Time Path Computation Considering Delays and Prohibitions for Intersection Movements. *Transportation Research B* 30: 359–67.

APPENDIX A

List of GIS Textbooks

Aronoff, S. 1989. *Geographic Information Systems: A Management Perspective.* Ottawa: WDL Publications.

Bernhardsen, T. 1999. *Geographic Information Systems: An Introduction.* 2d ed. New York: John Wiley & Sons.

Bonham-Carter, G.F. 1994. *Geographic Information Systems for Geoscientists: Modeling with GIS.* New York: Pergamon Press.

Booth, B. 1999. *Getting Started with ArcInfo.* Redlands, CA: ESRI Press.

Burrough, P.A., and R.A. McDonnell. 1998. *Principles of Geographical Information Systems.* Oxford: Oxford University Press.

Chrisman, N.R. 2002. *Exploring Geographic Information Systems.* 2d ed. New York: John Wiley & Sons.

Clarke, K.C. 2001. *Getting Started with Geographic Information Systems.* 3rd ed. Upper Saddle River, NJ: Prentice Hall.

DeMers, M.N. 2000. *Fundamentals of Geographic Information Systems.* 2d ed. New York: John Wiley & Sons.

ESRI Press. 1998. *Getting to Know ArcView GIS.* 3rd ed. Redlands, CA.

Heywood, I., S. Cornelius, and S. Carver. 1998. *An Introduction to Geographical Information Systems.* Upper Saddle River, NJ: Prentice Hall.

Huxhold, W.E. 1991. *An Introduction to Urban Geographic Information Systems.* New York: Oxford University Press.

Laurini, R., and D. Thompson. 1992. *Fundamentals of Spatial Information Systems.* London: Academic Press.

Lee, J., and D.W.S. Wong. 2001. *Statistical Analysis with ArcView GIS.* New York: John Wiley & Sons.

Longley, P.A., M.F. Goodchild, D.J. Maguire, and D.W. Rhind. 2001. *Geographic Information Systems and Science.* New York: John Wiley & Sons.

Longley, P.A., M.F. Goodchild, D.J. Maguire, and D.W. Rhind. 1999. *Geographical Information Systems: Principles, Techniques, Applications and Management.* 2d ed. New York: John Wiley & Sons.

MacDonald, A. 1999. *Building a Geodatabase.* Redlands, CA: ESRI Press.

Maguire, D.J., M.F. Goodchild, and D.W. Rhind. 1991. *Geographical Information Systems: Principles and Applications.* Harlow, Essex, England: Longman.

Martin, D.S. 1996. *Geographic Information Systems: Socioeconomic Applications.* 2d ed. London: Routledge.

Mitchell, A. 1999. *The ESRI Guide to GIS Analysis, Volume 1: Geographic Patterns & Relationships.* Redlands, CA: ESRI Press.

Ormsby T., and J. Alvi, 1999. *Extending ArcView GIS.* Redlands, CA: ESRI Press.

Peuquet, D.J., and D.F. Marble. 1990. *Introductory Readings in Geographic Information Systems.* London: Taylor and Francis.

Star J.L, and J.E. Estes. 1990. *Geographic Information Systems: An Introduction.* Englewood Cliffs, NJ: Prentice Hall.

Worboys, M.F. 1995. *GIS: A Computing Perspective.* London: Taylor and Francis.

Zeiler, M. 1999. *Modeling Our World: The ESRI Guide to Geodatabase Design.* Redlands, CA: ESRI Press.

APPENDIX B

MAGAZINES

GEO World

GEO Asia/Pacific

Geo Info Systems

GEO Europe

Business GEOgraphics

JOURNALS

*International Journal of Geographical
     Information Science*

*Cartography and Geographic Information
     Science*

Transactions in GIS

*Journal of Geographic Information and Decision
     Analysis (an electronic journal)* http://publish
     .uwo.ca/~jmalczew/gida.htm/

Geographical Systems

*Photogrammetric Engineering and Remote
     Sensing*

Remote Sensing of the Environment

List of State-level Websites for Existing GIS Data

The Geologic Survey of Alabama
http://www.gsa.state.al.us/gsa/GIS/clearinghouse.html/

Alaska State Geospatial Data Clearinghouse
http://www.asgdc.state.ak.us/

The University of Arizona Library
http://dizzy.library.arizona.edu/users/arawan/datamap.htm/

The Arkansas Geolibrary Catalog
http://cast.cast.uark.edu/local/isite/GeoLibraryCatalog.htm/

Stephen P. Teale Data Center [California]
http://www.gislab.teale.ca.gov/

Colorado Department of Natural Resources
http://dnr.state.co.us/

Map and Geographic Information Center, University of Connecticut
http://magic.lib.uconn.edu/

Delaware Geospatial Clearinghouse Node
http://gis.smith.udel.edu/fgdc/gateway/

Florida Department of Environmental Protection GIS
http://www.myflorida.com/myflorida/environment/learn/gis/index.html/

Georgia GIS Data Clearinghouse
http://gis.state.ga.us/

Hawaii Statewide GIS Program
http://www.state.hi.us/dbedt/gis/

INSIDE Idaho Geospatial Clearinghouse
http://inside.uidaho.edu/

Illinois Natural Resources Geospatial Data Clearinghouse
http://www.isgs.uiuc.edu/nsdihome/

Indiana GIS Initiative
http://www.state.in.us/ingisi/

Iowa Department of Natural Resources, Natural Resources Geographic Information System Library
http://samuel.igsb.uiowa.edu/nrgis/gishome.htm/

State of Kansas Geographic Information Systems Initiative's Data Access & Support Center
http://gisdasc.kgs.ukans.edu/dasc.html/

Office of Geographic Information, Kentucky
http://ogis.state.ky.us/

Louisiana Geographic Information Center
http://lagic.lsu.edu/

Maine Office of GIS
http://apollo.ogis.state.me.us/

Maryland State and Local Government Metadata Clearinghouse Node
http://www.msgic.state.md.us/

Massachusetts Geographic Information System
http://www.state.ma.us/mgis/massgis.htm/

Michigan GIS
http://www.state.mi.us/dmb/mic/gis/

Minnesota Department of Natural Resources
http://deli.dnr.state.mn.us/

Mississippi Automated Resource Information System
http://www.maris.state.ms.us/

Missouri Spatial Data Information Service
http://msdis.missouri.edu/

Montana Natural Resource Information System
http://nris.state.mt.us/

Nebraska Geospatial Data Clearinghouse
http://geodata.state.ne.us/

Nevada Bureau of Mines and Geology
http://www.nbmg.unr.edu/

New Hampshire GRANIT Web Site
http://www.granit.sr.unh.edu/

New Jersey Geographic Metadata Clearinghouse
http://njgeodata.rutgers.edu/

New Mexico Resource Geographic
Information System
http://rgis.unm.edu/

New York State GIS Clearinghouse
http://www.nysgis.state.ny.us/

North Carolina Geographic Data Clearinghouse
http://cgia.cgia.state.nc.us/ncgdc/

North Dakota Geological Survey,
Geographic Information Systems Center
http://www.state.nd.us/ndgs/ndgs_gis.htm/

Ohio Department of Administrative Services,
Geographic Information Systems Support Center
http://www.geodata.state.oh.us/

Spatial and Environmental Information
Clearinghouse, Oklahoma State University
http://www.seic.okstate.edu/

Oregon Geospatial Data Clearinghouse
http://www.sscgis.state.or.us/

Pennsylvania Spatial Data Access
http://www.pasda.psu.edu/

Rhode Island Geographic Information System
http://www.edc.uri.edu/rigis/

South Carolina Department of Natural Resources
GIS Data Clearinghouse
http://water.dnr.state.sc.us/gisdata/

South Dakota Geological Survey
http://www.sdgs.usd.edu/

The Map Library at the University of Tennessee,
Knoxville
http://www.lib.utk.edu/~cic/

Texas Natural Resources Information System
http://www.tnris.state.tx.us/

Utah AGRC State Geographic Information
Database
http://agrc.its.state.ut.us/

Vermont Geographic Information System
http://geo-vt.uvm.edu/

University of Virginia Library,
Geospatial & Statistical Data Center
http://fisher.lib.virginia.edu/

Washington State Geospatial Clearinghouse
http://wa-node.gis.washington.edu/

West Virginia GIS Technical Center
http://wvgis.wvu.edu/

Wisconsin Land Information Clearinghouse
http://badger.state.wi.us/agencies/wlib/sco/pages/

Wyoming Natural Resources Data Clearinghouse
http://www.sdvc.uwyo.edu/clearinghouse/

APPENDIX D

Metadata for a Map of Largest Cities in each Montana County

The following is a complete example of metadata downloaded from the Montana Natural Resource Information System website (http://nris.state.mt.us/). "Data about data" are absolutely important to GIS users. For example, the spatial reference information section describing the coverage's coordinate system and its parameters will be needed to re-project the coverage onto a different coordinate system.

IDENTIFICATION_INFORMATION:

Citation
    Originator: Montana State Library
    Publication_Date: 11/24/1992
    Title: Largest City in each Montana County
    Online_Linkage: http://nris.state.mt.us/nsdi/nris/e00/ct4.zip
    Online_Linkage: http://nris.state.mt.us/nsdi/nris/shape/ct4.zip
Description
    Abstract:
        Point locations for the largest city in each Montana county, selected from the U.S. Geological Survey Geographic Names Information System. These are the same as the county seats, except for Madison County, where Ennis, the largest town, is in the coverage, but Virginia City, the county seat, is not. There are two levels of annotation in the coverage — level One has a label for every city, level Two has a larger label for 13 larger cities. The locations of Circle and Choteau were incorrect in the GNIS data. These have been corrected.

Purpose:
    Small scale base map data.
Time_Period_of_Content
    Calendar_Date: 01/01/1988
Access_Constraints:
    None
Use_Constraints:
    Not for use in applications where cities should be perceived as areas rather than point locations.
Native_Data_Set_Environment:
    Arc/Info version 7.2.1 - SunOS version 5.7 and Windows NT Pathname = /montana/city2

DATA_QUALITY_INFORMATION:

Attribute_Accuracy_Report:
    City names are correct.   Latitude and Longitude locations are within the city limits.
Completeness_Report:
    Data set is complete.
Horizontal_Positional_Accuracy_Report:
    Point locations are within the corresponding city limits.
Vertical_Positional_Accuracy_Report:
    NONE
Lineage
    Source_Information
        Originator: U.S. Geological Survey National Mapping Division
        Publication_Date: 01/01/1992
        Title: Geographic Names Information System

336

Publication_Information
   Publication_Place: Reston, VA
   Publisher: U.S. Geologic Survey
Source_Scale_Denominator: 100000
Type_of_Source_Media: unknown
Source_Time_Period_of_Content
   Calendar_Date: 01/01/1992
Source_Citation_Abbreviation: gnis
Source_Contribution:
     The cities in the coverage were
     selected from this data source.
Process_Step
  Process_Description:
    Convert the GNIS file to an Arc/Info
    coverage.
  Source_Used_Citation_Abbreviation: gnis
  Process_Date: 08/21/1992
  Source_Produced_Citation_Abbreviation:
    imp1
Process_Step
  Process_Description:
    Select the cities from the GNIS data.
  Source_Used_Citation_Abbreviation:
    imp1
  Process_Date: 11/04/1992
  Source_Produced_Citation_Abbreviation:
    imp2
Process_Step
  Process_Description:
    Move the cities of Choteau and Circle to
    their correct locations.
  Source_Used_Citation_Abbreviation: imp2
  Process_Date: 11/24/1992
  Source_Produced_Citation_Abbreviation:
    fins

SPATIAL_DATA_ORGANIZATION_ INFORMATION:

Direct_Spatial_Reference_Method: Point
Point_and_Vector_Object_Information
  Number of Points: 56

SPATIAL_REFERENCE_INFORMATION:

Horizontal_Coordinate_System_Definition
  Grid_Coordinate_System_Name: State Plane
  Coordinate System 1983
    SPCS_Zone_Identifier: 2500
    Map_Projection_Name:
      Lambert Conformal Conic
    Standard_Parallel: 45
    Standard_Parallel: 49
    Longitude_of_Central_Meridian: -109.5
    Latitude_of_Projection_Origin: 44.25
    False_Easting: 600000
    False_Northing: 0
  Planar_Distance_Units: meters
  Geodetic_Model
    Horizontal_Datum_Name: North American
    Datum of 1983
  Altitude_Encoding_Method:
   Implicit coordinate

ENTITY_AND_ATTRIBUTE_ INFORMATION:

Entity_Type_Definition: Point Attribute Table
  Attribute_Label: NAME
    Attribute_Definition: Name of City
      Unrepresentable_Domain: Character
      field
    Beginning_Date_of_Attribute_Values:
    01/01/1992
  Attribute_Label: LONG
    Attribute_Definition: Longitude of city
    center
      Range_Domain
        Range_Domain_Minimum:
        -115.5550003051
        Range_Domain_Maximum:
        -104.1557998657
    Attribute_Units_of_Measure: Decimal
    Degrees
    Attribute_Measurement_Resolution:
    0.0000999999974
    Beginning_Date_of_Attribute_Values:
    01/27/1992

Attribute_Label: LAT
  Attribute_Definition: Latitude of city center
    Range_Domain
      Range_Domain_Minimum: 45
      Range_Domain_Maximum: 49
  Attribute_Units_of_Measure: Decimal
  Degrees
  Attribute_Measurement_Resolution:
  0.0000999999974
  Beginning_Date_of_Attribute_Values:
  01/27/1992

MONTANA STATE LIBRARY GIS DATA DISTRIBUTION INFORMATION:

Distributor
  Contact_Information
    Contact_Organization_Primary
      Contact_Organization: Montana State
      Library
      Contact_Person: Gerry Daumiller
    Contact_Position: Senior GIS Programmer
    Contact_Address
      Address_Type: mailing address
      Address: P.O. Box 201800
      City: Helena
      State_or_Province: Montana
      Postal_Code: 59620-1800
      Country: USA
    Contact_Voice_Telephone: (406) 444-5358
    Contact_Facsimile_Telephone: (406) 444-0581
    Contact_Electronic_Mail_Address:
    gdaumiller@state.mt.us

Hours_of_Service: Monday-Friday, 8-5,
    Mountain Time
Distribution_Liability: Users must assume
    responsibility to determine the usability of
    this data for their purposes.
Standard_Order_Process
    Non-digital_Form: Contact the State Library
      to see if the data is available as hardcopy
      maps.
Standard_Order_Process
    Digital_Form
      Digital_Transfer_Information
        Format_Name: Arc/Info Export
      Digital_Transfer_Option
        Offline_Option
          Offline_Media: CD-ROM
          Recording_Format   ISO 9660
    Digital_Form
      Digital_Transfer_Information
        Format_Name: ESRI Shapefile
      Digital_Transfer_Option
        Offline_Option
          Offline_Media: CD-ROM
          Recording_Format   ISO 9660
    Fees: For-profit organizations must pay our
      costs to reproduce the data.   Fees can be
      waived if doing Government work.

METADATA_REFERENCE_INFORMATION:

Metadata_Date: 01/04/1996
Metadata_Review_Date: 10/05/1996
Metadata_Contact: Gerry Daumiller

INDEX